Energy Democracy

Craig Morris • Arne Jungjohann

Energy Democracy

Germany's Energiewende to Renewables

palgrave
macmillan

Craig Morris
Petite Planète, Freiburg and Berlin,
Germany

Arne Jungjohann
Stuttgart, Germany

ISBN 978-3-319-31890-5   ISBN 978-3-319-31891-2   (eBook)
DOI 10.1007/978-3-319-31891-2

Library of Congress Control Number: 2016945578

Cover illustration: © VPC Travel Photo / Alamy Stock Photo

Printed on acid-free paper

This Palgrave Macmillan imprint is published by Springer Nature
The registered company is Springer International Publishing AG Switzerland

For Phillip and Pascale

Energy democracy, n. [ɛnərdʒi dɪmɑkrəsi] 1) when citizens and communities can make their own energy, even when it hurts energy corporations financially; 2) something currently mainly pursued in Denmark and Germany but that can spread around the world during the current window of opportunity; 3) the most often overlooked benefit of distributed renewables in the fight against climate change; 4) something to fight for as the path to better quality of life with stronger communities and better personal relationships.

Preface

This book is a history of Germany's energy transition—its *Energiewende*. It is not *the* history of the Energiewende. A book of this length couldn't be definitive or exhaustive. On the other hand, most international readers—and, no doubt, numerous Germans—who assume that the Energiewende started in 2011 after the nuclear accident in Fukushima are likely to be surprised to hear how old the grassroots movement is.

Since both of us have spent at least the past decade communicating German energy and climate policymaking for the international community, we were struck by the awkward perception of the Energiewende after Fukushima. Following the nuclear phaseout in 2011, the Energiewende drew attention around the world for being allegedly exceptional with its move from nuclear power to wind and solar. Yet, Germany is not exceptional for any of this. Austria, Belgium, Denmark, Italy, and Switzerland all have nuclear phaseout plans (some already completed), and France aims to reduce its reliance on nuclear. The rest of Europe either has no nuclear reactors or faces an unplanned phaseout as old plants reach the end of their lifetimes and country after country fails to build new reactors.

Germany isn't exceptional for wind or solar either. Italy and Greece have a higher share of solar power on their grids; China has more installed. Denmark, Portugal, Ireland, Spain, Romania, and Lithuania have a higher or similar share of wind power. And China and the US have more installed.

The Energiewende is nonetheless exceptional in one way too often overlooked. Apart from Denmark and, more recently, Scotland, Germany is the only country in the world where the switch to renewables is a switch to energy democracy. We wanted to remedy that oversight with this book.

Morris is an American writer based in Germany; Jungjohann, a former Washington-based policy expert fostering transatlantic cooperation on energy and climate matters. As contributing editor of *Renewables International*, Morris suddenly found himself thrust into the role of explaining the Energiewende in English in the wake of Fukushima, when few journalists working in English knew the full history. In 2012, we then joined forces to found the website EnergyTransition.de, first with an e-book and later with a blog, to help fill the gap. We are both often asked why a transition to renewables largely in the hands of citizens has been possible in Germany but not, say, in the US and the UK (at least not up to now).

With his years of experience in the *Deutscher Bundestag*, Jungjohann brought numerous insider stories and interview contacts to the table. Morris, in return, personally knew many of the people from Wyhl to Schönau—arguably, the region that gave birth to the Energiewende—because he had lived in the area since 1992. An American–German duo was also interesting given the topic and target audience. Perhaps ironically, Jungjohann sometimes felt uncomfortable with Morris's praise of Germany and criticism of the US, while Morris could not always agree with Jungjohann's praise of the US or criticism of Germany. Each author probably expects more of his home country than is reasonable. The result is a more balanced book than either author would have written alone.

Several of the chapters are based heavily on individual books, as our footnotes reveal. Readers may wonder why we did not use broader sources of information. They certainly exist: Paul Gipe has produced impressive work on the history of wind power, even with a focus on Germany; John Perlin has covered the history of solar masterfully. Gabrielle Hecht has also written authoritatively on the history of nuclear in France, and we do not cite her in our comparison either.

The reason is that all these books are already available in English. In using books only available in German and French, we aimed to at least make the gist of this untranslated literature accessible to a wider audience. It should be noted that the general findings and conclusions in the literature largely overlap; Hecht and Topçu come to very much the same conclusions about nuclear France, as do Gipe, Oelker, and Lobe for wind or Perlin and Janzing for solar. Later chapters also rely heavily on the fantastic German news archive at Energie-Chronik.de run by German journalist Udo Leuschner. As we move closer to the future, Morris's own online publications are referenced more frequently, as is our collaboration at EnergyTransition.de.

Our intent is to show international readers how German citizens got their government behind a policy that the public wanted. As such, our story downplays numerous aspects important for the German energy sector's history, but not necessarily for the energy transition as a grassroots movement. One example is the transformation of the energy sector in the former East Germany after the Fall of the Wall, which this book does not deal with in depth.

For the sake of brevity, this book does not tell the tale of every part of the grassroots movement worth mentioning either. For instance, Gorleben—the proposed site of (West) Germany's nuclear waste repository—played a huge role in the anti-nuclear movement; it thus deserves its own subchapter, which we unfortunately did not have space for.

Furthermore, in addition to the numerous people mentioned briefly, there were also countless environmental and renewable energy campaigns and trade shows driven by citizens. Numerous state and local governments also passed ambitious legislation and set up Energy Agencies, which have done excellent work in helping the public implement not only renewable energy technology, but also efficiency. Not all these helpful participants could be mentioned.

Many people helped shape this book. The authors would like to thank those who provided wise counsel and encouraging feedback at different stages of the project: energy analyst Toby Couture, journalist Osha Davidson, energy policy expert Jose Etcheverry, John Farrell of the ILSR, WWEA executive director Stefan Gsänger, Paul Gunther of Beyond

Nuclear, Bastian Hermisson of the Washington office of the Heinrich Böll Stiftung, author Mark Hertsgaard, Gisa Holzhausen of the GIZ, Tetsunari Iida of ISEP, Andreas Kraemer of Ecologic Institute, journalist Bernward Janzing, Anna Leidreiter of the World Future Council, Amory Lovins of the Rocky Mountain Institute, Georg Maue of the German Embassy Washington DC, Daniel Lerch of the Post Carbon Institute, energy analyst Hugo Lukas, journalist Gregory McDonald, English linguist Paul McPherron, Phyllis McPherron, Michael Mehling of MIT, Jennifer Morgan of Greenpeace International, Diane Moss of the 100% Renewables Institute, Klaus Müschen of the German Environmental Agency (UBA), Christ Nelder of the Rocky Mountain Institute, energy analyst Uwe Nestle, Alan Nogee of the Union of Concerned Scientists, Alexander Ochs of the Worldwatch Institute, Sara Peach of University of North Carolina, Josef Pesch of EWS, energy analyst Raffaele Piria, Anya Schoolman of the Committee Power Network, Joachim Seel of UC Berkeley, Dave Toke of the University of Aberdeen, energy analyst Sezin Topçu, energy analyst Dieter Seifried, Dirk Vansintjan of REscoop, former Lead Mission Operations Engineer at SpaceX Daniel Villani, and retired EU official James Wimberley. Arne's special thanks go to Verena for covering his back in an exciting and demanding time with a young family. We also thank our editors from Palgrave Macmillan, Rachael Ballard and Chloe Fitzsimmons, and Natascha Spörle for her support for comments and endnotes. Graphics for this book are available at www.EnergiewendeBook.de. The authors alone are solely responsible for any errors or omissions.

Finally, Paul Gipe deserves special recognition. In the 1970s, he began his career travelling around the mid-West collecting old wind power equipment from farmers. When California's wind boom happened in the early 1980s, he was there—and when it collapsed. In the 1990s, he became a rare US expert on Danish and German energy policy.

When the first wind turbines in the US started breaking in the 1990s, he called for them to be dismantled (as in Denmark and Germany) to keep America beautiful (many are still standing today). In 2008, when Ontario passed its Green Energy Act, Gipe was executive director of a Canadian NGO explaining to the Province's politicians what community energy is. In 2012, he told Jungjohann and Morris that they should meet.

This book is written in the spirit of everything Gipe has represented for the past 40 years: community empowerment and renewable energy—in other words, energy democracy. The authors continue to carry Paul's torch as he transitions to his own well-deserved (semi-)retirement.

Contents

List of Tables

List of Boxes

1

Energiewende: The Solution to More Problems Than Climate Change

On a sunny day in Berlin at the end of March 2015, Frank-Walter Steinmeier, Germany's Foreign Minister, took the stage at the Energy Transition Dialogue, the government's first international conference on the energy transition. He asked the crowd of some 900 people from more than 50 countries, "Do you know what the word for 'Energiewende' is in Indonesian? Or Spanish? I can tell you. It's 'Energiewende'." He speculated that the word could become a successor to "kindergarten" as the next German word taken up by the world.

If so, pronunciation is an obstacle (try: energhy-vend-uh). While the *Energie* part is self-explanatory, *Wende* is anything but. It can mean U-turn—but then, the energy transition is to take us on a new path, not back where we came from. But *Wende* can also mean a fundamental turn for the better. In fact, what we call the Fall of the Wall in English is *die Wende* for the Germans.

The main obstacle to becoming Germany's next major linguistic export is not language, however. It's partly target attainment. Germany must demonstrate that a heavily industrialized country can successfully move beyond nuclear power and fossil fuel, running its economy mainly on renewable resources. In 2010, the German government set several

C. Morris, A. Jungjohann, *Energy Democracy*,
DOI 10.1007/978-3-319-31891-2_1

1

targets for mid-century: at least 80 percent renewable electricity, at least 60 percent renewables in final energy (including heat and motor fuels), and a 50 percent cut in primary energy[1] consumption. In the process, carbon emissions from energy consumption are to be reduced by 80 to 95 percent.

Germany is an industrial powerhouse and one of the leading export nations in the world. It has to complete this transition while continuing to improve economic well-being. The transition to renewables was understood as a way of making German industry future-proof all along. When the Renewable Energy Act (EEG) was adopted in 2000, one of its coauthors, the late Hermann Scheer, emphasized the new law's contribution to development aid and industrial policy in a speech given at the Bundestag on the day the law was passed:

> Some types of renewable energy, such as photovoltaics, are still relatively expensive here but are already the cheaper alternative in areas of the Third World with no grid access. And if we want to do more there, we have to get the industry moving here ... We are responsible for this problem [of climate change] and should not shrug this responsibility by only thinking of ourselves.

In other words, the Energiewende is Germany's plan not only for a sustainable, affordable power supply, but an industry policy to ensure future exports—and a development aid policy to make good for previous emissions. Germany has helped bring down the cost of wind and solar power in particular for developing countries and the rest of the world.

By 2050, Germany aims to demonstrate that an industrial powerhouse can do without nuclear even as it draws down fossil fuel consumption, partly thanks to efficiency, to low levels. It should be a wonderful future for the Germans with well-paying jobs in their walkable cities surrounded by economically vibrant and clean rural communities. Germany would be surrounded by friendly neighboring countries no longer quibbling over scarce fossil resources (which by then will primarily be used as valu-

---

[1] A distinction is made between primary and final energy. Primary energy is what goes into the energy conversion process; final energy, what comes out of it. So a lump of coal is primary energy, while the electricity that comes out of the plant is final energy.

able raw materials anyway, not burned in inefficient processes). That, at least, has been the vision all along.

No Government Master Plan

So when did the Energiewende start? Did it kick off in the year 2000 with the passing of the EEG? Those behind the law often say so. But in reality, there is no single starting point to Germany's energy transition.

The EEG is unthinkable without its predecessor, the Feed-in Act of 1991. Inspired by previous Danish legislation and a US law called Public Utility Regulatory Policies Act (PURPA) from 1978, the Feed-in Act primarily opened the door to the energy sector for builders of wind farms. The legislation incentivized deployment—actually building things— rather than research and development after R&D subsidies had proven disappointing in the 1980s.

And we can go back even further. In 1980, a book entitled "Energiewende"[2] was published by the newly founded Öko-Institut, a research lab of independent analysts. The subtitle of the book is telling: "growth and prosperity without petroleum and uranium." The goal was not to phase out coal. On the contrary, the most ambitious scenario had 45 percent renewable energy and 55 percent coal. This target was for all energy, not just electricity; liquefied coal and renewable electricity would have powered mobility. Coal remains a blind spot in Germany's energy transition. In their defense, the authors point out that the result would have been a drastic cut in carbon emissions. But obviously, the focus was primarily on energy independence in the wake of two oil crises.

Back then, climate change did not play the prominent role it does today in public debate. Global warming would not take the foreground until 1986, when German news weekly *Der Spiegel* put on its cover an image of the cathedral in Cologne underwater. The then Chancellor Helmut Kohl spoke soon afterward in the German Parliament about the "grave threat

---

[2] Krause, Florentin, Hartmut Bossel, and Karl F. Müller-Reißmann. *Energie-Wende: Wachstum und Wohlstand ohne Erdöl und Uran; ein Alternativ-Bericht des Öko-Instituts/Freiburg.* 3. Aufl., 10.-12- Tsd. Frankfurt/M.: S. Fischer, 1981.

of global warming." But 1986 was also the year of the nuclear accident at Chernobyl. From that point on, reducing the risk of nuclear power and fighting climate change became central—and conflicting—goals of the Energiewende. While Chernobyl spoke against nuclear power, climate change spoke in favor of it. By the end of the 1980s, climate change had clearly taken center stage as the main goal of the Energiewende. But prior to Chernobyl, there was another goal: energy democracy.

The term "Energiewende" was coined in the 1970s, when a conservative rural community protested plans to industrialize the area with the construction of a new nuclear plant and attempts to attract new industry—such as a lead production plant—as buyers of all of the electricity that local families and businesses did not need. Locals fought against big business coming to town, changing the local surroundings, and reaping most of the profits while the community ran the risks. The battle of this rural German community against a nuclear reactor and a lead production plant very much resembles the fight of small-town America against fracking today. The Energiewende thus began as a grassroots movement for greater democracy in the energy sector and against privatizing profits and socializing risks.

Visitors who come to Germany to learn about its energy transition often ask how the government gets the public to play along, especially given the high price tag. But if you understand the Energiewende as a grassroots movement rather than a governmental master plan, you know the question should be: how did the Germans get their government to do what the public wanted even when it hurt big energy companies? The question is poignant today for people who also want to know how they can get money out of politics in order to make the government more responsive to popular will.

The Energiewende movement was a constant struggle. Its admirers sometimes depict it as a master plan of German engineering in combination with that magical German ability to reach a consensus across party lines and throughout society. Interestingly, the Energiewende's critics seem to agree that this combination exists, though they depict it as practically Soviet-style central planning based on a green ideology permeating most of German society. In reality, both of these views are slightly skewed; the Energiewende consensus has to be renegotiated continually. At certain times, it made progress by lurching forward, only to get stuck again for years as its proponents fought to prevent rollback. There never was a true master plan, and there is none today. And the pushback is real.

Whether we say the Energiewende started in the 1970s or in 1980, 1991, or 2000, one thing is certain—it did not start in 2011. Yet, that year the term became commonly known around the world—and closely associated with Chancellor Angela Merkel's reaction to the nuclear accident in Fukushima. Yet, as German author Udo Leuschner has pointed out, "Energiewende" became an official term for the government during the Social Democrat & Green coalition from 1998 to 2005. It then disappeared under Chancellor Merkel until 2011.[3] For instance, in 2007, the government first proposed its target of a 40 percent reduction in carbon emissions by 2020 relative to 1990. The proposal was called the "Integrated energy and climate package," and across 98 pages the document did not use the word "Energiewende" at all.[4] Three years later, the proposal became official policy, and the title reveals the calisthenics the government was willing to perform in order to avoid saying "Energiewende" (this time, across 40 pages): "Energy Concept for an Environmentally Friendly, Reliable and Affordable Energy Supply."[5] If "Energiewende" joins the ranks of "kindergarten" as the next German word given to the world, there is no doubt that this linguistic migration did not start until 2011.

---

[3] Leuschner, Udo. "Energiechronik: Kanzlerin will sich jetzt persönlich um die "Energiewende" kümmern." May 2012. Accessed February 4, 2016. http://www.energie-chronik.de/120502.htm.

[4] Bundesministerium für Wirtschaft und Technologie, and Bundesministerium für Umwelt, Naturschutz und Reaktorsicherheit. "Bericht zur Umsetzung der in der Kabinettsklausur am 23./24.08.2007 in Meseberg beschlossenen Eckpunkte für ein Integriertes Energie- und Klimaprogramm: Das Integrierte Energie- und Klimaprogramm der Bundesregierung." Accessed February 4, 2016. http://www.bmub.bund.de/fileadmin/bmu-import/files/pdfs/allgemein/application/pdf/gesamtbericht_iekp.pdf.

[5] Bundesregierung. "Energiekonzept für eine umweltschonende, zuverlässige und bezahlbare Energieversorgung." 28. September 2010. Accessed February 4, 2016. http://www.bundesregierung.de/ContentArchiv/DE/Archiv17/_Anlagen/2012/02/energiekonzept-final.pdf?__blob=publicationFile&v=5.

Fukushima Draws International Attention to the Energiewende

In March 2011, the nuclear accident in Fukushima put the global spotlight on Japan. The world looked on in horror at the devastation, but in great admiration at the Japanese people's ability to work together. But as the days and weeks dragged on, the Daiichi nuclear plant in the prefecture of Fukushima remained in the headlines. Top industry spokespeople and politicians had to step in front of the camera and assure the public that…well, actually, they didn't really know what was going on.

In a distant second place in the news and half a world away from the epicenter, Germany was an unlikely contender for attention. Its predicament was entirely man-made. Chancellor Angela Merkel, a Christian Democrat, reacted to the nuclear accident in Japan by shutting down eight of Germany's 17 nuclear plants 3 days after the tsunami. In doing so, she reversed her party's long-held position on nuclear power overnight. She handed down her decision like a decree, knowing full well that no one in her party who wanted a future in politics would dare challenge her. Public support for the phaseout was simply too high at around 80 percent. When the law came to a vote 3 months later in Parliament, it received nearly unanimous support. Merkel's conservative party, the Social Democrats, and the Green Party all voted in favor, despite their former bitter fights on the issue. The only party voting against the law was the successor of the communist party from East Germany, which said the phaseout did not go far enough.

Outside Germany, the decision was perceived as a sudden, even surprising move—not more or less the reinstitution of the nuclear phaseout agreement of 2000, which had only been repealed in late 2010. Reactions abroad broke down cleanly along people's stance toward nuclear. Germany was either hysterically overreacting or the only country that had the guts to do the right thing. That year, in 2011, the German government began talking a lot about the "Energiewende." Newscasters and politicians in Germany began conflating the 2011 nuclear phaseout with the Energiewende, and people began saying misleading things like "since the Energiewende began in 2011."

The sudden attention drawn to Merkel's nuclear phaseout unfortunately made the event seem more unusual than it was. In fact, numerous neighboring countries have stepped away from nuclear. Austria blocked the opening of a finished nuclear plant way back in 1978—a year *before* the accident at Three Mile Island in the US—and adopted a nuclear phaseout in 1997. Italy adopted one in 1985, the year *before* the accident at Chernobyl.[6] After the accident at Fukushima, Belgium also resolved to shut down its nuclear plants.[7] So did Switzerland, though the Swiss continue to debate the issue.[8] If the UK does not build a new nuclear reactor or extend the life spans of existing ones, only one will be in operation in the country by 2023, the first year without nuclear in Germany.[9] And even France aims to reduce the share of nuclear power from around 75 to 50 percent of supply by 2025. Germany is not an outlier with its nuclear phaseout; it lies in the midst of a European movement away from nuclear and toward renewables.

For three reasons, Germany's nuclear phaseout and the Energiewende drew so much attention around the world in 2011: the timing, the message, and the messenger. First, seen outside of its historical context, the 2011 phaseout decision did indeed look like a sudden, not well-prepared response to Fukushima with far-reaching consequences. How could international observers who do not follow daily German domestic politics know that the decision merely put into law more or less the initial nuclear phaseout deadline implemented by the former government 10 years before? How could they know that renewable energy goals for 2050 proposed in 2007 and adopted in 2010 were left unchanged in 2011?

---

[6] Morris, Craig. "Angst…that the Energiewende will work." Energy Transition. The German Energiewende. 24 July 2014. Accessed February 4, 2016. http://energytransition.de/2014/07/angst-that-the-energiewende-will-work/.

[7] Morris, Craig. "Belgium to phase out nuclear." Renewables International. The Magazine. 31 October 2011. Accessed February 4, 2016. http://www.renewablesinternational.net/belgium-to-phase-out-nuclear/150/537/32293/.

[8] Morris, Craig. "Nuclear phaseout in Japan and elsewhere." Renewables International. The Magazine. 30 March 2012. Accessed February 4, 2016. http://www.renewablesinternational.net/nuclear-phaseout-in-japan-and-elsewhere/150/537/33531/.

[9] World Nuclear Association. "Nuclear Power in the United Kingdom." 31 December 2015. Accessed February 4, 2016. http://www.world-nuclear.org/info/Country-Profiles/Countries-T-Z/United-Kingdom/.

The message was also bold—Germany aimed to be the first large, highly industrialized country to transition to a low-carbon economy with renewables but without nuclear power. In the UK, "decarbonization" explicitly includes nuclear, as does the US approach of "all of the above," in which everyone gets a piece of the pie. In contrast, the Energiewende makes enemies. Much of the international attention paid to Germany's energy transition has come from the pro-nuclear camp, which rightly feels challenged by Germany's aims. If successful, the Energiewende will prove that nuclear is not needed, not even for a low-carbon future.

The German experiment is also poised to demonstrate that going renewable is an attractive option for any country—a scary message for the oil, coal, and gas sectors. A look at their German peers tells utilities around the world that their classic business model is crumbling. The Big Four utilities' share of the German power market is shrinking as citizens and new players start making their own energy. By 2012, nearly half of investments in new solar, biomass, and wind power came from citizens and energy co-ops. Institutional investors such as banks, insurance companies, and municipal energy suppliers made up most of the other half; the Big Four, only 5.5 percent.[10] Finally, there was the messenger—Germany. To be certain, other countries have a greater share of renewable energy. Norway practically has 100 percent green electricity, but the example is impossible to follow; other countries cannot decide to be relatively small populations living along fjords with loads of hydropower. On the other hand, Denmark, Spain, and Portugal have far greater shares of wind power than Germany does, but the Germans draw more attention as a manufacturing powerhouse. If you can run a demanding industrial base like Germany's on wind and solar without traditional baseload power plants, the sentiment seems to be, it will more easily work elsewhere.

But let's be clear: no one worries about the German economy as much as the pro-fracking, pro-nuclear camp pretends to. At times, it sounds as though such critics might suffer from a bit of *angst* themselves—that the Germans might pull off their Energiewende.

---

[10] Morris, Craig. "Citizens own half of German renewable energy." Energy Transition. The German Energiewende. 29 October 2013. Accessed February 4, 2016. http://energytransition.de/2013/10/citizens-own-half-of-german-renewables/.

To be fair, only time will tell if the Germans can succeed. Despite their reputations for engineering wizardry, the Germans struggle with megaprojects as much as anyone else does—just look at the fiasco at Hamburg's *Elbphilharmonie* concert hall, which started off with a price tag of €77M but is now expected to cost 10 times as much, and the opening has been postponed from 2010 to 2016. Another example is Berlin's new airport, which is going to be years late and billions of euros over budget. The Energiewende is the mother of all megaprojects—"our man-on-the-moon project," as Steinmeier put it in 2015. The US Apollo Mission is remembered as a success, which it certainly was in the end. But it started off with Apollo 1, the vessel that caught fire on the launch pad, killing three astronauts and briefly casting doubt on the entire project. Megaprojects are not easy. The Energiewende might yet have its Apollo 1.

New Players Are Needed

How do you get utilities to close power plants that are working just fine in order to make space for renewable electricity? The challenge is financial, not technical, and the solution can only be political—like Germany's nuclear phaseout. The International Renewable Energy Agency (IRENA) says the "early retirement" of conventional capacity comes before the need for power storage.[11] We need to be asking questions about early retirement now. It requires a political breakthrough.

US author Bill McKibben says society should give utilities two options: "play a crucial role, or, at least, get out of the way."[12] The second option is the one pursued in Germany until recently. From England to the US, the debate focuses on what individuals (homeowners) can do, not what communities can do (note that Scotland is an exception; it has begun promoting community renewables). In Germany, communities and citizen energy cooperatives have told utilities, "step aside, we are

[11] International Renewable Energy Agency (IRENA). *REmap 2030. A Renewable Energy Raodmap.* Abu Dhabi: IRENA, 2014. Accessed February 4, 2016. http://www.irena.org/remap/IRENA_REmap_Report_June_2014.pdf.

[12] McKibben, Bill. "Power to the people." The New Yorker. 29 June 2015. Accessed 15 February 2016. http://www.newyorker.com/magazine/2015/06/29/power-to-the-people.

coming through" with the energy transition. In 2013, the British government did not seem to realize that citizens can build their own renewable energy projects; it merely welcomed their input when they said no to utility wind farms.[13] Germany allows its communities to say yes to wind farms—ones they built themselves.

Granted, there is a bright side to the US approach. Because Americans rely on their utilities to make the transition, some US utilities have more progressive business models (encouraging efficiency, for instance) and are ahead when it comes to things like smart meters. Why is German-style energy democracy not at the top of the agenda in America or England? Will other countries catch on to community renewable energy while the transition's window of opportunity is still open?

Relying on big conventional utilities for the transition is a dangerous tactic anyway. For instance, Florida Power & Light (FP&L) has been a major driver behind large wind projects across the US for many years now. In 2009, it was "America's no. 1 producer of wind power."[14] In 2014, it still claimed to be the "world's largest generator of renewable energy from wind and the sun."[15] But most of its investments in renewables were made in areas where it does not have competing assets: California, New Mexico, Nevada, New Jersey, and Ontario. At home in Florida, FP&L has recently joined the massive utility pushback against residential solar, helping ensure that the Sunshine State continues to get roughly 5/6th of its electricity from natural gas and coal—and less than 1 percent from solar. Citizens in Florida do not have the right to sell any electricity they generate, and firms offering third-party ownership—such as SolarCity—cannot do business in the state. "It's no secret we play an active role in public policy," an FP&L spokesperson is quoted in Rolling Stone.[16]

---

[13] Harvey, Fiona, and Peter Walker. "Residents to get more say over windfarms." The Guardian. 6 June 2013. Accessed February 4, 2016. http://www.theguardian.com/environment/2013/jun/06/residents-get-more-say-wind-farm.

[14] Gunther, Marc. "FPL: America's no. 1 wind power." Greenbiz. 21 September 2009. Accessed 15 February 2016. http://www.greenbiz.com/blog/2009/09/21/fpl-americas-no-1-wind-power.

[15] Via its subsidiary NextrEraEnergy, see website accessed on 15 February 2016. http://www.nexter-aenergy.com/company/factsheet.shtml.

[16] Dickinson, Tim. "The Koch brothers dirty war on solar power." Rolling Stone. 11 February 2016. Accessed 15 February 2016. http://www.rollingstone.com/politics/news/the-koch-brothers-dirty-war-on-solar-power-20160211?page=2

The lesson here should be clear: utilities often build renewables in new areas in order to enter new markets. Germany's Eon is behind a massive 782 MW wind farm in Roscoe, Texas—and also behind the world's largest offshore wind farm, the 630 MW London Array. Yet, the firm had only 213 MW of wind power capacity installed in all of Germany—its home market—at the end of 2014.[17] Germany's RWE and France's EDF are other examples of utilities that do an astonishing amount of business with renewables abroad relative to their investments in green electricity at home. Utilities are admittedly building giant wind and solar projects everywhere, but largely in competitor territory, not their own. And they see community projects built by citizens as a threat within their own turf. Corporate utilities don't want their own investments in renewables competing with their existing conventional energy assets, and they don't want yours either.

We may wake up a generation from now and realize that a window has closed—that corporations went renewable for everyone. Indeed, many Americans praise SolarCity, a firm listed on the stock exchange, for lowering the "soft cost" (the red tape) of installation for residential solar arrays. Yet, third-party ownership is not needed to bring down soft costs; the installed cost of solar in Germany, where citizen ownership is far more popular than corporate ownership, has historically been considerably lower than in the US and was 40 percent lower in 2014.[18] Germans prefer to own their own solar arrays, not have yet another listed corporation in between them and their solar roofs. A recent study found that third-party ownership—the SolarCity business model—hands most of the benefits to the solar middleman. Own the array yourself, and you get all the profits.[19]

---

[17] Eon homepage, access 15 February 2016. http://www.eon.com/en/sustainability/regional-activities/germany.html.

[18] Morris, Craig. "Solar twice as expensive in US as in Germany." Energy Transition. The German Energiewende. 11 May 2015. Accessed February 4, 2016. http://energytransition.de/2015/05/solar-twice-as-expensive-in-us-as-in-germany/.

[19] Farrell, John. "If You Can't Beat 'Em, Own 'Em—Utilities Muscle in to Rooftop Solar Market." Institute for Local Self-Reliance. 11 August 2015. Accessed February 4, 2016. http://ilsr.org/if-you-cant-beat-em-own-em-utilities-muscle-in-to-rooftop-solar-market/.

Most Americans cannot choose their power provider, whereas all Germans—indeed, all Europeans—have been able to for the past decade. Bill McKibben explains the US situation thus[20]:

Utilities are monopolies: since it would make no sense to have six sets of power poles and lines, utilities are granted exclusive rights to a territory. To keep the nation's utilities honest, they are typically regulated at the state level by a public-service commission that sets rates, evaluates performance, and enforces mandates, such as a requirement that a certain amount of power come from renewable sources.

Power lines may be a natural monopoly, but power plants are not. Europe has chosen to "unbundle" companies in the power sector; specifically, a firm that owns a power plant cannot own a high-voltage grid and vice versa. As our history explains, Germany even tried to do without a regulator for years, implementing one only when deregulation failed to work properly. The story shows that German policymakers are committed to market competition: power generators compete, and retail power providers compete. Furthermore, the power market is open to newcomers, especially those generating renewable energy.

Crucially, German utilities are not required to make any investments in renewable sources. In the US, Renewable Energy Portfolios often require utilities to provide a certain amount of green electricity (say, 33 percent by 2020 in California), and Renewable Obligation Certificates (ROCs) in the UK mainly benefited big players. In contrast, German policy incentivizes investments for everyone and forces grid operators to purchase the electricity. Germans are therefore free not only to pick their power provider but to make their own energy. You might say that the Germans have long enjoyed freedoms that Americans are just beginning to realize they lack.

---

[20] McKibben, Bill. "Power to the people." The New Yorker. 29 June 2015. Accessed 15 February 2016. http://www.newyorker.com/magazine/2015/06/29/power-to-the-people.

What Lessons Can We Draw From the Energiewende?

The history of the Energiewende reveals that there are brief windows of opportunity to change course. Once or twice a decade, German political activists and businesspeople were able to seize the moment and take a giant step forward. In-between these steps, they hunkered down so as not to lose ground. Partly, these giant steps were possible because utilities doubted for so long that solar and wind would not make much of a difference anyway. If you are looking for conclusions to draw from German experience, this insight will not help; utilities everywhere are very cognizant now of the fundamental threat that solar and wind pose to conventional business models. No utility today will repeat the mistake German utilities made and ignore renewables. But useful lessons can be drawn from the story of how Germans climbed through these temporary windows of opportunity. If you feel that, despite all your preparations for your endeavors, you are banging your head against a wall, maybe you are. Look for that open window.

The most important lesson to learn from Germany's grassroots Energiewende is one few outsiders expect: the energy transition represents a one-time window of opportunity to democratize the energy sector. Now that solar and wind are becoming competitive with fossil fuel (thanks partly to Germany's financial commitment while these renewable energy sources were still expensive), markets will increasingly push the transition. Up to a certain level, it will come one way or the other—in the hands of big business or with at least a partial shift in ownership that strengthens communities and provides greater competition on markets.

The history of the Energiewende clearly shows that it is a mistake to leave a fundamental transition up to those with vested interests. Large power companies have assets to defend. They will develop renewables at a pace that does not undercut those assets. Climate change mitigation cannot wait that long, but more is at stake. The year 2050 might be cleaner, but also less democratic.

The good news is that a low-carbon future is within reach. But the energy sector need not consist of a small number of large corporations. It can consist of numerous smaller firms truly competing for customers. It can also consist of more community groups and individual efforts.

Once our future new renewable energy infrastructure has been built—and it will be completed within a generation or two—it will not be dismantled just so we can have more competition, a level playing field between companies of different sizes, and more citizen involvement. When big projects go up, the companies behind them secure themselves a share of the market before newcomers can get involved.

The goal of this book is not to encourage people to start copying Germany's Energiewende. Rather, the German story provides a number of poignant lessons. You have the right to make your own energy. You have the right to do so profitably. The role of corporations can be smaller. Communities can be stronger. And a country can rally around a common goal. German energy policy is unique worldwide in an unusual respect: it has a brand, the Energiewende.

2

The Birth of a Movement: 1970s Protests for Democracy in Wyhl

Somewhere along German banks of the Rhine overlooking France lies one of the least visited monuments in the world. A modest stone at 246 km on the river is tucked away in some bushes next to a little used parking lot. Yet, when the monument was inaugurated in 2000, Germany's Undersecretary of the Environment attended. For this stone marks the spot where protesting citizens blocked construction of a nuclear plant for the first time in Germany—and in the world.

The campaign sparked a movement now known as the Energiewende. The protests in Wyhl inspired the first nonviolent direct-action group campaign against a planned nuclear plant in the US (in Seabrook, New Hampshire) in 1976, after two US activists, Randy Kehler and Betsy Corner, visited the German protest camp.

While some commemoration events are occasionally held at the monument, unsuspecting passers-by will hardly even notice the only three-foot-tall stone. Like the first solar panels and wind power generators built at the time, it seems quite homemade. Try to decipher the brief wording ("we said no"), and you may also have trouble, for it is written in the local dialect—instead of the standard German "wir haben nein gesagt," it reads "nai hämmer gsait" along with a date: 18 February 1975.

© The Editor(s) (if applicable) and The Author(s) 2016
C. Morris, A. Jungjohann, *Energy Democracy*,
DOI 10.1007/978-3-319-31891-2_2

It wasn't the first monument built. A much larger wooden structure was set up initially. Unknown parties—probably local proponents of the proposed nuclear plant—burned it down and threw the remaining charred pieces into a leading protester's yard.

Nuclear long divided the community and the country. The new monument may thus be an attempt to mark the grounds quietly. It is also robust enough to survive any attacks or attempts to remove it. Tapered toward the top, the stone is likely to slip out of a machine claw that grabs it from above. The stone's shape is thus a subtle nod to the "dragon's teeth"—triangular blocks of poured concrete to block the passage of military vehicles in war—used in later years to stop trains transporting nuclear waste.

Breisach—The First Site Proposed

Wyhl was not the first place proposed for the plant, nor was 18 February 1975 the start of the protest. In 1971, the government of the state of Baden-Württemberg announced that a nuclear plant would be built on the Rhine near Breisach, some 20 minutes south. Across the Rhine, the French began construction that year of a nuclear plant in Fessenheim, some 20 minutes further south again.[1]

Breisach is a historic town of some 15,000 inhabitants. Built atop a plateau, the town's medieval church with its two spires can be seen for miles around. You can also see the Fessenheim plant from the church when the weather is good.

Citizens of Breisach did not want the nuclear plant. The governor of Baden-Württemberg at the time was Hans Filbinger from Freiburg, the region's largest city some 20 minutes to the east. His state government wanted the plant. Filbinger expected power consumption to increase in

---

[1] Information about the history of Wyhl is partly taken from personal accounts with the authors. The following publication also proved to be a useful source of information: *Bürger gegen Kernkraftwerke*, Hans-Helmut Wüstenhagen, Rororo, 1975. Quotes are also taken from two video documentaries: (1) *Welcome to the Energiewende*, http://welcometotheenergiewende.blogspot.de/; and (2) *40 Jahre AKW-Widerstand: Wyhl*, SWR, https://www.youtube.com/watch?v=4ybq9MjL2fE.

his state by 70 percent, and four nuclear reactors were to be built in the 1970s alone.

Local politicians liked the idea of progress and jobs being created, as did many workers, but local winegrowers and farmers did not. Stretching some 60 km from the Black Forest in Germany to the Vosges Mountains in France, the Rhine Valley is the result of tectonic activity leaving behind a flat plain studded with a few volcanoes. The volcanoes have long been dormant, but moderate seismic activity is still common. The most recent major earthquake ravaged the Swiss city of Basel, some 20 min south of Fessenheim, back in 1356. Though no scientific measurements were possible then, estimates have put that quake at a magnitude of up to seven.

The locals in Breisgau were not concerned about earthquakes any more than the French were when building the two reactors in Fessenheim (with earthquake protection only up to a magnitude of 6.7). Rather, Breisach was a largely farming community, with plenty of orchards and vineyards along Tuni Hill and in Kaiserstuhl, a large dormant bed of volcanic rock covered by a type of soil called "loess," which is good for vineyards.

It may sound funny today, but the first history of events in Wyhl—published in 1975, when the movement was still underway—cites a flyer distributed by the protesters, which opened with the sentence: "one of the worst effects of the nuclear plant will be water vapor from the cooling towers."[2] Farmers feared that water vapor from the nuclear plant's cooling tower might change the local climate, detrimentally affecting their harvests. The German Weather Service estimated that local precipitation around an existing nuclear plant (Biblis in Hessen) had increased by 12 to 13 percent when the plant went into operation. Radioactivity was also mentioned at the time, though the concept was poorly understood among the general public (nuclear proponents would argue it still is today). When the plant was proposed for Breisach, protesters quickly organized a parade of 500 cars and tractors.[3]

---

[2] Wüstenhagen, Hans-Helmut, *Bürger gegen Kernkraftwerke: Wyhl, der Anfang?* Reinbek bei Hamburg: Rowohlt, 1975, p. 42.

[3] Wüstenhagen, p. 21

A local pharmacist and reserve army captain launched the campaign, collecting a whopping 65,000 signatures in the process.[4] Local politicians reacted by contacting Filbinger's office in Stuttgart. They asked if another location couldn't be found. One was—directly north of Kaiserstuhl in a village called Wyhl.

A Franco–German Alliance

The mayor of Wyhl thought the opportunity was too good to pass up. He contacted Stuttgart during the dispute in Breisach and asked if the project could be moved within his constituency. He did so without informing his citizens. At a meeting not open to the public, the city council offered to sell a plot of land suitable for a nuclear reactor to the state government. And so, on 19 July 1973 it was officially announced that the nuclear plant was to be built in Wyhl. Across the border in France, a buyer for the power was also found: Germany's *Chemiewerke München* was to build a new lead production plant at the same time after failing to find a community willing to have it in Germany.

Now, citizens on both sides of the Rhine reacted with alarm. Everyone felt that politicians out of touch with voters were turning rural communities into industrial landscapes. Jean-Jacques Rettig was a campaigner against the idea in France. "From Rotterdam to Basel, they wanted to connect industry along the entire Rhine," he remembers. In Germany, Siegfried Göppel, a mill owner outside Wyhl, was one of those spearheading the resistance. "The nuclear plants were just the beginning," he explains. "First, we need the energy. Then, industry will come. And the people who live here were to move to the foothills of the Black Forest in Germany or the Vosges Mountains in France"—in other words, 30 km away to the edge of the Rhine Valley.[5]

---

[4] Gensch, Goggo. "Wyhl? "Nai hämmer gsait." Der Widerstand gegen das Atomkraftwerk am Kaiserstuhl.": SWR-Geschichtsdokumentationen. 2013. Accessed February 4, 2016. http://www.swr.de/geschichte/wyhl-atomkraft-widerstand/-/id=100754/nid=100754/did=12047138/1x51213/index.html.

[5] Petite Planète. *Welcome to the Energiewende.* 2013. Accessed February 4, 2016. http://welcometotheenergiewende.blogspot.de/2013/10/full-movie.html.

For these rural communities, it seemed the upheaval never stopped. "This area was 80 % destroyed in World War II," Göppel remembers, "and people had been evacuated three times." What's more, the area had just undergone *Flurbereinigung*, a reallocation of property. Over generations, plots had increasingly been divided up, becoming smaller and smaller. As some people left farming, they would sell a plot to another farmer, but not necessarily the one with adjacent land. The result was a patchwork of disconnected farm plots. To make things worse, loess erodes easily. So during the difficult process of readjusting small plots to create single farms again, the government had provided funding to turn places like Tuni Hill into the terraces characteristic of the area today. The expensive terraces also allowed modern farming machines to be used. And now, everyone was to move away?

But the year was 1973, and the first oil crisis had struck. European countries turned their attention to domestic energy production. Nuclear appeared to be key; the fact that uranium needed to be imported was (and still is) downplayed. The next year, the French would launch their Messmer plan to build 170 nuclear plants by 2000. (Only a third were completed.) Worldwide, the 1970s were the decade of the greatest growth of nuclear power in the west. Germany was no exception. On the contrary, nuclear opponents demonstrated that Germany, Switzerland, and France were about to make the Upper Rhine the area with the largest concentration of nuclear plants in the world because each government was rushing to build so many reactors in the area—without consulting each other or coordinating plans.

The villagers seemed to have no chance resisting nuclear in the wake of the oil crisis. But then, something happened in France that changed everything—French citizens successfully blocked the lead production plant across the Rhine from Wyhl in Marckolsheim. Locals went to the construction site and squatted; German citizens joined them to ensure that the site was occupied day and night. This international, though still local, cooperation is all the more remarkable when we consider that, only a generation earlier, the people in these two regions had waged war against each other.

But environmental impacts do not respect political borders. The organization of local winegrowers calculated that lead oxide emissions from the German lead plant in France would have rained down on an area up to 15 km from the plant; the potential harvest losses were estimated at up to 15 million deutsche marks per year within a 5 km radius. "In preparation for the occupation of the construction site, we practiced being dragged away with the help of the local fire department," one French protester recollects. "The elderly sat in the holes dug into the ground for the first piles to be rammed in and refused to leave," Rettig recalls.[6] France had seen severe protests by long-haired students since 1968, but these folks were not hippies, and the police did not wish to be seen roughing up grandma and grandpa.[7]

When construction workers came to the site and saw staid citizens literally planted in the ground, many of them found the sight amusing. The mood was thus more relaxed than tense—and certainly far from the civil war atmosphere that was about to begin on the German side of the Rhine. A scary moment occurred on 20 September 1974, when Germans coming to relieve their French friends from their posts were prevented from entering France for six hours by border patrols. But the event did not escalate.[8]

Then, on 29 September 1974, Rettig received information leading him to believe that police were about to evacuate the site by force. The local church tower was turned into a lookout post, but the police did not come. Perhaps it was a tactic to test the courage of the protesters.

On 13 October 1974, 11 representatives on the city council of Marckolsheim stepped down because of the protests, forcing immediate elections. The citizen initiative opposing the lead plant simply declared itself a political party and won all 11 seats.[9]

The German firm that wanted the lead plant began calling protesters "anarchists and leftist extremists," foregrounding the completely inaccurate and disrespectful defamation of citizen protesters that would become commonplace in Germany. But the French government did not immediately react. Then, in mid-November, the French press incorrectly

---

[6] Goggo Gensch, 2013.

[7] Ibid.

[8] Wüstenhagen, p. 59.

[9] Wüstenhagen, p. 55.

reported that the government had withdrawn the construction permit; the protesters remained on the construction site and demanded evidence, which could not be provided. Perhaps the canard was also a ruse to get the occupiers to leave. Finally, on 25 February 1975—exactly a week after the date on the memorial stone in Wyhl—the French government put an end to the planned lead processing facility.

"Without the success of the French in Marckolsheim, the nuclear plant in Wyhl would have been built," Göppel believes today. It wasn't just the loss of a major power buyer that turned the tide. More than anything, the French had showed the Germans how to protest—and proved that civil disobedience on the brink of illegality could work.

The campaign now drew the French to protest in Germany. Culturally, the part of France in which Marckolsheim is located is called Alsace. The French word for German—*allemand*—is the proper name for Alemannic, the dialect spoken from Alsace in France, Baden in Germany, and large parts of Switzerland. In the twelfth century, the area produced some of the greatest literature written in Middle High German. Though Alsace was taken over by the French King Louis XIV in the mid-seventeenth century, rulers at the time did not care what language the peasantry spoke. It was not until Napoleon came to power that the idea of a modern republic was born—and with it, the requirement that citizens speak the government's language. From 1870 to 1945, the region also repeatedly passed between French and German hands. Today, young adults in Alsace are generally monolingual Francophones, but their parents who protested in Marckolsheim generally still spoke a dialect of German very similar to the one in Baden.

There were thus no communication problems between the Alsatians and the Badeners, and cooperation felt like overcoming an unnatural divide. Rettig remembers one Alsatian protester having a eureka moment: "Our parents shot at each other. But the next time Paris and Bonn [then still the capital of West Germany] have some stupid idea, we are not going to play along!"[10] This cross-border identity allowed protesters from France, Switzerland, and Germany to work together in a feeling of unity that had been overshadowed in 200 years of nationalism. The

---

[10] Petite Planéte, Welcome to the Energiewende, 2013.

Swiss began their own campaign against a nuclear plant to be built in Kaiseraugst, which was also blocked.

Protests Turn Violent

In Germany, men and women, young and old, all turned out—but Filbinger's government would have none of it. The protesters in Wyhl had tried the path taken in Breisach. They collected 90,000 signatures within seven months of the announced move of the nuclear plant from Breisgau to Wyhl.[11] But there were two big differences. First, their mayor was against them. And second, the nuclear plant had already been ordered for Breisach in the spring of 1974, and the delay cost money every day.

On 5 November 1974, the economics minister of the state of Baden-Württemberg announced that the permitting process had already begun. A busload of citizens from Kaiserstuhl went to the capital on 17 December 1974 to speak with the governor, but no member of his cabinet would meet with them, though a few parliamentarians did. The citizens were disappointed at the reception they were given; their bus was not allowed to park near the parliament building, and they were left waiting in the rain for quite a long time.[12]

In the end, the trip did no good anyway. The citizens wanted a referendum on the sale of the land to the nuclear firm. The mayor of Wyhl feared the nuclear opponents might win it, but the economics minister in the capital publicly said a referendum would not matter—he would exercise eminent domain and simply take the land if it went the wrong way.

Against this discouraging backdrop, a referendum was held on 12 January 1975 in Wyhl. In the months before the vote, the radio was full of pro-nuclear public service messages. Then, a public-awareness meeting was held. Experts invited by the government explained what nuclear plants were. Bernd Nössler remembers one particularly unconvincing presentation: "An engineer from the utility came, held up a tea pot and

---

[11] Goggo Gensch, 2013.
[12] Wüstenhagen, p. 61.

said, 'This is just like a nuclear plant. You pour water in, and steam comes out the top. And see, nothing else happens.'"[13]

But the protesters had already heard too many other opinions from their own experts to believe the teapot tale. Locals brought their own scientists to present opposing viewpoints at a public hearing in preparation for the referendum. Unfortunately, their experts were from out of town, and only citizens of Wyhl were allowed into the hearing. The protesters pointed out that the government's own experts and representatives were from even farther away, but to no avail. There were seven items on the agenda and 69 objections to the first item. The government presented its case for the first item, skipped all the objections, and attempted to move on to the second item. Pandemonium broke out. Clearly, the government wasn't interested in a debate; it had come to set the record straight. The activists and their experts were prevented from speaking; they left the hall out of protest. "These idiots must think we are idiots," one local expressed in dismay.[14]

The economics minister, who had attended the meeting, left in a rush as a crowd of protesters boycotting the meeting surrounded his car. To make a quick getaway, the driver smashed into two parked cars before driving across a field to get out of town. Although nearly 100 people saw what happened, police officers who witnessed the event refused to charge the minister's driver with hit and run.[15]

On 10 January 1975, Filbinger sent a letter to everyone in Wyhl praising the benefits of the proposed plant and asking for a vote of support in the upcoming referendum.[16] Two days later, the mayor stepped onto the upstairs balcony of City Hall to announce the result with a stone face: 692 votes against the nuclear plant (43 percent), 883 for it (55 percent). It was a clear victory for the mayor and the governor. And it did no good.

Theoretically, the land could now be sold to the nuclear plant builder. But technically, construction could not begin because citizens had filed suit against the permitting procedure. The Protestant church publicly

---

[13] Goggo Gensch, 2013
[14] Wüstenhagen, p. 61.
[15] Goggo Gensch, 2013.
[16] Ibid.

expressed its solidarity with the protest movement, calling for the discussion to be continued before construction began. It also criticized the "defamation and criminalization of nuclear opponents."[17]

The government had not only sent letters to citizens asking for support, but had also ensured that the media reported on the new nuclear plant with a positive spin. It used the police to intimidate protesters. In the months before the referendum, more than 100 people were interrogated and charged with disturbing the peace; if found guilty, they faced a maximum sentence of 13 years in jail. Telephones were tapped, homes were searched, and people were picked up for interrogation by police at their workplace as well—just to make sure that everyone knew they were criminals. On seeing how his congregation was being treated, one local pastor stated publicly, "there is no absolution on Ash Wednesday for what [the governor] is doing here in Wyhl."[18]

With the nuclear plant ordered, Filbinger gave the go-ahead for construction on 17 February 1975. It was time for a showdown, potentially a bloody one. That same day, bulldozers appeared on the construction site, and workers began cutting down the first trees. Women protesters wanted to go out to the site right away, but the men talked them out of it for the moment. Anyone setting foot on any part of the construction site, including parts of the forest that the machines had not yet gotten to, would be fined 200 deutsche marks.[19]

The next day, the first protest tents were set up, and the first shack was built along with a campfire. On 18 February 1975—the date on the memorial stone—180 opponents spent the night for the first time in what would become a nine-month occupation of the site. That first night, construction workers protected their equipment by putting a fence up around it. The firm explained that any delays caused by the protesters would result in damages of hundreds of thousands of marks per day. Governor Filbinger claimed that the protests were being "organized by extremists from all parts of Germany."[20]

---

[17] Wüstenhagen, p. 61.

[18] Wüstenhagen, p. 71.

[19] Wüstenhagen, p. 75.

[20] Wüstenhagen, p. 76.

Perhaps to prove that the resistance was not local, Filbinger ordered the police to forcefully remove the protesters two days later—and to focus on out-of-towners. Fifty-four people were arrested, but only eight of them were locals. Thirty-three were university students from nearby cities, and 12 were French people. A Dutch protester was also arrested. French protesters said they had never seen such brutality. People were dragged across the forest by their clothes and dumped into muddy puddles—in the freezing winter. Then, they were taken to police headquarters in their cold, soaking wet clothes for hours of interrogation.[21]

Bernd Nössler, the local baker, was up early as usual the morning when the police removed the first protesters. When he saw the riot control vehicles driving past his bakery, he quickly grabbed the phone to call everyone he knew so the camp could be warned. "But the phone was dead," he says, his voice still full of shock today. "The only call you could make was to the hospital. It was like living in a dictatorship." This news further incensed the public when it became known afterward.[22]

Police showed up with German shepherds. Some protesters got bitten. The community reacted with *Bauernschläue* (the cleverness of farmers), as winegrower Albert Helbing recalls: "We asked around in the community to see if anyone had a dog in heat. One was found. And when that dog reached the site, all those months of training the police dogs proved useless."[23]

Two local Christian Democrat politicians—the party of the governor—were dismayed at the police brutality and expressed their disapproval in writing to Stuttgart: "The state government needs to finally realize that the overwhelming majority of the population is behind the protest."[24]

Three days later—a Sunday—the protesters had indeed not left. The police set up a three-meter-tall barrier of razor wire and moved in with water cannons. The police chief from a nearby town, who usually commanded around 100 officers, was given 1000. Police boats patrolled the

---

[21] Wüstenhagen, p. 79.
[22] Goggo Gensch, 2013.
[23] Ibid.
[24] Wüstenhagen, p. 79.

Rhine, helicopters flew overhead, and a bus full of criminal prosecutors appeared on the scene. But it was a puny show of force against the 28,000 people (police estimate) who had joined the protest by then.[25]

The question was no longer whether a nuclear plant should be built. The public had seen police brutality against peaceful protesters, including women and the elderly. The show of force backfired miserably. The news drew nationwide attention and caused general outrage. From that point on, Wyhl was no longer just about industrializing a farming community, but also about governmental authoritarianism.

Increasingly, the social stress on police officers rose. A middle-aged woman among the protesters recognized her nephew in the police line and began shouting, "shame on you!" Father Gerhard Richter, a local pastor, went through the crowd on both sides of the barbed wire pleading for nonviolence. (Protesters had thrown stones at police officers, for instance.) Whenever he recognized someone from his congregation on either side of the razor wire, he asked them if they planned to come to church next Sunday—and if so, to make sure they did nothing between now and then to cause strife in the church.[26]

Since citizens couldn't rely on telephones, the villagers illegally listened in on police radio with homemade devices. They learned of a large police force waiting at a nearby rest area on the autobahn to clear the still occupied site. The villagers heard the officers pleading not to be sent in. At time, the pleas bordered on refusal.[27]

Shortly thereafter, the police chief went public. In a local newspaper, he published a long letter explaining the growing reluctance of police officers to arrest protesters. He made it clear that he was "not an opponent of nuclear power plants. On the contrary, I know that we need them." His problem was with the government using his police force for its own devices[28]:

---

[25] Wüstenhagen, p. 80.
[26] Personal communication with the authors.
[27] Wüstenhagen, p. 88.
[28] Wüstenhagen, pp. 87–88.

Most of us knew a bit about the motives of the occupiers, and most of these police officers were no longer certain that we were on the right side...Few of the protesters were offensive in their statements. Most of the demonstrators simply stated their case...I was not the only officer who learned a lot.

The police learned that both the governor and the economics minister were on the local utility's supervisory board, thus using their offices to do the company a favor. "We worked overtime on behalf of the state government—or perhaps I should say, on behalf of the utility," he wrote. He also learned that the governor was himself breaking the law by not respecting a court order to wait until its review of the construction permit was finished. The government had not told him any of this, and it was not easy to find this information in the media either. And then there was the referendum, which he learned was a complete farce: "If you do not voluntarily vote yes and sell the land, we will take it from you!" he summed up voters' options.[29]

Eventually, the police force backed down. Farmers then cut through the undefended razor wire, figured out how to drive the construction vehicles (which were not so different from their farming equipment), and emptied the site. But the government and the nuclear firm continued their shenanigans. When mill owner Siegfried Göppel made sure that the construction machinery was returned to its rightful owner and not kept as ransom (as some protesters proposed), he and 10 of his friends were sued for 150,000 marks because the machines they tried to protect and return had allegedly been damaged.[30]

Negotiations Replace Violence

With the police gone, the protesters settled in for an indefinite occupation of the site. The local hunting club had forked over its budget to have expert opinions written up—the ones that the government refused to discuss at public hearings. Now, legal expertise was needed to defend

---

[29] Wüstenhagen, p. 88.
[30] Wüstenhagen, p. 90.

protesters. A "friendship house" with space for up to 500 people was built to serve as a meeting place and as a retreat, so people didn't have to stand outdoors all the time. It included a kitchen and a restroom (the latter, basically a hole in the ground).[31]

The goal of meetings at the friendship house was not only to discuss expert reports and court cases. Events were also held to draw more people out to the protest regularly. A series of lectures led to the epithet *Volkshochschule*, the name for German adult education centers. Songs were sung and slides of Canada shown, just for entertainment. One of the songs had the following couplet:

> *"Atom und Blei,*
> *Und das Leben ist vorbei"*
> *("Nuclear and lead,*
> *and there's no life ahead.")*

Increasingly, the camp drew international attention. "Native Americans and Buddhists from Asia visited us as well," Rettig remembers.[32] And when the two American anti-nuclear activists from New England visited, they brought the documentary "Lovejoy's Nuclear War" with them, which tells the tale of a lone Massachusetts farmer protesting against a nuclear plant to be built.[33]

The adult education classes also offered a forum for the first nuclear experts. One prominent speaker was German-born US anti-nuclear activist Ernest Sternglass, who had already made a name for himself in the US anti-nuclear movement. Here, months after the protests in Wyhl had begun, many locals first learned of the risks of radioactivity. The government's experts initially viewed these laypeople with disdain for not knowing what "ions" were. Slowly, that was changing, but it wasn't the government that was educating the public. People were educating themselves. A learning curve had begun.

---

[31] Wüstenhagen, p. 91.
[32] Goggo Gensch, 2013.
[33] Personal communication with the authors.

In the 1960s, uranium was to be mined at Menzenschwand in the Black Forest, some 60 minutes from Wyhl. There, supporters argued partly that radon baths could be created in the process, making a radioactive spa area out of the sleepy region (radon is a decay product of uranium once falsely held to have health benefits). Opponents did not argue, however, that radon is in fact a major cause of cancer (as we now know); rather, they objected that the project would not only bring spa tourism, but also a mining industry to a rural community that basically enjoyed its peace and quiet—just as in Wyhl.

The *Volkshochschule* not only imparted expert knowledge about radiation to laypeople, but also brought local experts together. The lawsuits had made it clear that independent scientific expertise was needed. A group of these researchers founded Germany's Öko-Institut in the late 1970s in nearby Freiburg and Darmstadt. One of their first publications was the book entitled *Energiewende* from 1980.

Governor Filbinger continued to call the protesters "communists," but the nation saw images of middle-aged farmers—men and women, generations within a family line—on tractors. These folks were not calling for workers of the world to unite. They wanted arrogant government and big business out of their quiet rural lives.

The state government realized that it had to change its strategy and negotiate. Governmental utility representatives met with protesters under the direction of Father Gerhard Richter, the pastor who had helped prevent violence during the scariest moments. Richter was not originally a nuclear protester. In 1971, he welcomed the new nuclear plant a few hours north in Obrigheim as a clean alternative to fossil energy.[34] But Wyhl was different. "I was concerned about the way people who call this place home were being treated. We faced an arrogant economic and political power. In my opinion, it was more interested in profits than in serving people's needs."[35]

On 31 January 1976, the Offenburg Agreement Richter helped negotiate was signed. All charges were dropped against citizens for damages,

---

[34] Kiefer, Gerhard. "Günter Richter: die Stimme der Anderen." Badische Zeitung. 26 July 2014. Accessed February 4, 2016. https://www.badische-zeitung.de/suedwest-1/guenter-richter-die-stimme-der-anderenDOUBLEHYPHEN87992485.html.

[35] Petite Planéte, Welcome to the Energiewende, 2013.

and all fines were nullified. The government agreed to have additional independent reports conducted on controversial issues. And the state would wait until the public's concerns had been addressed before continuing construction—though it essentially had the leeway to declare that point had been reached whenever it saw fit.

In return, citizens agreed to limit their resistance to political means and courts. The occupation was over. This stipulation almost toppled the agreement when it was made public. In the end, however, there was a consensus that the site did not need to be occupied if no construction work was done, which was easy enough to verify.

From then on, legal experts and engineers fought things out. The court case quickly revolved around the safety of the reactor. The lower court in Freiburg ruled in favor of the protesters in 1977, but the appeals court in Mannheim overruled that decision in 1982. By that time, however, the plant had been postponed for six years already. Power demand had remained basically flat. The lights had not gone out, as Governor Filbinger promised they would if the nuclear plant was not built. Filbinger himself had since been replaced as governor by Lothar Späth, a much less authoritarian leader. He had already been involved in negotiations with local villagers on behalf of the state government in the 1970s, initially in secret; it was the first time the German government had ever negotiated directly with its citizens. Finally, the accident at Three Mile Island had also occurred in 1979, further casting doubt on the safety of nuclear power.

When the court ruled in favor of the nuclear plant in 1982, protesters from France, Switzerland, and Germany poured out to the construction site once again. From a makeshift stage with loudspeakers, Swiss campaign leader Martin Dierle told the crowd of thousands to be prepared for the worst brutality yet[36]:

> Friends, this is going to be very difficult. There will be no romantic evenings at the friendship house. We are not going to be sitting around nipping at our wine while we watch the moon go down over the mountains.

---

[36] Goggo Gensch, 2013.

The protesters stayed despite Dierle's warnings, but Governor Späth had no interest in sending in the troops. That week in Kaiserstuhl, Germany took a step out of its authoritarian past, and Späth was one of a long succession of modern political leaders lasting up to today who respect the public.

In short, the proponents of nuclear had won the court case but lost the argument. The reactor clearly was not needed, and the public had not changed its opinion about safety. The utility bought the property after winning the appeal in Mannheim, but the state government showed no interest in moving forward with the project. After more than a decade in limbo, the project was officially canceled in 1994.

The Violent Conflict Continues in Brokdorf and Elsewhere[37]

The success in Wyhl can be overstated. Similar protests were held at the same time against a nuclear plant at Brokdorf in northern Germany. In 1977, West Germany also began investigating Gorleben, a town on the border to then communist East Germany, as a final repository for nuclear waste. There, protesters also set up camp, but the government stuck to its plans until 2012. As of 2015, the search for a final repository in Germany was completely up in the air.

Protesters also managed to prevent construction of a nuclear waste reprocessing facility in Wackersdorf starting in the mid-1980s, but they failed to prevent two other reactors from going into operation, one in Grohnde and another in Brokdorf. At the latter, the violence reached greater proportions and lasted longer than in Wyhl.

---

[37] For more information on Brokdorf, see Link, Rainer. "Der Brokdorf-Komplex: Kampfhandlung in der Wilster Marsch." Deutschlandfunk. Hörspiel/Hintergrund Kultur. Redaktion: Hermann Theißen. Sendung: 29 January 2013. Accessed February 4, 2016. http://www.deutschlandfunk.de/der-brokdorf-komplex-kampfhandlungen-in-der-wilster-marsch.media.eff51ca73c52ecc8cf-25848762da380d.pdf and Hubert, Antje. *Das Ding am Deich: Vom Widerstand gegen ein Atomkraftwerk*. DVD documentary. 2012. Accessed February 4, 2016. http://www.dingamdeich.de/2-0-news.html.

In Brokdorf, conclusions were drawn from Wyhl on both sides of the issue. When the construction permit was granted in Brokdorf in late October 1976, trucks first moved in at 2:00 AM to begin setting up barbed wire around the perimeter. Protesters were not going to be allowed to squat. Inside, police dogs were everywhere. Nonetheless, it didn't take long for protesters to get into the grounds. "What you need is some carpet," some more experienced protesters from nearby Hamburg told locals just days later. Once the carpets had gone over the barbed wire, passage was relatively smooth sailing.

The police responded with helicopters and teargas. The experienced protesters told the others to start bringing swimming goggles to help against the teargas. But there wasn't much anyone could do against helicopters. And as in Wyhl, telephones were tapped.

When the confrontation became violent, politicians pleaded with leaders of the protest to file suit. Once the matter had gone to court, construction could be briefly suspended without anyone losing face. The court ruled in favor of the protesters, arguing that the nuclear plant operator did not know what to do with the waste—but politicians were prepared for this ruling and had another card up their sleeve. They changed the law to read that a repository was not needed as long as investigations into a final repository were being conducted. Construction could now continue, and there was no basis for a legal appeal.

On 28 February 1981, an estimated 100,000 people came out to protest renewed continuation of construction. Police set up roadblocks miles away to prevent, as they publicly explained, large guns from being brought in. "We brought teabags," one protester remembers, "and toilet paper." The campaign against the nuclear plant in Brokdorf failed; the reactor went online in October 1986, just months after the nuclear accident in Chernobyl.

Germany continued permitting new nuclear plants until 1982, when construction began on the last one, which went into operation in 1988 (Neckarwestheim II). But the banking sector largely abandoned nuclear in the 1970s.[38] Aside from France, nuclear plant construction hit a wall

---

[38] See "Nuclear power: the crisis in Europe and Japan," International Business. In Business Week. 25 December 1978, p. 44.

in Western countries in the 1980s based on cost. Protests against nuclear power nonetheless played a role by discouraging governments from subsidizing nuclear to an extent that would have suited bankers.

Citizens Fighting an Authoritarian Government

In April 2013, locals came together in Wyhl to celebrate some of the most prominent protesters from the 1970s. Father Richter and Jean-Jacques Rettig received an award for their commitment to the nonviolent protests. The ceremony took place during the town's open door day for businesses to recruit young workers. Youth unemployment is a mere 7 percent in the area (with general unemployment below 6 percent overall that year in Germany). Wyhl is doing well. But a rift remains.[39]

The mayor who tried to bring nuclear into his community remained in office and was reelected until 1992. As the plebiscite showed, there was a majority in favor of his plan within the village itself, which has a rather large population of workers who commute. In this respect, Wyhl is a bit of an exception; most other nearby villages consist predominantly of farmers and winegrowers—the group that initially opposed the nuclear plant the most.

In light of the referendum's quite clear outcome (55 to 43 percent), what sense does it make to depict the resistance in Kaiserstuhl as a grassroots democratic movement? Granted, the majority did not prevail. Then again, we rightly consider "elections" in dictatorships like North Korea to be undemocratic; voters simply have no real option. When the majority always gets its way, political scientists speak of the "tyranny of the majority"; democracy is about negotiating compromises, not giving majorities what they want all the time. In the case of Wyhl, citizens were intimidated and critics of the government not allowed to speak at public hearings. There was no real option to the nuclear plant. One of the protesters quite succinctly expressed a sentiment about the referendum that will sound familiar to citizens today: "money won."

---

[39] Petite Planéte, Welcome to the Energiewende, 2013.

The referendum was restricted to the community of Wyhl, which owned the land in question; neighboring communities did not have any input. Tens of thousands of people came to protest, and scores of thousands had signed petitions against the planned nuclear plant. Hundreds of thousands in neighboring communities would have been affected. But none of that mattered when a mere 883 people from Wyhl said yes to the idea.

The plebiscite thus revealed a crisis of democracy; citizens in Wyhl and the surrounding area overwhelmingly did not feel the process was legitimate. Their willingness to risk physical harm was courageous, but it could have gone wrong, as the example of Brokdorf illustrates. Nonviolent civil disobedience is rightly held to be an honorable form of protest, but it is only recommendable and effective if authorities are willing to refrain from an incommensurate use of force.

Germans largely drew lessons about the role of civil society and democracy—and the limits on state authoritarianism—from events in Wyhl and Brokdorf. It was not just about nuclear, and perhaps it was not primarily about nuclear.

One of the coauthors of the "Energiewende" study from 1980 stated in an interview recently that "our work did away with the claimed monopoly on technological competence that the energy industry used to assert: that there were no technically feasible alternatives to its ideas."[40] Wyhl was thus the beginning of the vast landscape of independent research institutes Germany now has.

German historians believe that Wyhl and other protests of that decade "turned former subjects into active citizens."[41] Some locals saw themselves in the tradition of protesters from the Farmers Wars from around 1500. The German public no longer accepts expert opinions at face value, but they also understand that they must educate themselves if they want to object. As Austrian Christine Lins, head of global renewables organiza-

---

[40] Morris, Craig. "Efficiency lacks a loud lobby": An interview with Florentin Krause." Energy Transition. The German Energiewende. 17 April 2013. Accessed February 4, 2016. http://energy-transition.de/2013/04/an-interview-with-florentin-krause/.

[41] Engels, Jens I. "Geschichte und Heimat der Widerstand gegen das Kernkraftwerk Wyhl." In *Wahrnehmung, Bewusstsein, Identifikation: Umweltprobleme und Umweltschutz als Triebfedern regionaler Entwicklung.* Edited by Kerstin Kretschmer, 103–30. Freiberg, 2003.

tion REN21, put it, the result is a uniquely informed public: "Where else but Germany can you talk about renewables with taxi drivers?" German officials also respect the knowledge of citizens more these days than they did in the 1970s. Manfred Konukiewitz of the German Development Ministry says that "40 million of the 80 million people in Germany are now energy experts. You hear people talking knowledgably about the energy transition in pubs."[42]

Without tax revenue from the nuclear reactor, Wyhl never got the indoor swimming pool that proponents of the nuclear plant had promised, but the town remained a thriving community with an identity. It seems that Germans are less interested in making a quick buck than in living a good life, which requires community. The Germans are not interested in the kind of boom towns going up in the US shale boom, such as in North Dakota, where there are "far, far too many men," according to one account.[43] There, for a lack of affordable housing, people sleep in cars. Entertainment mainly consists in getting drunk and going to strip joints. And the boom in a particular town "could easily last another 20 years"—a time frame that German villagers would consider too short to sacrifice the communities they call home for and simultaneously too long to be overrun by out-of-towners who can't behave themselves.

The Kaiserstuhlers might be criticized for being uninformed country bumpkins and NIMBYists who only care about something when it affects their backyards. This depiction is unfair, however, in light of the national movement that grew out of the protest; likewise, these laypeople educated themselves in order to speak eye to eye with experts. And then there is the saying, "think globally, act locally"—which the Kaiserstuhlers did. They didn't just oppose nuclear at home; after they learned more

[42] Morris, Craig. "REN21 releases new Global Statis Report." Renewables International. The Magazine. 28 November 2012. Accessed February 4, 2016. http://www.renewablesinternational. net/ren21-releases-new-global-status-report/150/537/58961/.

[43] Gottesdiener, Laura. "I Worked in a Strip Club in a North Dakota Fracking Boomtown." Motherjones. 14 October 2014. Accessed February 4, 2016. http://www.motherjones.com/environment/2014/10/inside-north-dakotas-crazy-oil-boom. Also see Sargent, Jonah, James Christenson, Eliot Popko, and Lewis Wolcox. "Running on Fumes in North Dakota." New York Times. Op-Docs: Season 3. 14 January 2014. Accessed February 4, 2016. http://www.nytimes.com/video/opinion/100000002648361/running-on-fumes-in-north-dakota.html.

about the technology, the slogan on posters read, "No nuclear plant in Wyhl—or anywhere!"

Most importantly, the struggle of these rural communities drew the attention of other parts of society. As different sectors of the public joined the movement, it gradually encompassed all of society as it went national. And the general public realized that protests alone were not enough; citizens would have to start coming up with solutions—because the experts had financial incentives not to find them.

3

Fledgling Wind Power: The Folly of Innovation Without Deployment

On 17 October 1983, the winds were blowing hard along the northwestern coast of Germany just an hour outside of Hamburg. In fact, they were blowing at 22 m/s. Modern wind turbines can handle such gusts. Still, 22 m/s represents strong gale force winds of around 50 mph.[1]

A reasonable person might conclude that such conditions are not ideal for the testing of a brand new wind turbine, certainly not one 10 times larger than the prototype it was based on. There was no dearth of reasonable people on hand that day either. The engineers who built the unit were there, but so were top governmental officials who had signed the research funding for the project. Executives from RWE, a utility entrusted with the project, were also on hand. And they had driven a long way from their headquarters in Germany's coal heartland to see the fruits of their labor. The engineers would have been willing to wait for calmer conditions, but the executives were busy people, and the project was already a year behind schedule.

---

[1] Oelker and Hinsch, p. 53.

© The Editor(s) (if applicable) and The Author(s) 2016
C. Morris, A. Jungjohann, *Energy Democracy*,
DOI 10.1007/978-3-319-31891-2_3

The signal was given to let the wind turbine's rotor start spinning in the wind. It was not hooked up to the grid, so the purpose was to demonstrate proper mechanical functioning. Without a grid connection, however, a bit of resistance was missing. It was a disaster in the making: strong winds and no load to push.

The turbine started to turn and quickly reached its maximum rotation speed, setting off the emergency brakes. The functional demonstration had become a crash test. The unit was designed to undergo only 35 such emergency stops over its entire service life, and it was inaugurated with one.

Why the rush? Perhaps the experts from the conventional energy sector simply were not used to waiting for better weather when coal and nuclear plants were inaugurated. But something else was at work as well, for it is hard to imagine German engineers opening a large hydropower facility during a flood.

No, the executives had not come to test the turbine. They had come to break it. On 28 February 1982, some 18 months earlier, an RWE executive stated at a board meeting (later leaked to a newspaper), "We need Growian to prove that wind power won't work."[2]

This giant prototype was called Growian. The name itself is revealing, for it rhymes with the German word *Grobian*, meaning an uncouth person. "Growian" actually stands for "*gr*osse *Wind*an*lage*," or "big wind turbine," but the clearly negative connotation will not have escaped the people who named the machine they were building in order to prove that "it wouldn't work." In the Renaissance book *Ship of Fools* that invented the name, Saint Grobian is the (fictional) patron saint of those who lack tact and good manners. Perhaps the decision-makers wished to plant in people's minds an association with something unpleasant from the very beginning.

---

[2] Oelker and Hinsch, p. 51.

Early German Wind Power Generators

The large group of engineers was aghast at the sabotage. One of them was Erich Hau. Originally an engineer from the aviation sector, he was asked to join the project by German mechanical engineering firm MAN, which was to build Growian in a research project completely funded by the German government. RWE was involved to help handle grid and other utility issues, along with the nuclear research center in Jülich. But it was the local utility from Hamburg, HEW, that managed the project overall, even though the firm had made it clear that it was not interested. After the government ensured that the firms would only have to cover just under 5 percent of the initial budget, with all overruns also paid by the government, everyone eventually said yes.[3]

As a German manufacturer of utility vehicles, MAN had no experience with wind power. Indeed, it wasn't particularly interested in the project either, but who can say no to a government handout? One of the company's executives explained his commitment to a later project thus: "If we get a research and development contract with 100 percent funding, of course we do it."[4]

To get Growian going, MAN first needed some expertise. In the 1970s, when the project began, Germany had little, but company executives figured that aviation must be close enough. After all, it was wind, right? Hau was therefore hired, but he remained realistic: "We didn't know anything about wind power at the time." He thought it was "crazy" to be asked to build the world's largest wind turbine "and we have never even seen a windmill to begin with."[5]

The first thing Hau therefore wanted to do was have a look at one— but where? Over in Stuttgart lived one of the world's leading wind power experts: Ulrich Hütter. He had published his dissertation at the University of Vienna in 1942 on the "Fundamentals of wind power plants." In the 1930s, he and his brother had experimented with gliders, with some success. But Hütter was retired by the time Growian came around.

---

[3] Oelker and Hinsch, p. 50.
[4] Oelker and Hinsch, p. 55.
[5] Oelker and Hinsch, p. 46.

In his dissertation, Hütter investigated the Betz Limit. The law of physics drawn up by and named after Albert Betz put the theoretical maximum efficiency of a wind turbine at 59.3 percent—in 1919, when wind generators were crude machines nailed together by farmers out of scrap metal and boards.

As a constructor of early wind generators, Hütter set out to test the Betz limit, so he focused largely on blades and aerodynamics. Hütter began making wind turbines in the late 1940s for a midsize firm outside Stuttgart called Allgaier. The firm hired him after seeing his single-blade turbine from 1946 and a later three-blade model with an output of 1.3 kW. It was used on a chicken farm.

Allgaier had bigger plans with Hütter. His first model for the company still only had an output of 1.3 kW and a rotor diameter of 8 m, but that experience was immediately used to build a 7.2 kW turbine with three blades 11.28 m across. It seemed like a good size for farms, and it would have been used mainly as a source of heat and for mechanical energy, such as pumping water; synchronizing the direct current that the generator produced with the alternating current from the grid posed a whole new set of obstacles that an aerodynamics expert like Hütter was ill-equipped to address.

This particular model was then marketed as the WE 10. Around 200 units were sold both in Germany and abroad. For instance, eight of them were used to drain rainwater from behind a levy that tended to fill up. In 1952, he was even asked to connect one to the grid and see what would happen. When the experiment basically worked, an entirely new market seemed to have opened up.[6]

In 1953, a competition was then held by the Chamber of Commerce in Stuttgart for the construction of a 100-kW wind turbine. Hütter's WE 10 had been delivered with outputs ranging from 6 to 10 kW. The new goal was an order of magnitude higher.

Hütter won the competition. Another participant had submitted a design with a single blade, a model Hütter had already tried. At the time, no one could say how many blades a wind turbine should have. A single blade saved on material, and hence cost, but it turned out to have fatal

---

drawbacks. First, it had to rotate much faster to get the same energy out of the "swept area" (essentially, the circle that the rotating blades pass through). Initially, it was (correctly) thought that faster blade rotation would generate more power. But the greater speed increased noise significantly and also added to wear and tear. The uneven load on the hub from the single blade also proved troublesome. Indeed, this imbalance is a main reason why you may never see a single-blade wind turbine in your life, but you may have a similar design in your pocket right now. Our cell phones vibrate using a piece of metal on one side of a tiny rotating shaft.

In the competition, Hütter beat the single-blade design not with his three-blade model, but with a two-blade rotor he wanted to try out. This option also uses less material than a three-blade design and is not as noisy as a single-blade unit, but it has its own special drawback: the blades are arranged like wings, but instead of flapping in sync, they teeter. Wind velocities increase with height, so a blade is exposed to faster winds when it is on top than when it is at the bottom. Hütter wanted to fix the problem with a "teetering hub," as the solution is known today in the world of helicopters (it is practically unknown now in the wind sector).

Still, 100 kW was a big project—too big, in fact, for Allgaier. So the firm got all the big names from mechanical engineering in the region to participate. Voith made the gearbox, Mannesmann the tower, Porsche the nacelle (the "chassis" on top of the tower that houses the gearbox and generator), and AEG the generator and electronics. The new turbine would have a wingspan of 34 m, three times longer than the WE 10.

In late 1957, the W-34 (the number generally refers to the rotor diameter) went into operation. Each 17-meter-long rotor blade was reinforced with fiberglass—a novel technology at the time. In several respects, this turbine was a success. It came closer to the Betz limit than any previous model. The fiberglass blades were a breakthrough and set the stage for blade development decades later. But for the time being, there would be no further development. The turbine was Hütter's last major project.[7]

In 1959, he became a professor of aviation construction at the Technical University of Stuttgart. That year, Allgaier backed out of the wind power sector altogether. In 1968, the W-34 was dismantled, and only one of its

---

[7] Oelker and Hinsch, p. 23.

blades was preserved at Hütter's university. Near the end of his career, NASA invited him to come to the US as an advisor. In the wake of the 1973 oil crisis, the US had launched a Federal Wind Energy Program—and they wanted to learn as much about the W-34 as possible. The machine no longer existed as such, but the technical drawings were found and sold to NASA for US$55,000. But Hütter had seen enough of the stop-and-go wind sector and was not interested in working closely with NASA: "I'm not going back into the wind power kindergarten," he told an associate.[8]

Hütter was also asked to do some advising in Germany. In 1974, he was summoned to the Research Ministry in Bonn to talk to some people in the department of "non-fossil, non-nuclear energy." (Later, German renewable energy advocate and parliamentarian Hermann Scheer would say of that department's name: "the double negation really said it all.")[9] The ministry wanted to start the program that would produce Growian, and Hütter was the country's biggest expert on wind power. He was asked how big a wind turbine could be.

The Research Ministry essentially needed a machine that would convince critics within the conventional energy sector. Size was everything. Based on the W-34, Hütter proposed a swept area of 1 ha, which would require a rotor blade diameter of 112.8 m in length—3½ times longer. He then went out on a limb and said he thought 3 MW was feasible—30 times the size of his W-34. "What, so little?" the official responded.[10]

Though the ministry official was not impressed, Hütter was Germany's best man for wind power. So Growian was built based on the W-34, with a teetering hub and two blades.

Growian—A Wind Turbine Made Big to Fail

The turbine was not only big in relation to previous competitors, but also relative to other things that had been put up in the air. The tower itself was just over 100 m tall, putting the hub height at 102 m. It was a size that the wind sector would not reach in serial production until 2002.

---

[8] Oelker and Hinsch, p. 23.
[9] Personal communication.
[10] Bauer and Lobe, p. 45.

The problem was that, in the early 1980s, no construction crane was available for such heights. The tallest one reached up 80 m. The obvious solution was to lower the tower, but officials from the Research Ministry would have nothing of it. The engineers now faced quite a dilemma. How could they hoist a 400-ton nacelle into the air without a crane?

Their solution was similar to the way giant church bells were pulled up to the top of spires centuries ago. In the case of Growian, the tower itself was to serve as the crane, and everything in the nacelle—in particular, the gearbox and generator—would have to go up from inside the tower.

This solution also had a fatal flaw—it meant that all parts would have to be small enough to fit within the tower, with larger sections welded together at the top. Germanic Lloyd, which had been asked to monitor the project, immediately expressed its concerns. The firm set forth standards in the shipbuilding industry but had no experience with wind turbines (no underwriters did). Still, welding seams were known to be a problem for ships, especially under changing loads.[11]

In the end, a small mathematical mistake led to a miscalculation of the peak load, so the welding seams proved to be the weak point in the system despite all the attention paid to the issue. As Hau remembers with regret, "The mistake that caused Growian to break so soon was a simple one that would've been easy to avoid."[12]

In 1984, Growian was connected to the grid, and despite the improper welding seams, it managed to withstand 38 emergency stops over its short life, three more than it was designed to withstand. In the end, though, it ran for a mere 420 operating hours, roughly "one for every ton that the nacelle weighs," as an expert from Germanic Lloyd put it.[13] Little remains of the giant turbine today aside from one of its blades. It can be found pointing straight up in the air atop an automobile museum in the southern German town of Sinsheim.

The engineers had not given up, however. The research leading up to Growian had led to other products as well. Like NASA, the Germans first set out to look for one of Hütter's old designs. Only one blade of the

---

[11] Oelker and Hinsch, p. 52.
[12] Oelker and Hinsch, p. 52.
[13] Oelker and Hinsch, p. 53.

W-34 was still around, and none of the WE 10s from the 1950s was still in operation—but one of them was found lying around "in pieces in a shed." Erich Hau and his colleagues put it back together and installed it on the roof of the MAN building to see how it worked. But when corporate management at the truck-building firm saw the three-blade turbine, it reminded them of the Mercedes emblem, and they ordered it taken down and relocated.[14]

By 1979, Hau and his team had come up with a prototype based on the WE 10 called Aeroman. With a capacity of 10 kW, it was essentially just a rebuild of the WE 10, not a step in the direction of Growian in terms of size. Nonetheless, MAN sales staff jumped on the opportunity to sell it when it proved to work more or less. The engineers working on the project were not happy; after all, the unit was still just a prototype. Nonetheless, one was sold to New Zealand, with others shipped to Australia and Korea in 1980. "It was crazy, really ridiculous," one team member commented.[15]

A slightly larger version of the Aeroman indicated that incremental progress would lead to greater success than increasing turbine sizes by leaps and bounds, as attempted with Growian. In the early 1980s, around the time that Growian was failing, a 30-kW version of the Aeroman became something of a hot export item, with around 300 sold to the Tehachapi wind farm in southern California alone. "The product had a good reputation among Tehachapi operators," remembers US wind power expert Paul Gipe, who worked for one of the firms in Tehachapi during that era.[16] When policy support was withdrawn in California in 1984, however, the state's wind market collapsed, sending the fledgling global market into a tailspin. Numerous German and Danish firms went bankrupt.

If incremental progress was the ticket, stable policy support was clearly needed. The Germans and the Danes both learned this lesson, creating a reliable policy environment in the 1990s, and both were wind power giants by 2000. In contrast, the US wind power sector went nowhere

---

[14] Oelker and Hinsch, p. 46.

[15] Oelker and Hinsch, p. 47.

[16] Gipe, Paul. "Photos of Aeroman Wind Turbines." Wind Works. 5 October 2013. Accessed February 5, 2016. http://www.wind-works.org/cms/index.php?id=531.

under an allegedly environmental Vice President Al Gore. After Clinton and Gore left office, the most wind power growth in the country resulted from a 1999 state law in Texas. The Governor at the time was a man named George W. Bush. But even today, the US wind power industry suffers from stop-and-go policies at the federal level.

Another lesson from the experience in California was "the Danes are coming." A Danish firm already had a 50-kW turbine to compete with the 30-kW Aeroman, and there was hardly a difference in price. By the end of the 1980s, there was further evidence that the Danes were doing something better than everyone else. After the failure of Growian, research funding focused on comparisons of the wide range of small wind turbines that had cropped up that decade in the Netherlands, Germany, and Denmark. On the German island of Pellworm in the North Sea, a wind farm had been created to test different models under similar conditions. The winner—indeed, the only one that ran properly—was a model called the Windmatic from Denmark. It had been built to be robust like a tractor, not lightweight and aerodynamic like an airplane. Indeed, unlike the US and Germany, Denmark has no major aerospace industry. What seemed to be a drawback initially turned out to be a crucial benefit.[17]

Developers of wind turbines were gradually coming to realize that the sector had as much, if not more, to do with heavy farming equipment than with aerospace. For instance, airplanes are regularly serviced, whereas wind turbines are expected to continue working across a wide temperature range and in conditions wet and dry with as little maintenance as possible. A wind turbine needs to be robust; any savings from the use of less material can quickly prove to be a bad investment if lightweight design leads to extended downtime. Not surprisingly, the world's largest wind turbine manufacturer—Denmark's Vestas—started off as a manufacturer of farming equipment. John Deere might have fared better with wind turbine development than NASA and Boeing did. It could hardly have done worse.

---

[17] Oelker and Hinsch, p. 47.

Inspiration from Denmark

Unlike the Danes, American researchers entrusted wind turbine development to aerospace experts, but like the Germans the Americans wanted to grow by leaps and bounds. When it got the technical drawings of the WE-34 from Germany, NASA contracted Westinghouse to develop a 100-kW turbine. MOD-0 (model zero) was based on the German design. It went into operation in 1975 and was inspiring enough for General Electric (GE) to be contracted to produce a 2-MW version—20 times more powerful. It was called MOD-1. Constructed in 1979, it was briefly the largest wind turbine in the world. It also "never generated enough electricity to be logged," as Paul Gipe remembers.[18]

Nonetheless, the next year Boeing completed a 2.5-MW version with a rotor diameter of 91 m. Called MOD-2, this turbine set up in Washington State revealed once and for all that development was going too fast. All these models produced far too expensive electricity, and they were unreliable to boot. GE backed out of the wind sector as a result in 1983 (during the boom in California), but research money remained available. By 1988, a 3.2-MW MOD-5B was set up in windy Oahu, Hawaii. And once again, it failed to meet expectations.[19]

Other countries also took a stab at big wind at this early stage and failed. In 1982, the Swedes tried their hand at a 3-MW turbine—the same generator size as Growian, though the rotor blade diameter of the Swedish unit was considerably smaller at 78 m. The Swedes had reacted to the accident at Three Mile Island with a nuclear phaseout plan adopted in 1980. But Swedish attempts to build big wind turbines from scratch—rather than incrementally starting from small units—proved to be fruitless endeavors.[20]

Although the Danes had a healthy grassroots movement of wind generator tinkerers, the Danish government also did not wish to be left behind, so in the early 1980s research money focused on the construc-

---

[18] Gipe, Paul. *Wind energy comes of age.* Wiley series in sustainable design. New York: Wiley, 1995, p. 113.

[19] Oelker and Hinsch, p. 50.

[20] Oelker and Hinsch, p. 50.

tion of two nearly identical 600-kW turbines with one major difference between them. The first turbine had "stall control," meaning that the shape of the blades caused turbulence at high wind velocities, thereby acting as a kind of natural brake system. This design was a tried-and-true Danish standard by the 1980s, having first been demonstrated on a three-blade turbine constructed by Danish wind power pioneer Johannes Juul in 1957—only two years before Hütter left the private sector for a post as a university professor.[21]

The other of the "Nibe twins" (the two turbines were named after the town they were installed near) used a different concept called "pitch control." Here, the blades pivot on their axis, turning into the wind at slow velocities and out of it at high velocities. Pitch control is used almost exclusively today. The test in Nibe allowed the Danes to challenge one of their main assumptions: that stall control was a solution for the future. While other countries simply rushed to scale up any prototype that looked half successful, the Danes remained open to the idea that they had not perfected the technology yet.

Still, the Danish foray into larger turbines was short-lived. The Nibe Twins provided insights, including the tough lesson that engineers simply were not yet ready for 600 kW. But note another difference in approaches between the Danish project and those in other countries: the fledgling wind generator industry in Denmark was already producing the largest turbines worldwide in serial production, but their attempt to build a big turbine nonetheless remained quite small in comparison—the Germans, Swedes, and Americans tried their hand at units 4 to 5 times larger starting from much smaller examples.

Where Does Innovation Come from?

These examples show how experts share with the general public an exaggerated fascination with breakthrough innovations. In the case of wind power, researchers faced a much wider range of options than discussed above. In addition to questions of stall or pitch control and the number

---

[21] Oelker and Hinsch, p. 22.

of blades, there is also the question of whether a turbine should face the wind or have the rotor placed downwind, that is, behind the tower relative to incoming wind.

An even bigger question is whether the rotor axis should be horizontal, as in all the models discussed above, or vertical. There are numerous design options for vertical-axis wind turbines, one of which is named after French aeronautic engineer Darrieus—who realized immediately what the drawbacks of the design were and never bothered to build one. This turbine type looks a bit like a giant egg beater sticking up in the landscape. Because it does not face any particular wind direction, it was hoped that this design would handle changing wind directions better than horizontal-axis turbines, which have to track the wind. But this benefit is minor compared to the drawback of not being able to start up independently when the wind starts blowing; Darrieus turbines require a motor to start turning.

They are also quite heavy relative to power output. One finds them today installed on the ground or on rooftops in urban environments, where turbulence from buildings would make horizontal-axis turbines a bit useless. For instance, three giant vertical-axis turbines are installed on the roof of Greenpeace's headquarters in Hamburg. "Two of them work quite well," says one staff member.[22] But the largest horizontal-axis turbines, such as the 7.5 MW E-126 made by Germany's Enercon, rest atop a tower roughly 140 m tall. The largest Darrieus turbine to date was probably a 200 kW system (0.2 MW) that was only 27 m tall. Constructed in Michigan, it ran for only a few hours.[23]

Nonetheless, the vertical-axis design and others continued to be repeatedly proposed and celebrated as potential breakthroughs by journalists and bloggers obviously not sufficiently familiar with the history of wind turbine development.

All too often, people tend to think of innovators as geniuses coming up with breakthroughs in the lab. In reality, innovations largely fall into two categories: haphazard discoveries and exhaustive comparisons.

---

[22] Personal communication.

[23] McLaren, Noa. "Wind turbine spins briefly in Ishpeming." Upper Michigan Source. 10 June 2011. Accessed February 5, 2016. http://uppermichiganssource.com/news/local/wind-turbine-spins-briefly-in-ishpeming?id=628563.

Thomas Edison is a good example of the latter. When he set out to make a better light bulb, he tried out numerous materials, including human hair, before settling on a carbon filament—something that had already been invented and therefore led to patent disputes. Edison's contribution lay in demonstrating that this option was the best in a wide range of possibilities, and he did so with a certain amount of entrepreneurial risk—financially, he needed to make the technology work.

A lot of innovation also comes from chance discoveries, with penicillin probably being the most famous example. Photovoltaics is another good example. The first solar cell was made in 1883 by American inventor Charles Fritts, who used selenium and gold, two materials not widely used in photovoltaics today. A decade earlier, selenium had proven to be photoconductive during tests for submarine telegraph cables—meaning that this discovery was accidental. Some 70 years later, three scientists at Bell Labs in New Jersey were working on improving the low efficiency of selenium solar cells when one of them proposed that silicon should be tried instead. A few years earlier, another Bell Labs researcher had—again, accidentally—discovered that impurities in a silicon crystal ("doped silicon") made the material more efficient for photovoltaics than selenium.

More than 90 percent of the solar cells we now have worldwide are the results of decades of gradual improvements made to these accidental discoveries of doped silicon solar cells. Likewise, three-bladed wind turbines now dominate the global wind power market, but they are not direct descendants of Growian, with its two blades and teetering rotor, neither of which are common today. They are direct further developments of small turbines whose design proved most valuable compared with other options tested just as rigorously. More importantly, these advances in the wind turbine sector came from people who—like Edison—ran an entrepreneurial risk and needed to make the technology work. In contrast, those who reluctantly signed a contract for a government research handout for a technology that competed with their current assets did not advance wind power nearly as much.

Major advances did not come from the conventional energy sector, either. Today, one manufacturer—Enercon—accounts for nearly 40 percent of all turbines installed in Germany, and the firm started off as a garage operation in the 1980s. Just as coal giant RWE was reluctant to take nuclear power seriously as a competitive technology to its assets (see Chap. 12), Germany's incumbent utilities did not drive wind turbine development, nor did Westinghouse or GE in the US. After leaving the wind sector in 1983, GE returned in 1997 by acquiring struggling German manufacturer Tacke; in 2002, it also took over bankrupt Enron's wind assets. From the mid-1980s to the mid-1990s, wind turbines made significant advances, and GE played no role. Nor did German engineering giant Siemens.

A turbine specialist in the gas and hydropower sectors, Siemens hopped on the wind power bandwagon at a very late stage by taking over Denmark's Bonus Energy in 2004. In the 1980s, Enercon, Tacke, Bonus, Vestas, and others were relatively small companies determined to make wind power a success. Their hearts were in it.

Repeatedly, we see examples of how conventional energy firms are not truly devoted to renewables. Though hard to quantify, their lack of commitment makes a great difference. Both BP and Shell were giants in the solar sector at the beginning of this century, but they left just in time for the photovoltaics boom that started in 2014 and is likely to be sustained. If anyone had the deep pockets required to withstand the consolidation phase in the photovoltaics sector from around 2010 to 2014, two of the world's largest oil firms did. They opted out.

They were not more devoted to biofuels. Indeed, though BP has officially been called "Beyond Petroleum" since 2000, there is currently little to suggest the name still fits. Likewise, Richard Branson of Virgin Airlines dabbled in biofuels research to much acclaim for a few years, but there is no sign he has made good on his word to devote his company's profits to making biofuels work in the aviation sector in order to prevent climate change.

It is clearly a mistake to expect conventional energy firms to herald in the age of renewables. Repeatedly, they argue that what proponents of renewables want simply won't work. Wilfried Steuers, the former head of a small German utility and president of the German Atomic Forum in the late 1990s, explained that "we put up two wind turbines, first to show that we are not against wind power…and second, to tell schoolchildren who come by to visit on a field trip that the things have not run this winter." (In the same interview, he also states, "Uranium cannot be so dangerous; otherwise, God wouldn't have created it.")[24]

---

[24] Bauer and Lobe, p. 78 ff.

Germany is fortunate enough to have held this debate for decades, and much progress has been made, so the argument that "renewables won't work" is now thoroughly discredited (while Americans such as Bill Gates continue to claim that breakthrough technologies are needed). Elsewhere, such as in the United States, citizens and their politicians continue to expect utilities to handle the energy transition for them, such as by requiring these utilities to provide a certain share of green electricity by a certain year—as though there were no conflict between renewable energy projects and these firms' stranded assets.

The history of wind power development suggests that innovation often comes from incremental steps brought about in combination with deployment—meaning that you have to build different designs, compare them, and try to learn from comparisons and mistakes when taking prototypes to the next level. Once a basic set of design types have proven successful, you then need a reliable market so that further advances can come from refinements during the manufacturing process and from economies of scale. The market, not a small coterie of corporate executives and governmental bureaucrats, should decide what designs are the best. But first, you need that market. Germany (and Denmark) succeeded in the 1990s because it implemented policies that allowed for trial and error and created a market for wind power. By the end of that decade, that market had settled on the small range of turbine types currently built.

An excellent example of the focus on innovation without deployment, Growian was clearly a step in the wrong direction, but it was not a complete failure. Werner Kleinkauf, who developed the control system for Growian, later cofounded German inverter manufacturer SMA, the largest supplier of photovoltaics inverters in the world. More importantly for wind power, the German Research Ministry decided to take a different approach based on the failure of Growian.

By the end of the 1980s, funding for gigantic wind power projects had eaten up three quarters of all state funding for wind power research, and there was no product on the market to show for it. Erich Hau summed up the bitter lesson from Growian quite well: "Without experience, not even large firms with a lot of engineers were able to develop wind turbines from scratch."[25]

---

[25] Oelker and Hinsch, p. 59.

Instead of trying to go too big too quickly with a single design, research funding at the end of the 1980s began focusing on comparisons of the wide range of small wind generators that tinkerers and do-it-yourselfers were selling at the time. These test wind farms led to the crucial insight that funding focusing more on deployment than on R&D might lead to even greater advances in innovation.

4

German Wind Pioneers Fighting Power Monopolies in the 1980s

In George Orwell's *Animal Farm*, published in 1945, the animals take over the farm. Their first major project—the one that the book centers on—is the construction of a windmill. But the animals do not wish to use it to grind grain. As Snowball, the brainy pig leader, explains, electricity would make life a lot easier for the animals on the farm.

The idea of a homemade windmill generating electricity on a farm was familiar to Orwell's readers at the time. Indeed, until the rural electrification plan expanded the grid in the US in the 1930s, most farms in North America either had such a primitive windmill or no electricity at all. The first windmill built not to pump water or grind grain is generally acknowledged to be the one built by American Charles F. Brush in 1888. It had an output of 12 kW and stood some 20 m tall. Consisting of 144 wooden planks, the rotor did not have proper "blades" as we know them today, but more closely resembled the "western windmills" that used to dot the landscape of North America, with a large number of blades covering the entire swept area. This option is not used today in modern wind turbines because the design provides more torque than speed, meaning that it is a bit like driving in first gear. You won't go very fast, but you can

C. Morris, A. Jungjohann, *Energy Democracy*,
DOI 10.1007/978-3-319-31891-2_4

tow just about anything. (Western windmills were mainly used to pump water, so that torque came in handy for heavy lifting.)

Hooked up to an array of batteries, Brush's wind power generator powered his home for 20 years—the same time frame that manufacturers offer for service warranties today. Brush also invented an early kind of streetlight called the arc lamp. In 1894, Norwegian explorer Fridtjof Nansen took a trip through the Northeast Passage on a sailboat with two of Brush's inventions: an arc lamp and a wind generator to power it. His arc lamp did not fare as well, although it was an improvement on a concept discovered decades before. Though Brush's design sold well in the 1880s, it gradually disappeared after that. The problem was partly that the arc was a good way of lighting a fire. It was also easier to manufacture large versions; a small version for bedtime reading proved to be a challenge. Imagine a giant streetlight giving off sparks on your bed stand, and you begin to understand why Edison wanted an incandescent light bulb.

The Germans were not pioneers in wind power generation at the time, but their landscape was nonetheless also studded with windmills providing mechanical energy. An estimated 20,000 windmills were in operation in the country around 1850, compared with 9000 in the Netherlands—and some 200,000 across Europe.[1]

In the first half of the twentieth century, German farms also frequently had to make their own electricity or do without entirely. Growian was constructed in a rural community at the mouth of the Elbe River called Kaiser Wilhelm Koog. The county was not connected to the grid until 1941. At the time, each farm had its own wind power generator. The mayor of Kaiser Wilhelm Koog when Growian was built was a child on one of those farms when the village got its grid connection. He had a 5-kW homemade wind power generator in his yard at the time.[2]

By the 1970s, of course, the grid had been expanded to even the smallest settlements in Germany. Nonetheless, people continued to hammer together their own makeshift wind power generators. One of them was

---

[1] Bauer and Lobe, p. 40.
[2] Oelker and Hinsch, p. 253.

Dietrich Koch. He had a small grid connection but wanted to become independent of the "power dictatorship" of the monopolist utility in his region, RWE.

In 1975, he had moved to a house on the outskirts of Mettingen, a town with only 12,000 inhabitants. The heating system consisted of a primitive stove he fired manually. There were also only two sockets in the entire building. The home was on the grid, alright, but only just barely. When he called RWE asking for a full-fledged modern grid connection so he could wire up the entire house, he was told, "that will cost 50,000 marks"—the price of five Volkswagen Golfs.[3]

With his schoolteacher's salary, Koch didn't have that kind of money. But on a summer vacation to Denmark, he drove past the Tvind project, where young people from various countries were working with their teachers and a few volunteer engineers to set up an absolutely gigantic wind turbine. Construction began in 1975, and the project was completed in 1978. The windmill has (it is still running in 2016) a rotor diameter of 54 m and a 2-MW generator, though it runs at a throttled output of 900 kW. It is roughly half as tall and a third as powerful as Growian. But where Germany's best engineers failed, a bunch of dedicated kids, teachers, and volunteers in Denmark succeeded. They built the world's largest wind turbine that worked.

Back home, Koch was determined to repeat the success of wind power in Germany, though not necessarily as a school project. Instead, in a casual conversation at a filling station, he learned that a Dutchman just across the border had recently also built a turbine. He drove over, and even from the autobahn he could see the name Lagerwey written across the nacelle up in the sky.[4]

Henk Lagerwey had just completed his master's thesis on "wind turbines with inverters and variable rotor speeds." The inverter was the electrical device that converted the turbine's direct current into the alternating current needed for a grid connection. Lagerwey was getting a degree in electrical engineering. The variable-speed aspect is therefore all the more

---

[3] Oelker and Hinsch, p. 208 ff.
[4] Oelker and Hinsch, p. 221 ff.

interesting because it is not obviously related to electronics. Lagerwey was not happy with the stall control used on turbines at the time; remember, this approach simply created turbulence around the blades at high speeds, preventing them from going faster. The young Dutchman found a way to use centrifugal force to pivot each blade on its axis, turning it out of the wind at high speed and into the wind at low speed. This is the "pitch control" approach used almost exclusively today. When the Danes compared pitch control to stall control with the Nibe Twins, they were essentially testing their own tried-and-true stall control against the new-comer pitch control from the Netherlands, which was simply proving to be too successful.

With Lagerwey, the Netherlands might have become an equally strong wind turbine powerhouse as Denmark did. Instead, the wind sector became an example of the "Dutch Disease"—when a country's fossil resources prevent it from developing new technologies. One might trace the Dutch failure to develop a major wind turbine manufacturer back to the booming gas sector in the country since the 1980s, when wind power was taking off elsewhere.

The Netherlands have the world's tenth largest gas field, but the Danes are also oil exporters from their North Sea resources and will remain so for the foreseeable future. Society occasionally pushes for change in one direction or the other. The Danes rallied around the Tvind windmill and eventually became global wind power leaders. The Dutch gathered in the district of Amsterdam called De Pijp and took back their streets, demanding that children should have the right of way over cars. The result was an equally admirable, but fundamentally different outcome: the culture of cycling now associated with the Dutch.[5]

But drives for societal change are not a zero-sum game; multiple campaigns can be successful simultaneously. Copenhagen, Denmark, is now generally recognized as the most cycling-friendly city in the world; clearly,

---

[5] Shahan, Zachary. "How did bicycling take over the Netherlands?" Treehugger. 19 March 2014. Accessed February 5, 2016. http://www.treehugger.com/bikes/how-did-bicycling-take-over-neth-erlands.html.

the Danes were successful at both developing wind power and promoting cycling.

Another big difference between the Netherlands and Denmark— indeed, between the Netherlands and Germany as well—when it comes to the success of wind power is more obvious: the lack of sustained policy support. The Dutch government never saw fit to provide start-ups like Lagerwey with a reliable domestic market year after year. The grassroots movement in Denmark and Germany led politicians in those countries to provide a reliable policy framework in the 1990s for wind power. The right policies might not have been adopted, and certainly not sustained, without strong, widespread popular support.

Koch was an early customer for Lagerwey, who had not even found his turbine manufacturing company yet. But when Koch purchased a 20-kW wind power generator made by young Henk and two of his in-laws on their farm, the German's troubles were just starting. The turbine needed a permit, but it did not fit into any particular category. Koch cleverly got around this potentially extended state of limbo by telling officials he wanted to build a nuclear bunker—the kind Americans were putting up in their backyards. Because so few Germans were concerned about nuclear war at the time, few citizens had constructed their own bunkers, so the government had implemented an official policy encouraging permitting for any such plants. Koch told local authorities that he needed the wind generator so that he would have electricity in his bunker, and the officials liked the idea. He then set up the turbine and reported back to the authorities that his land turned out to be filled with hard rock underground, making construction of a bunker much more expensive than originally planned—so he would not build one. "Officially, I only noticed the rocky underground after construction started," he admitted with a telling smile.[6]

But even when the permit had been granted, it wasn't easy getting the turbine into the country. In the 1970s, only a generation after World War II, Europe still had strict borders. Traffic regulations proved to be a problem; the tower extended out of the truck an arm's length too far. Since

---

[6] Oelker and Hinsch, p. 221 ff.

he had already reached the border, the Dutch officials reluctantly waived him on, but the Germans would not allow him to continue. "You will need a special police escort," he was told—and that arrangement could be set up in a couple of weeks.

Koch had already ordered and paid for a crane for the next day, so he could not wait. He went to the next phone booth and called the local TV station. After hearing Koch's story, the reporter on the other end said, "Don't worry, we'll be over in a bit." An hour later, a TV crew began reporting live, and the embarrassed border officials suddenly decided that an arm's length wasn't so bad for such a short trip of only 20 minutes. The haggling at the border turned out to be a blessing in disguise, for the TV crew kept up with the story over the next few days as the first wind generator owned by a private citizen was hooked up to the grid. This time, the revolution was televised.

Of all utilities, coal giant RWE was the one that had the first citizen-owned wind turbine hooked up to its grid. The firm initially reacted the way it would for several decades to come: with condescension. Its experts had originally claimed that the turbine was a stupid idea and would not even generate a single kilowatt-hour per day. "In its first year of operation, my little unit produced more electricity than the three-megawatt Growian, which cost more than 100 million marks in state funding—and I paid for my unit out completely of my own pocket," Koch says with glee.[7]

But RWE was not about to take Koch's success sitting down. It did not wish to purchase any electricity from its customers, so it merely paid 2 Pf for Koch's excess wind power. "They were charging me 28 pfennigs at the time! I was pissed," Koch remembers. "So I started heating the swimming pool, I built a greenhouse, I did everything just so I wouldn't have to sell my electricity to RWE. I didn't want them to use my windmill to finance their nuclear power."

Amidst all the media attention, Koch and his windmill quickly became quite an attraction. One person he inspired was Heinrich Bartelt. A resident of a nearby village, Bartelt had also already worked with some friends to construct their own little turbine with three blades and a rotor

---

[7] Oelker and Hinsch, p. 221 ff.

diameter of 3.6 m in 1979. He was 24 at the time but had been thinking about how to use the force of wind since 1973—not only because of the oil crisis, but because some buildings on his family's farm had been destroyed in a storm that year.

In 1982, Koch and Bartelt met at a meeting of one of the fledgling wind energy organizations that would later merge to become the German Wind Energy Association (BWE). Bartelt already knew the man from the radio. And while everyone else was talking about various technical problems, these two young men focused more on politics.[8]

How German Power Providers Became Monopolies

Since 1935, Germany had been broken up into the kind of regulated utility monopoly zones still common in most parts of the United States today— or, as various pro-renewables publications from the 1980s in Germany put it, "power dictatorships." In essence, the Energy Management Act of 1935 simply specified in law what was already common practice. The area served by large utilities like RWE often bordered on smaller municipal utilities. In order to grow geographically, they needed to take over their competitors. To do so, executives at big utilities came up with the idea of a "concession fee" for municipalities. In return for control of local grids and power supply, municipals would receive a cut of the sales. The concession fee still exists today; in 2014, it made up around 7 percent of the retail rate. Cities do not provide the utility that pays the fee any service; the utility simply pays it in return for the privilege of being a monopoly power provider—out of gratitude to the city for not having its own independent municipal utility. And of course, utilities don't actually pay the fee; they pass it on to ratepayers in full. Germany is one of the few countries that taxes power consumption in order to cross-subsidize such local services as public swimming pools and public transport.

---

[8] Oelker and Hinsch, p. 222 ff.

The 1935 law codified this arrangement. City officials were happy because revenue still went into their local budgets from power sales. Utilities like RWE were happy because they could keep the areas they had expanded into. As one veteran campaigner for renewable energy, Johannes Lackmann, once put it, the concession fee essentially means that losses from public transportation are socialized while profits from the power sector are privatized.[9]

Ratepayers were the losers because they now had to pay for big utilities to be profitable and for local governments to have revenue—but no one represented their interests in the 1930s. Granted, the law explicitly spoke of "making energy supply as reliable and inexpensive as possible," but the real objective was different. The Nazis had been in power for two years, and their main objective was to get the power sector ready for war.[10]

Municipal utilities had set up quite a large number of relatively small power generators, an arrangement that had benefits from a military perspective. Large plants would be much easier targets. In return, the Act specified that power lines are a natural monopoly. The large utilities were happy to hear this because it essentially meant the country was now carved up into monopolies by grid zone (similar to the situation in the US), so the big boys would not have to compete with each other. In fact, the Act of 1935 explicitly speaks of the "detrimental effects of competition on the national economy." Within these monopoly zones, the big utilities were able to buy out or buy into most small utilities and municipals. Relations between RWE and municipal governments became so entwined that in 2014 municipalities held roughly a quarter of the shares in the company. That year, the firm posted its first loss (directly as a result of its failure to take the energy transition seriously) since World War II, and the public suddenly realized that the demise of the country's big utilities was going to have a severe impact on municipal budgets; if RWE goes down, it's taking the public swimming pool with it.

---

[9] Oelker and Hinsch, p. 351.

[10] Also see Becker, Peter. *Aufstieg und Krise der deutschen Stromkonzerne: Zugleich ein Beitrag zur Entwicklung des Energierechts*. Bochum: Ponte Press, 2011.

Haggling with Utilities Over Grid Connections

What bothered early wind pioneers like Bartelt and Koch most was their inability to sell power to the grid at a profit. Under their leadership in 1985, their wind energy group brought together a few dozen people to create a wind farm interest group (called IWB), which may have defined what community ownership of wind power means in Germany for the first time. There were three criteria: minimized visual impact on landscapes, distributed supply, and integration in municipal energy supply concepts. In layman's terms, they didn't want to have a lot of wind turbines in a small area, especially when owned by out-of-town corporations. They also specifically were rejecting the "wind wall" concept that had been briefly experimented with during the boom that had just ended in California. There, wind farm planners believed it was a good idea to put as many wind turbines as possible side-by-side and behind each other on a windy slope. The visual result was a hill literally plastered with machines—exactly the nightmare vision that wind power opponents paint today (and that no wind farm has tried to replicate since 1984).[11]

This group of wind pioneers thus had strict criteria for their own projects. But their main opponent was not the public—it was still their local utility, RWE, and the lack of clarity in the permitting process. The latter was often a Catch-22 situation. Permits might only be granted in a particular state if the wind turbine was clearly on the property of a farm. But some counties required wind turbines to be 200 m from the nearest building, which often pushed the turbine off of the farmer's own land.

And then there was the meeting that Bartelt had with an RWE representative. The utility didn't mind connecting turbines; it simply wanted 100,000 marks for the hookup. Bartelt held the paper under his counterparty's nose and said, "This is what we mean by an abuse of monopoly power—you want a fortune for a job that takes just a few minutes!" The money was not the only problem; as another wind pioneer once put it, "Applying for a grid connection was about as complicated as trying to get out of communist East Germany back then."[12]

---

[11] Oelker and Hinsch, p. 222.
[12] Oelker and Hinsch, p. 223 ff.

The fundamental question in the 1980s was whether the Energy Management Act of 1935—or anything else since—required the monopoly utilities to hook up privately owned power generators at all. In the case of wind power, the pioneers had taken the matter to court, and a ruling from 1983 stipulated that wind turbines could be built as long as no other public concerns were detrimentally affected. German lawyer Ivo Dane fought hard for, as he put it, "the civil right to install wind turbines and receive proper compensation for the electricity" in the case. Unfortunately, in supporting Dane's view, the court spoke of "turbines exporting all of their power to the grid." In reality, few of them did—most systems at the time produced power used directly, with only the excess exported to the grid. Various local officials interpreted the word "all" to be material or immaterial, so the legal uncertainty continued.[13]

Bartelt's own experience in December 1987 illustrates the problem well. Heinrich and his brother Josef were surprised how easy it was to receive a permit for a 30-meter-tall unit they wanted to put up less than 100 m from their shed. The local official said he thought wind power was great, called the minimum distance a bunch of garbage, and added that the height was not a problem either.[14]

In December 1988, Heinrich Bartelt finished setting up the system after he had spent most of that year trying to get RWE to accept his submission for a grid connection based on the construction permit. He had simply grown tired of waiting and decided to hook everything up himself. RWE may have been tipped off, or maybe it was just happenstance, but the very day when Bartelt wanted to make the connection, a delegation representing the utility paid an unexpected visit to the farm. Bartelt was already out back working on the machine, "but my mother knew what she needed to do." First, she invited the gentlemen into the kitchen and offered them pie and schnapps. When Heinrich finally came in, the real argument got going. "You can't simply run the meter backwards. That would constitute compensation at the full retail level," the delegation argued. "It's the only fair solution," Heinrich answered. Two hours of haggling and several rounds of schnapps later, the delegation

---

[13] Oelker and Hinsch, p. 317.
[14] Oelker and Hinsch, p. 223 ff.

left, the turbine went into operation, and the Bartelt family farm had its first turbine.[15]

And then there was the story of Norbert Giese. In 1987, he wrote his master's thesis in geology on the potential of wind power in Schleswig-Holstein, the German state north of Hamburg bordering on Denmark. If wind power does not seem a natural fit for geology, keep in mind that there were few classes on wind power at all at the time. Students of mechanical engineering, for instance, spent only 90 minutes covering both wind and solar technology during their entire studies.[16]

In his master's thesis, Giese demonstrated at the time that the state would be able to get 10 percent of its electricity from wind power. Some 25 years later, the rural state with some of the best wind conditions in Germany was already producing nearly as much renewable electricity (including solar and biomass) as it consumed in total power supply—meaning that renewables covered the equivalent of 90 percent of demand (the state trades electricity with its neighbors). It currently has a long-term goal of being 300 percent renewable in the power sector, so it will be a huge supplier.

Naturally, the notion of 10 percent wind power seemed extravagant in the late 1980s. To get an idea of the mood, consider the following text from an ad published by the "nuclear energy information circle" in German weekly *Die Zeit* from 1990:[17]

The Danes are European leaders in the use of wind energy: in 1988, Denmark got almost 1/100th of its electricity from wind—equivalent to 0.9 percent of total power consumption. Such an intensive use of wind power is not possible in Germany because of the different climate conditions. In 1989, wind power only made up 0.03 percent of total power consumption [in Germany]. We will therefore continue to rely on other environmentally friendly types of power generation, such as nuclear power, which currently makes up 40 percent of power production.

---

[15] Oelker and Hinsch, p. 224.

[16] Oelker and Hinsch, p. 162 ff.

[17] Morris, Craig. "Germany cannot get 0.9 percent of its electricity from wind power." Renewables International. The Magazine. 7 January 2015. Accessed February 5, 2016. http://www.renewables international.net/germany-cannot-get-09-percent-of-its-electricity-from-wind-power/150/435/82718/.

Three years later, a group of utilities—including RWE and some of the predecessor firms that later became Eon—unabashedly signed their name to an ad claiming that "renewables—such as solar, water, and wind—cannot cover more than four percent of our power demand, not even in the long term." Germany already had 3 percent hydropower, so the ad states that an additional 1 percent cannot come from solar and wind. In 2015, the two collectively made up around 17 percent.

Giese's interest in wind power began in the early 1980s. He was one of the protesters against the nuclear plant in Brokdorf. "But I was always one of the people who not only wanted to say no to nuclear, but also yes to something else."

In December 1989, he began selling turbines for Denmark's Bonus. Like so many of the pioneers, he was less concerned about money and more concerned about doing the right thing: "If I consume power, then I am responsible for how it is generated." At the time Giese joined the Danish firm, its 150-kW machine was a hot item, but the company was already getting started with a 450-kW unit, the largest serially produced turbine in the world at the time. With a price tag of 870,000 marks, it was also more than an individual investor could be expected to have. Giese thus joined the wind sector right at the time when it moved out of its beginnings in the community of do-it-yourselfers and started becoming a truly modern industry.

The German farming community also knew a little bit about pooling resources, so they continued to be the driver of investments. Around 1990, the head of a Husum shipbuilding firm wanted to bring people together in a fund to construct 50 wind turbines with a total of 12.5 MW, an average of 250 kW per turbine. This project was to go up within a stone's throw of the Danish border in Friedrich Wilhelm Lübke Koog, some 90 minutes north of Kaiser Wilhelm Koog, where Growian had been built. The biggest investors were institutional ones from southern Germany, common practice in the shipbuilding sector, as was allowing smaller local investors to get involved. In this case, some local residents invested, and when the project went well, they began to wonder why they didn't just do things on their own rather than chip into a project that would mainly benefit large investors a thousand km to the south.

On 15 January 1991, an open town hall meeting was therefore held to discuss the idea. It wasn't easy going. As one of the initiators put it, "If you want to get three farmers to agree to something, you have to kill two of them." But in the end, 33 people chipped in at least the minimum stake of 3000 marks at the first meeting. As usual with such projects afterwards, more and more people would want to get involved later. But for the time being, this group had come together with enough equity to leverage millions in bank loans and upfront governmental subsidies for a wind farm.[18]

Enercon's Breakthrough

These laypeople knew nothing about wind turbines, but the 450-kW machine from Bonus seemed interesting, and they also looked into a 300-kW unit from Germany's Enercon. Here, Giese dropped the baton as the representative of Bonus in Germany. "It was one of my worst days ever," he says in retrospect. The farmers came by to see Giese personally, and he immediately realized they had no idea what they were talking about except that they wanted some wind turbines. He did not take them seriously.

The competition did. At the time, Enercon consisted of only two people. They received a request for a price on a sheet of paper ripped out of a school notebook, with the chads still dangling on it. How much would 22 of the 300-kW turbines cost? It was the biggest order Enercon had ever received. The boss, Aloys Wobben, took the question seriously and invited the group over.[19]

Wobben drove them around to visit a couple of satisfied customers. Enercon had the good fortune of being founded in the state of Lower Saxony, which was the first to offer financial support for wind turbines in 1987. The local utility was also generally interested in wind power and ordered an early single-blade unit called the Monopteros with 250 kW to test alongside Enercon's 300-kW unit. It was a very early attempt by a utility to test different options already on the market. But in a way,

---

[18] Oelker and Hinsch, p. 303 ff.
[19] Oelker and Hinsch, p. 304 ff.

the test failed when the manufacturer of the Monopteros was unable to deliver a unit. But the utility, EWE, was quite pleased with Enercon's product and ordered 10 additional units in 1989—even though the hydraulics had failed in the original unit tested.

This hydraulics failure got Wobben to thinking. "You can't stop a gearbox from leaking," he had decided in 1988. He then set out on a bold idea: throw out the gearbox, the hydraulics, and everything else that could leak. And in March 1922, the prototype of a 450-kW turbine from Enercon went up without a gearbox. Called "direct drive," the replacement design became a hot item in Germany, and demand exploded by 1994. During development, a lot of work needed to be done on electronics, so the E-40 (as the unit was called) had a unique ability to stabilize the grid, making the turbine an especially attractive export item for countries with weaker grids than Germany's.[20]

None of this development would have been possible without Enercon's first buyers, however, such as EWE. As of 2014, the northern German utility was the fifth largest in the country, though at a good distance behind southern Germany's EnBW in fourth position. It also made a name for itself as a utility that embraced Germany's energy transition. That year, roughly a quarter of the electricity it generated came from renewable energy, on par with the national average but roughly 6 times more than the share of renewables in RWE's supply, for instance.

The first big order Enercon received from this farming community on the Danish border made a huge difference as well. The farmers were not just in it for the money, however. Institutional investors actually fared better with such projects because they could cash in on tax incentives. (This drawback still plagues the energy sector in the US, where tax incentives continue to favor big business financially over committed communities, who often cannot even benefit from tax credits.) But after discussing the matter at their first meeting, the German farmers decided to go ahead with the idea. "If we don't do something for the environment, who will?" one of them summed up the general consensus. The director of Enercon's first branch office also remembers an underlying wish to be independent:

---

[20] Oelker and Hinsch, p. 132.

"70 percent of our first customers were farmers, and they were interested in the technology—not only about money, but also about doing something different."[21]

The price tag was a whopping 16 million marks.[22] Funding from the State of Schleswig-Holstein covered 25 percent of the purchase price for the first seven units. The next seven fell under the federal government's new 250 MW Program, an extension of the successful 100 MW Program launched in the late 1980s by two Christian Democrat politicians. It was the first federal program that, in the wake of Growian's failure, redirected state funding from R&D to deployment. In 1984, a survey taken by government officials revealed that 450 to 500 small wind turbines had already been built by 1982, and a later study revealed that do-it-yourselfers had put up 58 percent of them. The experts found that most of the concepts were not good, but 59 percent of the owners said they were happy with their products, which were mainly used to provide heat; almost none of the turbines were connected to the grid. When the first applications for the 100 MW Program came in, the Research Ministry realized that a lot of the submissions came from tinkerers. Not all of them included design documentation, and some of the experimenters had put a unit together from scrap.

A local cooperative bank provided the loan to cover the rest of the money needed in Lübke Koog, and the locals came up with 2.3 million in equity, roughly 70,000 per person. But the money did not have to be liquid; the farmers owned property that could be mortgaged. Still, the situation was new for them financially. "We didn't know what to wear when you sign a 16 million mark contract," one of them remembers.[23]

The wind farm in Lübke Koog was not the first one owned by a group of citizens. Three months earlier, a smaller project had begun in a nearby village. But it was the Lübke Koog wind farm that drew everyone's attention and showed other Germans how a small rural community could build industrial infrastructure.

---

[21] Oelker and Hinsch, p. 190.
[22] Oelker and Hinsch, p. 299.
[23] Oelker and Hinsch, p. 305.

Wobben's "direct-drive" turbine was not the first of its kind. In a way, he reinvented the wheel. Way back in the summer of 1973, Walter Schönball, a population-growth expert at the UN in Geneva, had come together with some engineering friends in their spare time and built a 70-kW turbine on the island of Sylt off the coast of Germany just across from Lübke Koog. It first implemented the direct-drive concept. Schönball and his friends would later found the Wind Energy Research and Application Association (VWFA), the first predecessor of the country's Wind Energy Association (BWE). They lacked something that Wobben later had: a market with which to further develop direct-drive turbines.[24]

These early developers were critical of the government's initial focus on big projects, with research conducted by large established corporations. These beginnings still characterize the BWE today. While the American Wind Energy Association, for instance, represents big industrial firms, the BWE does so only indirectly. It has an industry committee for turbine manufacturers and other corporations to express their interests. Mainly, the BWE continues to represent the interests of Germany's community-owned wind farms.

Early Wind Power in Eastern Germany

Hermann Honnef also indirectly played a role in the development of direct-drive turbines. In the 1930s, he dreamt of a 30-MW turbine with 175-meter propellers reaching up 300 m into the sky. It was far too early for any such thing to even be attempted; indeed, the largest wind turbine attempted to date is only around half that size. But his example inspired Gerd Otto, born in 1928.[25]

Otto could also have gone down in history as a pioneer in the German wind sector, but he worked in communist East Germany. Initially, Otto was an engineer who designed trains used to transport lignite, the dirtiest type of coal—and Germany happens to have the largest reserves in

---

[24] Oelker and Hinsch, p. 64.
[25] Oelker and Hinsch, p. 275 ff.

the world. Otto was looking for a way of replacing lignite, but he wasn't happy about the prospect of nuclear either, so he poured over Honnef's old ideas. In 1986, he quit his job as an engineer out of frustration and became a beekeeper—a clear sign of protest against his government. Then, he began writing.

In 1989, he published a book on wind power. It contained a drawing that became popular in the German renewable energy community starting in the 1990s. Next to a wind turbine is a pile of coal of the size needed to generate as much electricity as the turbine would produce. The pile of coal is roughly as tall as the turbine, but many times wider at the base. His book also contained drawings of turbines reaching up to 30 and even 50 MW. And he calculated that direct-drive turbines—the ones that Wobben was just starting to think about designing—would be a cheaper option for units between 200 and 500 kW.

Otto was already retired in the 1990s, when the German wind power market began growing sustainably, but just before Reunification he impacted events one last time. East Germany had done little with wind power. In 1946, a wind turbine was built near the Baltic using spare parts in war-torn Rostock; the generator was a motor from a tank, and the blades were taken from a helicopter. The project did not lead to further development, but 40 years later, perhaps inspired by research being conducted in West Germany, the East German Science Ministry had a 20-kW turbine built on the Baltic Sea in the 1980s as a part of a heat supply system. And once again, the technology was not developed further.

Then, Otto became active in the late 1980s. The turbine he got the government to build showed the special challenges that East German engineers faced. Otto had petitioned the government to build one of Honnef's 30-MW turbines from the designs created in the 1930s. In 1985, he even had the idea patented. The goal was to power Berlin entirely with wind turbines.

At the time, East Germany had a special submission process to encourage its citizens to come forward with their ideas about how to improve the country. If you filled in the paperwork correctly, you were guaranteed an answer within four weeks. Gerd Otto's idea was handed over to Otto Jörn, who was to respond.

Jörn found that 30 MW (10 times the size of the failed Growian) might be a bit large to start with but agreed that the government should devote some research attention to wind power. He was then contracted to put together a team that would build a modern wind turbine to power a heating system that dried timber. But the country was so broke that equipment and materials were hard to come by. Practically everything made was shipped abroad to ensure an inflow of foreign currency. East German generators, ironically, were therefore used in early wind turbines sold by Danish pioneer Vestas, but Jörn had a hard time getting one himself—even though he was conducting a project for the Ministry of Coal and Energy. When he wanted to review the first sparse publications on wind power generators from the West, he had to receive special permission from the Stasi, East Germany's dreaded secret police.[26]

The tower for the turbine wasn't easy to come by either. There was some fresh steel on hand, but in order to get 16 tons of it, Jörn covered the metal with rust by hand so it would be classified as scrap metal officially. That way, he could keep it in the country. The ruse worked.

In May 1989, the turbine was hooked up to the grid and exported its first power. The experiment had succeeded. A few months later, Jörn was even able to set up a 200-kW Vestas unit along the coast. It was 3 October 1989, just a month before the Berlin Wall would come down—an event that, however, no one saw coming at the time.

After the wall came down, everything was in limbo for a while. But on 8 June 1990, the first post-Reunification wind turbine was set up in the former East Germany—an 80-kW unit made by Enercon. All the officials from the days of communism were still in office, but suddenly people were relaxed, open, and embracing change. It took less than two months for the turbine to be built once the permit had been granted.[27]

Sensing a unique business opportunity, Enercon financed the turbine itself. Wobben himself drove over. There were no hotels around, so he slept in a sleeping bag on the living room floor of the East German man coordinating the project locally, Karl Hartung. His pastor had come back

---

[26] Oelker and Hinsch, p. 227.
[27] Oelker and Hinsch, p. 279.

from his first visit to West Germany a few months before with a wind power magazine and stories about what was being done with turbines just across the border.

Enercon mainly wanted to be first in the East. Although both parties were German, communication barriers were significant. Hartung had no phone and no car, so twice a week he would hitchhike to a phone booth in West Germany (international calls would have been more expensive) to coordinate details.

The Protestant Church in East Germany was, of course, instrumental in the peaceful protests that led to the Fall of the Wall in 1989. In the West, the church was heavily involved in wind power. Most of these farming communities building their own wind farms were not groups of green hippies, but churchgoing conservatives. One West German congregation provided the funding for a 20-kW system built by a partner church community in East Germany. The turbine was set up on 9 November 1988—the same day the Berlin Wall fell. "I missed everything that day because of wind power," one of the East German wind pioneers involved in the project stated.

Needless to say, the Fall of the Wall changed everything—except the role of the church. Both in the East and the West, the Protestant Church in particular continues to be a major driving force for greater citizen involvement, healthier democracy, and stronger communities. The story of the peaceful revolution in East Germany is relatively well-known. Major church leaders were at the head of some of the protests, which were often held in churches. But the involvement of churches in the West is less known. There, the revolution was perhaps less spectacular, stretching as it did across decades. But it was always about the same call for healthier democracy, stronger communities, and greater citizen input. And it took place largely pertaining to energy issues.

German citizens realized they would need to come together to fight incumbent monopoly utilities. In the 1970s, a church group came together to provide a million marks for the protest against the nuclear plant in Brokdorf. Part of the money was used to set up a 75-kW Lagerwey wind turbine, built within sight of the plant. Since then, church groups have remained a major player in the Energiewende. There is even a campaign of "solar churches"—congregations putting solar roofs on church buildings. In southern Germany, where there is less wind than along the northern coast, church groups were also crucial, but merely in a different part of the energy sector. There, a church group helped spearhead a campaign for a community to buy back its grid.[28]

---

[28] Oelker and Hinsch, p. 303.

5

The Power Rebels of Schönau

At the beginning of the 1990s, Schönau, a small town with a population of 2500 in the Black Forest, began collecting donations for victims of the nuclear accident in Chernobyl, Ukraine. In 1993, a church group brought over 20 schoolchildren to spend a few weeks at lower levels of radiation in southwest Germany. In 1995, the retreat was repeated, and lots of the children returned, but one girl didn't. Her story upset people in Schönau. She had died of leukemia. Even worse for some locals, her death was not included in official statistics of Chernobyl victims.

People not only collected money, but also sold household items at a flea market. One of the bigger objects donated was a crib, and the mother had a peculiar stipulation. "Don't tell anyone this was Joachim's."[1] In the 1980s, her son had played for SC Freiburg, the nearby football team. With 81 goals, he remains the team's top scorer even today. Joachim went on to an impressive career as a coach. The world knows him better by

---

This chapter is largely based on personal talks and Janzing, Bernward, and Dieter Seifried. *Störfall mit Charme: Die Schönauer Stromrebellen im Widerstand gegen die Atomkraft; wie eine Elterninitiative, die sich nach Tschernobyl gründet, zu einem bundesweiten Stromversorger wird.* Vöhrenbach: Dold, 2008.

---

[1] Janzing and Seifried, p. 18.

© The Editor(s) (if applicable) and The Author(s) 2016
C. Morris, A. Jungjohann, *Energy Democracy*,
DOI 10.1007/978-3-319-31891-2_5

his nickname, Jogi. As head coach, Joachim Löw led Germany's national team to the World Cup in 2014.

Radiation from Chernobyl Reaches the Black Forest

The town's reaction to Chernobyl depended on the weather. The accident occurred on 26 April 1986, but the Soviet Union made no announcement. Two days later, however, sensors at a Swedish nuclear reactor detected higher levels of radiation, which the Swedes could not explain. The Soviets were embarrassed into admitting the accident on 29 April 1986.

The next morning, the regional newspaper *Badische Zeitung* summed up the event in a mere 24 lines under the heading "Accident at nuclear plant." The initial reaction was hardly a panic, but officials soon began confiscating produce grown outdoors and telling people to keep their children indoors. To make things worse, May 1 is Labor Day outside the US (where the holiday was shifted to September 1 in an effort to ensure that workers of the world do not unite annually). It was a Thursday that year, so children normally would have been at school. Instead, they sat at home, and their parents told them they could not go outside and play for more than a half an hour. The country's top health officials said so.[2]

The weather was fantastic that day, so the children had trouble understanding why they couldn't go out—and their parents had trouble explaining the reason. It was a danger no one could see, smell, or feel. That day, a radioactive cloud from Chernobyl—1600 km away as the bird flies—reached the area, and measurements of strange-sounding elements few laypeople had heard of before skyrocketed at a measurement station atop the nearby Schauinsland Mountain. Weather reports began including radiation levels along with temperatures and precipitation.

---

[2] Janzing and Seifried, p. 5.

One of those parents was Ursula Sladek. A mother of five, she was laid up with a broken leg from a skiing accident that week. Her husband, Michael, was the town's general practitioner and was therefore one of the people not able to take off work that day. He had his hands full explaining to everyone what the risk was and what they could do about it. He wasn't exactly sure of the former, having only had a rudimentary introduction to the health effects of radioactivity at medical school; and he certainly didn't know the latter—like so many other public figures, he felt helpless when asked what his specific recommendations were.

Bedridden, Ursula was not in much of a position that day to run after five children out enjoying the splendid weather that day, and Michael was too busy to help. So their kids stayed out playing longer than health officials were recommending. The couple now felt even more helpless, and they wanted to do something.

They were not alone. Wolf Dieter Drescher was a local forester. He and his wife put an ad in the town's paper: "Who is worried about the future of their children after Chernobyl; who wants to do something and doesn't know what?" Seven people answered, including Ursula and Michael. The group got together and called itself "Parents against Nuclear Power." It was only one of numerous such informal groups that came together that month across the country, but most of them focused on setting up help networks in case of a nuclear disaster. The one in Schönau was different. It set out to make the town less reliant on nuclear power.[3]

First, the group reasoned that less power should be consumed. They took the idea to their local utility, KWR. Though the firm was investor-owned, it prided itself on having close contact with the community—at least in publicity. Nonetheless, when the firm was awarded a 20-year concession to run the town's grid in 1974, it had one condition: small, historic hydropower plants it did not own would have to be shut down entirely.[4]

[3] Janzing and Seifried, p. 8.
[4] Janzing and Seifried, p. 11.

"Treated Like Retards Who Deserved Pity"

Historically, small hydropower was the electricity source that the region started with. In 1905, Schönau purchased land for the first hydropower plant, which had an output of 48 horsepower—smaller than most car engines today. The system generated direct current, which was stored in batteries, and a gasoline engine served as a backup generator. When a 28-horsepower hydropower plant was built nearby in 1930, the new plant became the "ancillary plant," while the old one was the "main plant."[5]

The new plant was built because electricity consumption had increased, but for many years the main plant ran at very low capacity, with electricity demand initially only being a third of maximum output. The focus was therefore on increasing power demand. When the ancillary plant was built a quarter of a century later, decision-makers were therefore careful to keep the new unit small lest capacity once again greatly outstrip demand, making the new investment unprofitable.

At the end of World War II, the town was still running on direct current, aside from two butchers and the local hospital, which had machines requiring alternating current. But in the 1950s, Germany underwent its *Wirtschaftswunder* (economic miracle), and power consumption in the 1950s grew rapidly—in Schönau by 12 percent annually. At the time, economic growth was synonymous with greater energy consumption.

The town completely switched to alternating current in the 1950s, and by the end of the decade Schönau was consuming twice as much electricity as it was generating from its own resources; it had begun importing electricity. In 1973, the city council resolved to commission the municipal grid to the local utility KWR, which paid 600,000 marks for the deal. In closing the main and ancillary plants as a part of the agreement, Schönau became completely dependent on KWR's power plants.[6]

The mind-set at the utility was still old-school in the 1980s, despite the first clear signs that economic growth could outstrip the increase in energy consumption. So when the group of parents approached the utility

---

[5] Janzing and Seifried, p. 38 ff.
[6] Janzing and Seifried, p. 43.

to explain their idea about conserving energy, they were hoping for the kind of open mind that the company stressed in its public relations work—but they met with absolute rejection and ridicule. One of the people in the citizen initiative was the local notary, and he summed up the meeting thus: "We were treated like retards who deserved pity."

But the Parents against Nuclear were not about to give up. After the disappointing meeting with utility representatives, they regrouped at the Four Lions Hotel's restaurant in Schönau. In addition to the notary and the forester, the local policeman was a part of the group, as were a few schoolteachers. The Sladeks were also still there, and Michael complained about how the Energy Management Act of 1935 was still in force. Large utilities were buying up small competitors, especially municipals, but the big firms didn't want a showdown between each other. Michael Sladek in particular wanted to find a way to get around that law. But how? Looking back today, he concludes, "Sometimes, you have to change the law."

But first, the Parents against Nuclear—or the Power Rebels, as they would later become known nationwide—took their case to the 22nd Conference of the Protestant Church in Frankfurt. In June 1987, it called on the German government to revise the law. Nothing happened. So Michael Sladek made a decision: "we have to be willing to take things into our own hands." "When we moved here in 1977, we weren't particularly political or ecological," his wife Ursula remembers. "We were just unbelievably naïve. After Chernobyl, we thought politicians would react, but the only thing that happened was that the levels of admissible radioactivity were increased."[7]

The Power Rebels held a power conservation competition in town. Drescher, the forester, used to work at the utility and knew that the firm had a lot of handheld power meters kept in stock. They could be given to households so that people could measure how much electricity particular appliances consume. Once again, the Power Rebels approached KWR with the idea, but at least they were not surprised this time when they were thrown out of the building.

Drescher contacted another utility and got what he wanted. "If KWR had cooperated, we would never have launched a bigger campaign," he

---

[7] Janzing and Seifried, p. 11.

says in retrospect, "but we really had to react to the way they disrespected us. If they had been a bit cleverer, they could have easily driven our group into the ground."[8]

By January 1987, the campaign for power conservation was ready to be rolled out. People measuring consumption at home was just the beginning. Local shops started offering discounts on efficient household appliances. The local savings and loan bank offered low-interest loans for investments in efficiency. Along with the hospital and high school, 140 households took part in the competition, which was surprisingly successful in its outreach mainly because the Power Rebels turned out to have a natural ability to make things enjoyable. Michael Sladek's catchy slogan for the campaign is a case in point: "Make your relationship with your power meter a loving one. Visit it every day."

The participants reduced their power consumption by 20 percent, with the winners cutting theirs in half. The press started to become interested—nationwide. Women's fashion magazine *Brigitte* covered the initiative, explaining, "If everyone did that, we wouldn't have so many environmental problems."

Only KWR was upset. Its CEO threatened to sue the campaigners for foregone losses. But instead of scaring the Power Rebels into submission, a large number of citizens just learning about the campaign were aghast at the utility's arrogance. It was the first in a long succession of communication missteps the company's management would make. Indeed, for the next 25 years the German power sector's top executives will repeatedly pay dearly for failing to take the Power Rebels seriously, not only in terms of charm, but also technical savvy.

Countering Legal Challenges with Increasing Expertise

KWR had already tried the stick. In August 1990, it opted for the carrot: management told city officials that it was willing to increase the concession fee, which went directly into the town's budget. The 20-year concession signed in 1974 would expire in 1994. At that point, the city could

---

[8] Janzing and Seifried, p. 14 ff.

renegotiate the concession fee anyway. The concession from 1974 put the fee at 3 percent of the retail rate, whereas 5 percent was the going rate for new concessions. So KWR offered to pay 5 percent starting as soon as the city signed a contract for a renewal of the concession.[9]

In November 1990, the Power Rebels reacted by founding two companies, one for distributed energy systems, and another to buy back the local grid. The former company did not go very far initially, given the utility's monopoly; it would eventually set up the first wind turbine in the region, reactivate an old hydropower plant, and build new biogas and cogeneration units (which produce both power and useful heat). But the second company was a challenge to that monopoly. It is here that the real showdown took place.

The first thing the Power Rebels had to address was the concession fee. If the mayor did not accept the utility's offer to start paying a higher concession the city fee three years early, it would lose an estimated 32,000 marks. At a meeting in the Four Lions Hotel, which was increasingly becoming an informal headquarters, the Power Rebels therefore decided that the first step would be to offer the city the equivalent amount of money it would receive from the increased concession fee up to 1990. That way, they would at least buy some time; the city would face no foregone losses from waiting while the Power Rebels planned their next step.

Within six weeks, 282 people pledged to cover the full amount. If the Power Rebels managed to buy back the grid, the donors would get the money back with interest and a return for the risk. Otherwise, they would lose their money. Still, City Hall had to play along. The Power Rebels had yet to make any serious business proposal; they had only stated their intent to buy back the grid. And why should the city believe that a policeman, a general practitioner, a forester, and some school teachers have the expertise needed in the power sector?

On 28 January 1991, the city council heard the Power Rebels out and accepted their offer. After all, the 282 pledgers represented more than 10 percent of the community's population. The campaigners were given three months to explain what exactly they wanted. If it hadn't occurred to them earlier, the Power Rebels now realized that they needed help. None of them knew how to propose a business plan for a grid takeover.

---

[9] Janzing and Seifried, p. 27 ff.

They began making calls. A friend of a friend recommended an engineer from Aachen, a town on the edge of West German coal country—but also a hotbed for the fledgling renewable energy movement. At the time, a local campaign had begun calling for "cost-covering remuneration" for solar power made by citizens, and the Aachen Model—as the policy would become known—led in 2000 to the country's Renewable Energy Act. Photovoltaics also partly has its beginnings in Aachen; it was here that Photon Magazine was founded in the mid-1990s.

The engineer's name was Wolfgang Zander. When a delegation of the Power Rebels met him, they liked each other at first sight. The Power Rebels had become accustomed to meeting with suits and ties from executive boards, but they always came in their normal attire. Michael Sladek is a bear of a man, with a beard that would allow him to double as Santa Claus. Zander showed up to the meeting dressed casually himself and with long hair. "They looked like me," he thought at the time.

The Power Rebels couldn't have found anyone better than Zander, not only in terms of the dress code. First, he wasn't in it for the money—which was good, because the folks from Schönau didn't have much. More importantly, he was able to put together a 500-page feasibility study by the deadline with input from some other crucial experts. The State Government of Baden-Württemberg, where Schönau is located, had to approve all municipal takeovers, so the study was submitted to the capitol in Stuttgart. It so impressed state officials that they recommended that Schönau's city council turn down KWR's offer and go with the Power Rebels. The utility was caught with its pants down. It was no longer dealing with a bunch of idealistic activists. It was now dealing with a bunch of idealistic activists and at least one guy who knew exactly what he was doing.[10]

This small victory led to even greater press coverage. An article entitled "A village under tension" published on 7 June 1991 in German weekly *Die Zeit* was penned by the young Fritz Vorholz, who would become one of the country's best journalists covering the Energiewende. Vorholz made Germans aware of how widespread expert support for the Power

---

[10] Janzing and Seifried, p. 32 ff.

Rebels had become.[11] The Christian Democrat politician who had headed the German government's Inquiry Commission on Protecting the Earth's Atmosphere was quoted saying, "What more can we wish for than such commitment from locals?" The climatologist who was chairman of the government's Climate Committee at the time said it was an example of "how local actions can solve global problems." And one of the country's top financial experts was quoted in the article as saying that the grid buyback was a "good project for private investors." He even offered to provide enough funding to cover any shortfall that the citizens could not muster themselves—not as a donation, but as an investment. In an initial indication of how much KWR must have been raking in the whole time, he estimated the return at "not below 10 percent in a conservative, prudent calculation."

Vorholz listed all the good things the citizens were trying to do and lamented that the "Black Forest village is at the mercy of an energy provider who thinks these plans are completely unrealistic." And he pointed out that KWR was a 100 percent subsidiary of a Swiss power firm that ran nuclear plants.

Taking Back the Grid

A month later, the city council of Schönau took a vote based on Zander's business plan. It went along party lines. As in Wyhl, the mayor of Schönau was in favor of the utility and skeptical of his own citizens. He also had his party, the Christian Democratic Union (CDU), in line. There were 13 members of the council, and the CDU had six of them, including the mayor. Only one more was needed for a majority. The four council members who did not belong to a party, one of which was Michael Sladek, were expected to vote in favor of the grid buyback. The remaining three were Social Democrats (SPD), and their vote was split—two in favor of the buyback, one against. The Power Rebels lost 7–6.

---

[11] Vorholz, Fritz. "Ein Dorf unter Spannung." Die Zeit Online. 7 June 1991. Accessed February 5, 2016. http://www.zeit.de/1991/24/ein-dorf-unter-spannung.

The outcome was foreseeable, though, and the Power Rebels had planned ahead. As soon as the gavel fell, finalizing the defeated resolution, one of the independent council members made another motion, this time for a plebiscite on the matter. No one was aware of this minor detail in municipal law, for it had never been used before; the city must conduct a plebiscite if 10 percent of its citizens sign a petition calling for one. Sometimes, it pays to have a notary on your team.

Within a month, twice as many signatures as necessary were collected. The city official who accepted the petition couldn't help but comment, "There were a couple of pretty surprising names on the list."

A plebiscite would be a battle for hearts and minds, but also a home game for the Power Rebels. Georg Thoma, a local and a winner of an Olympic medal in skiing, agreed to go mountain biking with supporters. Gingerbread hearts with "*ja*" (yes) written in sugar on top were handed out. Special events were held for both children and the elderly.

Michael Sladek was keen on making sure the issue remained nonpartisan. The Green Party visited town to show its support, but no one came to greet them. Schönau is conservative, and the Greens get fewer votes there than the national average.

On a radio talk show that reached the southwestern quarter of Germany, Michael Sladek and Wolfgang Zander debated the mayor and a KWR executive. It was largely a technical discussion. Sladek called for small, distributed cogeneration units; the mayor said local pollution would increase as a result. The winner was clear in terms of popularity, with the moderator calling Schönau "an example of living democracy."

The debate on the radio show also focused on whether the Undersecretary at the Environmental Ministry in Stuttgart, a CDU member, broke party lines when he recommended that Schönau sell to the citizen movement instead of the utility—against the wishes of Schönau's CDU mayor. The pressure increased on the Undersecretary until he eventually sided with the mayor. When word leaked that the Undersecretary had explained his turnaround off the record by saying, "I like what you are doing, but I would also be against it if I were your mayor," the campaigners were dismayed. "I learned that politicians often say one thing to the public and something else behind closed doors," one of them stated.[12]

---

[12] Janzing and Seifried, p. 35 ff.

On 27 October 1991, the plebiscite was finally held. Voter turnout was quite high at 74.3 percent, and the yays defeated the nays 729 to 579. The people clearly expressed their will for the grid buyback. The CDU party whip on the city council summed up the minority's disbelief succinctly: "That can't be. I don't believe it."

The campaigners had just two more hurdles to jump. First, the Economics Ministry had to confirm that the Power Rebels met the legal requirements to become a power provider. The city government was not going to take over the utility itself; rather, the Power Rebels would as a new cooperative. But the campaigners had no electrical expertise. Did they even know who needs to be hired for what? And second, they had to actually purchase the grid commission. What price would the utility demand, and how would they come up with the money? Twenty years earlier, the city had commissioned the grid to KWR for 600,000 marks. The price would certainly be much higher now.

The Winding Road to Victory

The victory in the plebiscite once again drew national attention. On Germany's Channel One, reporter Franz Alt covered the topic on 12 November 1991. Alt himself would later become a card-carrying supporter of renewable energy.

One of the people watching Alt's show was Walter Bolz. He was a tax advisor for the nearby city of Rottweil, which is a leader in cogeneration units. "They need me," he told his girlfriend next to him on the sofa. The next day, he called Michael Sladek, who surprised him by saying, "I've been waiting for your call." An official from Rottweil had also seen the show and called up the Sladeks to recommend their tax advisor.

Bolz helped the Power Rebels get assistance from other municipal utilities to find out what specifically needed to be done. It wasn't always easy. At times, a utility Bolz contacted needed to ask its CDU mayor for permission to help a group of citizens go against the wishes of their CDU mayor in Schönau. But gradually, the Power Rebels began to design their cooperative energy provider.

City council elections were held in 1994, and the winners would be the ones that sign the new concession, either with the old utility or the new cooperative. The current mayor announced he would not run for reelection. The independents managed to take a seat from the CDU in the elections, thereby tipping the scale. So in November 1995, the new city council officially resolved not to give a commission to KWR, but to EWS, as the Power Rebels call the energy cooperative they created.

But then, the CDU turned the tables, calling for its own plebiscite. The party doubted that the cooperative had the required expertise. This time around, the question posed to the people is reversed: "Should the city council resolution to sign a contract with EWS be disregarded and the concession with KWR extended?" Those who voted yes for the Power Rebels in the last plebiscite now have to vote no. The tactic assumed that people were too stupid to switch from "yes" to "no."

A rift increasingly divided the town's inhabitants, as also happened in Wyhl. People stopped shopping at particular stores if the owner had expressed an opposing opinion on the issue. A local marmalade producer began selling jars with a single word written on them: no. The city's biggest company—indeed, the only true industry in town—responded with its own campaign. The firm made brushes, everything from toothbrushes to special brushes for industry. It was typical of Germany's *Mittelstand*—midsize, often family-owned firms that would be considered too small to be global players in most countries and would hence be taken over by a large corporation. In Germany, they find their market niche and defend their independence.

Like most German industrial firms, the brush maker in Schönau didn't like the idea of a bunch of laypeople buying back the grid. On toothbrush boxes sold locally, it started writing a single word: "yes." The company took out a full-page ad in the paper, explaining that "we would have to reconsider whether it makes sense at all to reinvest in Schönau" if locals took charge of the grid.

KWR stepped up its own campaign, claiming that EWS would post a loss of 30,000 marks annually and that people would also be laid off in the region. The utility began showing off all its equipment—from backup generators to heavy tractors needed to restore power lines in the snowy forest—so that people would begin to wonder how a citizen initiative could ever do all that. The awareness-raising campaign was worth 30,000 marks to the firm, exactly the amount it said the cooperative would lose each year. Obviously, a lot of money was being made from the grid.[13]

The Power Rebels didn't have the money to keep up, but they were more numerous. They spent time going door-to-door answering everyone's questions. The visits were carefully organized. Teachers did not knock on the doors of their pupils, for instance. EWS also now had its own offices, which became a meeting place for regular events on everything from insulation to solar energy and unrelated community issues. The events were reminiscent of those held at the protesters' camp in Wyhl some 20 years before.

On 25 October 1994, KWR announced the sales price for the grid: 8.7 million marks, more than 14 times more than it paid the city for the concession 20 years ago. Michael Sladek took it in stride: "Schönau apparently has the most valuable grid in the entire country."[14]

EWS had already conducted its own estimate of the grid's value and came up with a much different figure: 3.9 million marks. The Power Rebels had also already started collecting money. Citizens could buy a stake in the new cooperative for 5000 marks per share. Locals alone purchased 480 such shares, bringing in 2.4 million marks. An additional 1.7 million came from shares sold to people across the country. The Power Rebels already had more money than they thought they needed—just not as much as the utility wanted.

The institutional investor who originally put the project's potential return "prudently" at above 10 percent was willing to fill the gap, but the idea went against the campaign's spirit. Why fight so hard only to hand a majority of the project to someone from out of town? The GLS Bank, which focuses on sustainability in investments, had been advising the Power Rebels and did not believe the institutional investor was a good idea. "Every fund we have created up to now has filled up completely," they assure EWS.[15]

---

[13] Janzing and Seifried, p. 45 ff.
[14] Janzing and Seifried, p. 56 ff.
[15] Janzing and Seifried, p. 55.

The strategy, at least, was clear. Pay the money so that the concession, which continued indefinitely in the interim, could be switched over. Then, EWS would challenge the purchase price in court to get part of their money back.

How "I Am a Failure" Became a Winning Slogan

Wolfgang Zander conducted EWS's audit of the local grid to determine its value. He was the one who put the figure at 3.9 million. And he was sure he was right. "Established utilities are not used to facing competent criticism," he reassured EWS.

In reaching his assessment, he took a close look at everything, even walking out to physically measure the length of underground cables. He found 21 kilometers. The utility asked a German auditing firm (since taken over by PriceWaterhouseCoopers) to produce an audit that would "stand up in court." These savvy auditors were not about to walk around in fields, however. They based their estimates on all their experience in the energy sector throughout the country, using assumptions they held to be applicable averages for utilities in Germany. Apparently, no one had ever audited the auditors, who put the length of underground cables in Schönau at 33 km. It was the most egregious error, but not the only one. The auditors also listed every household power meter at its full new price without any depreciation—as though every single unit would have to be replaced as a part of the handover.

A two-man engineering firm from Aachen had showed up one of the country's biggest consultancies in the energy sector. When the facts were revealed, the utility's auditors admitted that Zander was correct, but the utility stuck to its price nonetheless—for good reason. The utility knew what the campaign's strategy was by now: buy first, sue later. But if such a high price were paid, the Economics Ministry, which still had to approve the deal, would argue that the price is too high, so the cooperative would not be profitable. It was a cynical tactic by the utility: charge a price it knows is too high in order to block the sale.

The Power Rebels decided to collect the missing money as donations. That way, it did not count as equity and would not be used in calculating profitability. All EWS needed was a million people out of a country of 80 million to donate five marks, the cost of a pizza. The GLS Bank's communications expert said outreach was needed. He contacted the country's 50 largest ad agencies asking for ideas to promote the campaign nationwide. There were only two conditions: the campaign cannot cost anything, and the agency must not have any connections to the energy sector. When he heard the idea, Michael Sladek didn't believe it would work. But 15 agencies actually responded with proposals—for free. "Whenever I have trouble imagining that something will work, I like to think of this episode," Ursula Sladek says today.[16]

The idea eventually chosen was one not even meant seriously when first proposed after a long round of frustrating brainstorming. Not surprisingly for a slogan that was successful, the particular wording has so many connotations that it is hard to translate. *Ich bin ein Störfall* means "I am a failure" in a word-by-word translation, though a better translation would be "I'm bringing down the system." In technology, *Störfall* is an accident (or, as the nuclear sector prefers to say, an incident), but unlike its English equivalents the German word does not suggest a chance event. Instead, *Stör-* comes from the verb *stören*, meaning "to bother." In other words, *Ich bin ein Störfall* meant something like "I'm going to bother you until you fail" in Schönau. The ads displayed normal people in a black and white photo behind the slogan on a red banner. At the bottom, a text explained who the people were:

- "Kristin Schreier (23). Student, Scandinavia fan, and *Störfall*."
- "Hanna Kück (9 months). Baby, teddy bear collector, and *Störfall*."
- "Peter Jacobs (45). Entrepreneur, father, and *Störfall*."

The GLS wanted to run the ad campaign on the tenth anniversary of the Chernobyl accident (or incident, depending). It got one of the country's leading philosophers, Carl Amery, to spearhead the ad campaign. Nonetheless, the campaign almost did not get going. Michael

---

[16] Janzing and Seifried, p. 59.

Sladek opposed: "I am not a *Störfall*." (His wife Ursula disagrees.)[17] In the end, the GLS put its foot down—"*Störfall* or we are out." Thus, on 10 September 1996, a bit late for Chernobyl's anniversary, the campaign was launched in newspapers across the country. The response was so positive that the ad agency's top management overcame its skepticism and proposed to run TV ads in addition.

The ads went wild. Cinemas ran it for free, as did TV stations. A version was done for radio. The biggest environmental NGOs in the country called on their members to donate: Greenpeace, the WWF, and the German chapter of Friends of the Earth. Across the country, people told friends not to give them a present for their birthday, but donate to the campaign instead. When a woman from France donated 25,000 marks, the Power Rebels thanked her in a personal postcard. She responded with another 25,000 marks.[18]

One of the biggest donors was probably chocolate maker Alfred Ritter. He began investing in solar and electric mobility after Chernobyl ruined plantations in Turkey where he sourced his hazelnuts. "That really got me thinking," he explains expanding into sectors unrelated to chocolate. Michael Sladek arranged a meeting with Ritter, who pledged 200,000 marks. Within six weeks, the first million had been collected.[19]

In March 1997, KWR lowered its offer from 8.7 to 5.7 million euros. In April, EWS had that much. They offered to take over the grid on July 1, and once again KWR laughed at them, saying that the cooperative did not even have its own transformer stations, which would take at least three months to deliver, "as everyone in the industry knows." Walter Bolz, who would have handled the matter, had passed away earlier that year, but his replacement called up a colleague from his old utility and asked if he could have a couple of transformer stations from the ones it kept in reserve. He could. By mid-June, the transformers were ready to go, and everything else was in place—two weeks before the proposed date that KWR thought impossible.

---

[17] Janzing and Seifried, p. 60.

[18] Janzing and Seifried, p. 62.

[19] Janzing and Seifried, p. 63.

Now, all the late Bolz' preparations were crucial. The handover was proposed to the Economics Ministry in Stuttgart. EWS had only hired three people, and Stuttgart wanted contracts signed with a neighboring utility or local electricians to ensure that emergency staff would be available in case of illness or emergencies. That was done, but the clock was still ticking, and there was no further word from Stuttgart. Then, just days before July 1, one of Michael Sladek's patients came into his office and reported what he had just heard on the radio: Stuttgart had just put the letter in the mail.

Everyone breathed a sigh of relief. In a few days, EWS would own the grid after more than a decade of campaigning. But there was one last stage in the fight: getting part of the money back in court.[20]

Peter Becker, a lawyer specializing in defending the rights of municipalities to own their own infrastructure, helped out EWS in court. In the early 1990s, he had made a name for himself defending towns in the former East Germany from West German utilities, which tried to move into the country as quickly as possible after reunification. Though a West German, Becker managed to bring together East German mayors and convince them to defend themselves. In many cases, they did so successfully.

In July 1997, EWS and KWR sign a contract stipulating that the payment of 5,683,458 marks is made contingent upon a subsequent legal review. Becker formulates the clause himself.

Out of court, the two parties agree to have a third audit of the grid's value conducted, and this one will count; the auditor has been jointly chosen. This report comes to a conclusion that surprises everyone—the grid is only worth 3.5 million marks, 400,000 less than EWS's Wolfgang Zander previously estimated. The utility has lost this war in every conceivable way: hearts, minds, data, facts, analyses, you name it. KWR immediately pays back 2.3 million marks including interest and chalks the whole story up to experience.

---

[20] Janzing and Seifried, p. 66.

Waiting for Liberalization Was Not an Option

On 1 April 1998, the German power market was liberalized. Mainly at the behest of the European Commission, Germany revised its Energy Management Act of 1935—the one Michael Sladek fought to get rid of. Since then, there have been no power monopolies. German customers are free to choose a provider from hundreds of competitors. Only the country's high-voltage transmission grid is still divided up into four zones as natural monopolies.

In a way, the Power Rebels could have saved themselves a lot of trouble by just waiting for liberalization. Their campaign succeeded only nine months earlier. But they would then have only been able to switch providers as customers. KWR still might have been the local provider. After all, even medium-voltage and low-voltage power lines represent a "natural monopoly"; competing power lines are not set up in the name of greater competition, as is done, say, with telecommunication towers for cell phones. In that respect, the Power Rebels' campaign was useful because liberalization of the power market would not have automatically brought about a change in grid ownership.

More importantly, the Power Rebels gradually appeared to be the only adults in a room full of whiners. Repeatedly, their estimates and analyses turned out to be more reliable than those from energy sector experts. And of course, power reliability has not dropped in the town since EWS took over in 1997. In 2007, for instance, the town experienced only three minutes of grid outages for the entire year, even less than the national average of around 15 minutes annually. This level, incidentally, is reached each month in the US, where minutes of downtime range from around 100 to 400 minutes annually, depending on the state. The British perform only slightly better at around 50 to 60 minutes of grid downtime annually.[21]

The brush manufacturer, Frisetta, thus never had a reason to leave town, as it threatened to do before the citizen takeover of the grid.

---

[21] Morris, Craig. "German grid keeps getting more reliable." Renewables International. The Magazine. 21 August 2015. Accessed February 5, 2016. http://www.renewablesinternational.net/german-grid-keeps-getting-more-reliable/150/537/89595/.

Perhaps that's why it stayed in town until 2003, when its polymer division relocated, reducing the town's power consumption by around 15 percent in the process. The brush division remains in Schönau, however, though an American firm took it over in 2013. Still, locals remain proud of their Brush Valley, as they sometimes joke over a glass of wine from Kaiserstuhl.[22]

The institutional investor that wanted to get involved in EWS also benefited from the campaign in Schönau. After liberalization, it founded Lichtblick, which quickly became Germany's biggest provider of 100 percent green electricity, serving some 600,000 customers in Germany in mid-2014. It is a competitor of EWS in that respect, and it is not surprising that EWS is smaller, for it was never the intention of the Power Rebels to grow beyond their own town's borders. They do, however, still provide assistance to other municipalities looking to buy back their grids, most recently in the successful campaign in Hamburg and the failed campaign in Berlin, both in 2014.[23]

Increasingly, Ursula Sladek has taken the foreground. Her crowning moment came in 2011, when she gave a speech in English on receiving the Goldman Environmental Prize, arguably the most prestigious such award in the world, at a ceremony in San Francisco. And once again, she was on crutches, this time with a broken ankle. She met with Barack Obama in the Oval Office. The President wanted to know more about EWS.

She turned out to be perfect for TV talk shows. She displays an unwavering ability to simplify complex findings so that a TV audience can understand them. Instead of using technical terms, such as "distributed energy," she says, "The Sun not only shines on RWE and Eon. It shines over the whole country."

She is quick-witted. When the CEO of RWE asked her in 2009 on live TV why EWS's customers paid 250 euros a year more than the cheapest

---

[22] Sattelberger, Dirk. "Schönau: Ranir übernimmt Zahnbürstenhersteller Frisetta." Badische Zeitung. 25 May 2013. Accessed February 5, 2016. http://www.badische-zeitung.de/schoenau/schoenau-ranir-uebernimmt-zahnbuerstenhersteller-frisettaDOUBLEHYPHEN72221230.html.

[23] Morris, Craig. "Vattenfall leaves Hamburg." Renewables International. The Magazine. 14 January 2014. Accessed February 5, 2016. http://www.renewablesinternational.net/vattenfall-leaves-hamburg/150/537/76082/.

competitors, she answered that rates at EWS were no higher than the big utility with the most nuclear and added, "is your company cheaper than mine?" (It wasn't.) On another show, she was asked whether her company would raise its rates next year as well. "Yes, but not as much as my colleague," she answered, pointing to the CEO of a conventional utility. "Maybe we were a bit cleverer in procurement," she added—as always, with an irresistible smile.

When pointing out the devastating environmental impacts of uranium mining, she asked, "As Christians, are we just supposed to say that doesn't concern us?" And when asked whether the nuclear phaseout wouldn't increase carbon emissions, she answered that there were other ways of reducing emissions (such as efficiency and a focus outside the power sector)—and that the main effect of extending nuclear plant commissions would be "reinforcing monopolistic structures in the power market." The studio audience roared in approval.

And KWR? In 1998, it took advantage of the liberalization of the power market to join forces with another firm in creating NaturEnergie, one of the first providers of green electricity in Germany, which currently serves more than 35,000 customers—compared with around 150,000 customers at EWS. Proponents of renewables warn consumers not to choose green power providers like NaturEnergie, however, because they simply sell at a premium the renewable electricity (generally hydropower) they already had, so those green power customers simply cross-subsidize the nuclear and fossil assets at these utilities.

In Schönau's elections of June 2015, the Independents won two seats (down from five), compared with five for the Christian Democrats (unchanged). The SPD rose out of the ashes from only two seats to seven. The Greens still have none.[24] The rift over the construction of a nuclear power plant that divided Wyhl lasted for decades, but it seems that the wounds have healed better in Schönau. The mayoral race was described as "heavy petting" in the press; no mud fight here. The Power Rebels are no longer a bone of contention either. In 2007, only 29 households in the town were not customers of EWS.

---

[24] http://www.stadt-schoenau.de/pb/,Lde/262749.html.

The story of EWS's struggle to become a community grid operator shows how important it is for citizens to develop their own expertise: in politics, technology, and public relations. But without charismatic leaders like Ursula and Michael Sladek, little would happen. Whereas Wyhl helped make the German government less authoritarian, the Power Rebels of Schönau helped make energy experts less arrogant.

The issue was not partisan; citizens wanted the right to make their own energy so they could decide how it is made. Schönau is a good example of a town where conservatives are also conservationists.

6

Renewable Energy in Conservative Communities

Drive out to Larrieden, and you are truly in rural Bavaria. The hamlet is located in-between rolling hills of lush green. You will need to go by car for a lack of public transportation. "There is only the school bus to Feuchtwangen," says Stefan Bayerlein, who heads a local community energy project. "So if you need to do some shopping and don't have a car, you leave in the early morning with the kids and come back in the early afternoon with them," he says, describing the inconvenience.

Reliance on cars doesn't exactly sound renewable, but Larrieden is nonetheless a major success story—one of a growing number of villages and small towns going 100 percent renewable for electricity and heat. "Saving the planet was not the main objective," Bayerlein explains, "saving the community was." Once a stable farming village some five kilometers from the small town of Feuchtwangen, Larrieden saw its population dwindle in recent decades. Shops began to close as people increasingly drove into town. "Property is cheap here, but practically every adult needs their own car, so the savings are relative," says Bayerlein. Today, Larrieden has no grocery store, no bakery, no shops at all. Like so many other places in Germany, the hamlet that once had basic local infrastructure has become a dead suburb. Next stop: ghost town.

© The Editor(s) (if applicable) and The Author(s) 2016   **95**
C. Morris, A. Jungjohann, *Energy Democracy*,
DOI 10.1007/978-3-319-31891-2_6

Headed by folks like Bayerlein, locals are fighting to keep their community alive. As you approach the village, you begin to see a wind turbine here and there. On an old farm building, a solar thermal array that has seen better years indicates a long tradition and early commitment to renewables. On two edges, Larrieden has a few structures that look a bit like circus tents: biogas units, as the gigantic piles of biomass under rain protection indicate. A few of the buildings, though not too many, also have photovoltaic roofs.

A number of coincidences came together to turn this conservative rural community into a renewable energy showcase. "For decades, we have had a technology research center in Feuchtwangen," explains Michael Köhlein, who manages a few of the biogas facilities today. "They worked on things like solar heat a long time ago, so we had such expertise at our fingertips." Then, a farmer working completely on his own figured out that energy crops would be a better way for him to make a living than animal husbandry. Whenever he got sick, he still had to tend to his farm animals, but energy crops allowed him to take a day or two off. Nearby, there was also a buyer of large amounts of heat: a residential complex for refugees.

That initial biogas project then got other villagers thinking. Bayerlein pulls out a map of Larrieden in his home office. It shows that there is not a single district heat network, but three. "You can see how this grew organically," he says. The village is strung along two main roads, making it cross-shaped. "It is far more efficient to pipe the biogas than the heat from the biogas unit," explains Köhlein. "By the time you have piped heat a kilometer, the losses are significant," he says, pointing to the position on the map where the three separate piped heat networks could converge but don't.

On his computer screen in his home office, Bayerlein displays a slideshow he often gives of the project; practically everyone who manages a community energy firm spends a considerable amount of time explaining the project to visitors from nearby towns and such faraway places as the United States and Japan. One of his main points is that going green was a side issue, not the main driver. "The villagers increasingly lived in isolation. Without local shops, cafés, and restaurants, there was no community life. People mainly saw each other when they drove by in cars, but there was no communication."

Then, some new people moved to the village and started talking to everyone about community renewable energy. Bayerlein says their newcomer status helped them bring everyone together. "All of the talk about what needed to be done opened up old wounds." Like the village of Wyhl, Larrieden had seen its agricultural property renegotiated as a part of the *Flurbereinigung* (land reform) campaign decades ago. During that process, the one or other landowner had been forced to give up a few square meters here or there. Negotiations about whose property would now have to be dug through for the district heat network reminded folks of those old wounds. "Without the newcomers, these talks might have ended in endless quarrel and gone nowhere," Bayerlein says. "With them, we were able to keep our eye on the common goal."

Sorting through his computer presentation, Bayerlein stops at a photo showing locals digging up a yard with shovels. To keep costs down, they did as much of the work as they could themselves. Digging doesn't exactly sound like a lot of fun, but the people in the photo look like they are having a good time. Bayerlein says the photo shows the point where everyone began to appreciate how the project was turning what had become an anonymous suburb back into a community.

Köhlein and Bayerlein are careful about perceptions as well. They know that they must avoid being perceived to be doing exactly what big energy companies do—reaping profits off of local resources, while the locals themselves have to deal with the impact. For instance, Köhlein regularly drives biomass through town with his tractor. After a few days, it starts to create a real mess on the roads. So once a week, he goes through with a sweeper attached to his tractor to clean up the streets. "It's not enough really," he admits, "the streets are still dirty for days at a time. But the public perception is tremendous—people really appreciate the effort. They know we don't have to do it. We do it because we care." Bayerlein concurs: "It's as much about the gesture as it is about the result."

Later, we visit a wind turbine on a hill overlooking the village. It, too, was built with investments from locals. And again, the locals did what they could to keep costs down. Bayerlein and Köhlein have to laugh about one aspect of the story. "When people heard that we wanted 20,000 euros for planning, they thought they were getting put through the ringer." He then points over to four turbines visible a few kilometers away on another hilltop. "But later, when wind planning firm Juwi put those up, they charged 100,000 euros per unit for the same services."

Juwi is widely heralded as exactly the kind of SME the Energiewende produces. Founded in the mid-1990s, Juwi develops turnkey wind farms, which it then either operates itself or sells to investors—institutional ones or communities. As such, it and others like it—Windwärts and Belelectric come to mind—performed a crucial role; at a time when major utilities were uninterested in renewables, these planning firms filled that gap. But hard-core community projects like the one in Larrieden still tend to see these big planning firms as out-of-towners who come in, build, sell, and leave. If renewable energy is food, big utilities are red meat, utility-scale renewable energy is organic meat, Juwi is vegetarian, and Larrieden is vegan.

Yet, the villagers in Larrieden are not leftist at all. The town is a conservative community within a county that votes Christian Union. Its representative in the Bundestag is Christian Union politician Josef Göppel, arguably the main proponent of community energy in German Parliament today. For instance, Göppel spearheaded the founding of the federal community energy organization BEEN in Berlin in 2013; up to that point, community projects had no nationwide lobby group. While Göppel was not a driving force in Larrieden, "He is very proud of us," Bayerlein says. But at one point, Köhlein adds, Göppel played a crucial role. The Bavarian government is the most skeptical of wind power of the 16 German states, probably the only one that could be called anti-wind. "When the local planning officials in Feuchtwangen began to give us trouble for the wind turbine permit, Göppel came in and gave them a real talking to," Köhlein remembers. "His intervention was crucial," Bayerlein adds.

Why Sometimes 100 Percent Renewables Is Not Enough

In the past few years, numerous German communities like Larrieden have begun to make international headlines for going 100 percent renewable. Feldheim is perhaps the most prominent. Unlike Larrieden, the town of a mere 150 people has the good fortune of being close to Berlin and hence quickly accessible to visiting journalists and energy experts. Other villages and rural communities are equally impressive, though a bit harder to reach. Located near Munich, Wildpoldsried has also received some press coverage. Jühnde, in contrast, is not close to a major city; it claims to be Germany's first "bioenergy village." Those who visit Freiburg—which calls itself Green City Freiburg—are likely to be carted off to see the rural community of Freiamt.

These success stories are everywhere in Germany. Add them up, and you begin to get impressive statewide statistics.[1] For instance, Mecklenburg-Vorpommern—the state bordering the Baltic—was already 120 percent renewable for electricity in 2013, meaning that its renewable power generation was 20 percent greater than its power demand.[2] Neighboring Schleswig-Holstein (bordering the North Sea, the Baltic, and Denmark) is also now 100 percent renewable in a net calculation for electricity—and has a target of 300 percent, as the state's (and Germany's first and only) Energiewende Minister Robert Habeck points out.[3] The state thus plans to be a major power exporter.

"I believe you can do 300 percent renewable power, but what will you do with the other 200 percent?" a skeptical Environmental Minister Peter Altmaier asked the governor of Schleswig-Holstein in 2012.[4] While in office as Environmental Minister, Altmaier calculated that the cumulative targets for Germany's 16 states actually exceeded the federal government's target. Was someone going to have to cut back?

At the heart of Altmaier's criticism is a simple fact: all these numbers are net calculations. Mecklenburg-Vorpommern produces more renew-

---

[1] For an overview of villages, towns, and regions in Germany that are going 100 percent renewable, see http://www.100-ee.de/.

[2] Morris, Craig. "German state already has 120 percent renewable power." Renewables International. The Magazine. 24 June 2014. Accessed February 5, 2016. http://www.renewablesinternational.net/german-state-already-has-120-percent-renewable-power/150/537/79680/.

[3] Hockenos, Paul. "Robert Habeck: Germany's First and Only Minister for the Energiewende." Energy Transition. The German Energiewende. 20 December 2013. Accessed February 5, 2016. http://energytransition.de/2013/12/robert-habeck-germanys-first-and-only-minister-for-the-energiewende/.

[4] Morris, Craig. "I know you can do 300 % renewable." Renewables International. The Magazine. 18 September 2012. Accessed February 5, 2016. http://www.renewablesinternational.net/i-know-you-can-do-300-renewable/150/505/56639/.

able electricity than it consumes over the year, but it is not self-sufficient. It needs to sell electricity to neighboring regions when it has too much, and it also generates and imports conventional power when it does not have enough renewable electricity.

The same holds true for tiny Feldheim, which has repeatedly been called "off grid." In reality, Feldheim is a hamlet stuck onto a gigantic wind farm. Though "household" is not a term in physics, journalists like to avoid such technical terms as "megawatts," explaining instead that a particular wind farm can serve a certain number of households. The 74 megawatts of wind turbines installed in Feldheim could cover the power demand of up to 100,000 households. Feldheim doesn't even have a hundred.

The wind farm (and the village) most definitely has a grid connection—how else would it sell 99 percent of the power it generates? Still, let's not belittle this success; like Jühnde and Freiamt, Feldheim has combined individual and group citizen projects not only to be 100 percent renewable net for electricity, but also for heat from local biomass. To go 100 percent renewable for heat, a district heat network had to be built. The next step is to further reduce reliance on the grid by means of battery storage.[5]

Such success stories are initially possible in small rural areas, but can large cities also go 100 percent renewable? The short answer is: they won't have to. The City of Frankfurt's plan is illustrative in this respect. It aims to be 100 percent renewable for electricity and heat by 2050 "at the latest."[6] (Transportation remains a hard nut to crack; Frankfurt's plan does not include Germany's largest airport, for instance.) With an incorporated population of nearly a million people and more than two million in the surrounding area, Frankfurt stands little chance of generating all its energy within its own area. The plan therefore focuses on efficiency (especially for heat) and energy imports from the rural surrounding area. German cities will encircle themselves with Feldheims. The hamlets will not go off-grid because conglomerations will need energy imports.

Feldheim is a messier story in another respect as well. The Germans say that *Bürgerenergie* (literally, "citizen energy," though we use the term "community energy" in this book) is the driving force behind their energy

---

[5] See http://www.neue-energien-forum-feldheim.de/index.php/energieautarkes-dorf.
[6] See http://frankfurt.de/sixcms/detail.php?id=4576&_ffmpar[_id_inhalt]=15931823.

transition. Feldheim is certainly one example, and there is no doubt that the community came together in the project. But as elsewhere, the whole story consists of individual efforts (such as solar roofs on homes) and outsider participation. How else could a mere 150 people leverage the capital for 74 megawatts of wind turbines when the price tag easily reaches 100 million euros?

Anyone who believes community renewable energy must be small and cannot move fast enough should obviously take a closer look at the German story, which tells a different tale. In 2015, a project in Medelby near the Danish border saw 360 German citizens leverage 127 million euros to build 82.3 megawatts spread across 24 wind turbines.[7] (For what it's worth, the project claims it will serve 59,000 households—far more than are in the local area.[8])

In Feldheim, the situation was a bit different. The firm that manages the wind farm, Energiequelle, has its headquarters some 60 kilometers from the town and is open to all investors. In contrast, you practically had to live within two kilometers of Medelby to invest. Is a project still "community energy" if out-of-towners are behind it? What if one of the big utilities simply allows locals to buy shares in the project? If a municipal utility is behind a renewable energy project without further citizen investors, is that a community project as well? And for that matter, citizens have always been able to buy shares in listed corporations, so haven't we had community energy all along?

So What Is "Bürgerenergie"?

First, corporations do not serve community needs today. They used to, however. When the CEO of General Motors famously stated in 1953 that "what was good for our country was good for General Motors, and vice versa," he spoke as the head of the firm that focused on stakeholder value, not just shareholder value. Communities are stakeholders in corporations

---

[7] Morris, Craig. "How big can a community wind farm be?" Renewables International. The Magazine. 11 May 2015. Accessed February 5, 2016. http://www.renewablesinternational.net/how-big-can-a-community-wind-farm-be/150/435/87457/.

[8] See http://www.sonnewindwaerme.de/schleswig-holsteins-groesster-buergerwindpark-eroeffnet.

even when they don't own stock because corporate activities affect communities. In the 1950s, the potential conflict between stakeholder value and shareholder value was still acknowledged, and both were given equal status. But by the 1990s, shareholder value reigned supreme, and the corporate world no longer acknowledged the potential for conflict. As long as this situation persists, corporations cannot be understood as potentially serving communities regardless of shareholder membership, which can change at any time.[9]

In contrast to anonymous shareholding, Germany has long had the principle of *Eigentum verpflichtet,* which could be translated as "with property comes responsibility." The term was used in the Weimar Constitution of 1919 (Germany's first democracy) and reappears in the current Basic Law (the German Constitution). The concept that property ownership entails a responsibility toward other stakeholders and the community has thus been well-established in Germany for a century; there is no clear equivalent legal term in the Anglo world.

Although German citizens have clearly been far more involved in the energy transition than normal people have in other countries (aside from Denmark), there is no clear definition of what *Bürgerenergie* means. One question is whether individual investments—such as residential solar roofs—qualify, or whether "community" always means a group must be involved. The Germans generally view individual solar roofs by citizens as community energy, especially because the investors in such projects are often involved in other group projects. When a small town talks about going renewable, it includes solar roofs.

Another distinction needs to be made between municipal and community. In general, municipal projects are not considered to be *Bürgerenergie.* For instance, Germany has roughly 900 municipal utilities, the smallest of which are staffed by a single person. These entities purchase electricity from big utilities on behalf of local governments. Such utilities often do not have the capacity to take the lead on community renewable energy projects, which are instead usually spearheaded by local citizens. When parents at a local public school launch a campaign to put a solar roof on a school building—

---

[9] Ho, Karen Z. *Liquidated: An ethnography of Wall Street.* A John Hope Franklin Center book. Durham: Duke Univ. Press, 2009.

for which they need the city's permission—the project is considered community. A city government putting solar on municipal buildings might not be.

Larger municipals, such as the one from Munich, make investments all over the world. Stadtwerke München is behind offshore wind farms in the German North Sea and Liverpool Bay (UK) and also has onshore wind investments across Germany—but also in France and Sweden. Here, this municipal utility hardly differs from RWE and Eon, Germany's two biggest utilities, which are listed corporations. They prefer to invest in renewables abroad, where the investments do not conflict with their existing power generation assets.

From the American perspective, a distinction also needs to be made between German and US cooperatives. In the US, rural cooperatives were created as a kind of municipal for areas without a major conglomeration as a part of the rural electrification program of the 1930s. Like German municipals, many of these US cooperatives invested in coal and nuclear over the decades. Today, some of them are therefore as outright hostile to renewable energy as conventional utilities are.[10]

In Germany, citizen energy cooperatives can be broken down into two groups. First, we have cooperatives in the strict legal sense of the term *Genossenschaften*. These groups have per-person voting rights regardless of the individual's investment. In contrast, numerous community energy projects have gone into business as limited liability companies; here, voting rights are granted relative to the individual's investment stake. The earliest community wind farms were launched as limited liability companies, a legal form that remained attractive until 2006, when the government changed the law to facilitate the founding of renewable energy cooperatives in order to encourage the founding of such entities.

We thus face some potential confusion:

- German cooperatives are not synonymous with US cooperatives;
- German municipals are not considered cooperatives;
- In Germany, community renewable energy projects can include individual efforts; and
- The term *Bürgerenergie* is used for community projects that are limited liability firms and co-ops in the legal sense (*Genossenschaften*).

---

[10] See http://cherrylandelectric.coop/2015/01/lessons-germany/.

The lack of a clear definition means that the German government cannot have a specific goal for the share of citizen ownership. As of 2012, community projects made up roughly half of investments in renewable energy in Germany, but that amount is shrinking now for two reasons: first, offshore wind is finally growing strongly, and there is no community ownership of offshore wind in Germany (in contrast to Denmark, where a solution for citizen participation was found); and second, the policy transition from feed-in tariffs to auctions since 2015 will also reduce community involvement. In addition, the municipal utility of Mannheim acquired Juwi in 2015. That planning firm will therefore not be developing so many projects in order to sell them to citizen co-ops—the utility will probably want to retain ownership. In contrast, Scotland has a clear target for "community and locally owned renewable energy" of 500 megawatts by 2020, and to do so it has clearly defined the term.[11]

If community energy begins to shrink in Germany, public support for the Energiewende may shrink as the options for citizens to take part become more restricted. The government is aware of what is at stake, but as of 2016 it did not seem to perceive the shift in renewable energy investments from citizens back to conventional utilities as a major problem.

Citizen Participation Raises Acceptance

Germany has the second-highest power rates in the EU after Denmark, a fact that has drawn a lot of attention abroad. Foreign onlookers frequently assume that the Germans must be up in arms about the high cost of the Energiewende. The foreigners are then surprised to hear that Energiewende demonstrations have been held, but that they were all by citizens defending their right to make their own energy. There has been no demonstration about higher electricity prices.

Citizen participation has been crucial toward raising acceptance levels for the energy transition in Germany. In other countries, locals are told that they live in, say, an area with great wind resources, so the utility is going to come in and put up scores of wind turbines. A natural reaction is

---

[11] See http://www.gov.scot/Topics/Business-Industry/Energy/Energy-sources/19185/Communities.

for people to complain about privatized profits and socialized impacts—
the utility will get most of the money, but local people will have to put
up with the changed landscape.

If people cannot build renewables, politicians can only allow citizens
to block projects—exactly what is happening right now in England,
where top energy politicians are pushing to allow citizens to "say no
to wind farms."[12] German experience suggests that citizens should be
allowed to say yes to wind farms—and that they are more likely to do
so if they set up the wind turbines themselves.

The wind sector in the US and the UK has to deal with talk of noise
and blight from wind turbines, concerns that are hard to view objectively.
After all, anyone who tries to record the noise from a wind turbine
may have trouble doing so if a car drives by, and whether a wind turbine
is uglier than other objects in built environments (power pylons, billboards, roads, buildings, etc.) is a matter of taste. When such concerns
are belittled, people feel that the experts do not take them seriously, that
they are powerless, and that big money will get what it wants anyway.

Communication experts who try to help utilities navigate what experts
take for NIMBYism therefore face a dilemma when they assume the
experts are right and the public is wrong. Here is one passage from the
2012 edition of *Responding to community outrage* by Peter Sandman (originally published in 1993), just after explaining that his intended reader is
utility representatives:[13]

> I am assuming that people are very upset about some risk, or likely to
> become very upset about it; that you do not believe their level of concern
> is technically justified; and that you are looking for ways to understand and
> respond better.

---

[12] See Hope, Christopher. "No more windfarms unless local people say yes, says new Energy
Secretary." The Telegraph. 17 May 2015. Accessed February 5, 2016. http://www.telegraph.co.uk/
news/earth/energy/windpower/11611050/Amber-Rudd-No-more-windfarms-unless-local-people-
say-yes.html and Harvey, Fiona, and Peter Walker. "Residents to get more say over windfarms." The
Guardian. 6 June 2013. Accessed February 4, 2016. http://www.theguardian.com/environ-
ment/2013/jun/06/residents-get-more-say-wind-farm.
[13] Sandman, Peter M. *Responding to community outrage: Strategies for effective risk communication.* 5.
print. Fairfax Va.: American Industrial Hygiene Association, 2003.

Sandman only includes one other option, which he says his book does not investigate: when "officials are trying to break through public denial or apathy about a serious hazard," such as when "agencies try to persuade homeowners to test for radon." In both cases, the experts are right and the public is wrong. He does not even consider what the German public has experienced during the Energiewende, in which energy experts and top politicians promise to the public for decades that wind power would not work, solar power would remain too expensive, the lights would go out without nuclear, and so on. We must consider cases in which the public is right and the experts are wrong.[14]

Treat people like they are wrong and you are right, and it is hard not to seem arrogant. In the case of noise and light from wind turbines (where there is often no clear wrong and right), the best approach is to let wind farms grow at a level the public accepts and with as much public participation as possible.

Vauban and Freiamt: Energy Democracy in Practice

Freiamt, Germany, is another good example of energy democracy in practice. This rural community began its path toward 100 percent renewable energy when Ernst Leimer, a local mechanic and a Christian Union member of the town council, proposed putting up a few wind turbines. Not everyone was happy. "If I had said I wanted to put up ten right away, the idea would have gone nowhere," he says. Instead, he found local investors and got enough local acceptance to put up a few turbines. When everyone could see what they look like in the landscape (not bad) and could experience how hard they were to hear, the focus turned to what a good investment they had been. Another few were added, and a third round brought the total up to ten. Add on all the biogas units and photovoltaics roofs in the community, and Freiamt is a net power exporter and gets most of its heat from renewables.

---

[14] For a more enlightened approach to Sandman, see Devine-Wright, Patrick, ed. *Renewable energy and the public: From NIMBY to participation*. London: Earthscan, 2011.

Leimer did all this in his spare time. "Some people play golf; I build renewable energy systems," he laughs. And distributed energy can progress quickly when so many communities move simultaneously, even when you take each step carefully to ensure public acceptance. In 2015, Germany had more solar installed than any other country in the world and was third in wind power behind only China and the US. Nearly half of those investments had been made by individuals by 2012, with almost another 40 percent coming from new players (largely SMEs and institutional investors).[15] People who worry that distributed energy might not move fast enough to prevent climate change need not worry; the German government is quite concerned about how quickly wind and solar grew in the hands of the public.

Of course, experts are still needed; no one is suggesting that laypeople know enough to design a future energy infrastructure. Chilean architect Alejandro Aravena perhaps put it best when he won the prestigious Pritzker Prize in 2016. Aravena engages with communities before he builds: "What we're trying to do by asking people to participate is envision what is the question, not what is the answer. There's nothing worse than answering the wrong questions well."[16]

He could have been talking about the Vauban neighborhood in Freiburg, Germany. The project has drawn global attention for one particular aspect: it is considered a car-free zone. It deserves greater attention for how that result came about—through energy democracy.

Visitors to the neighborhood—and they come several times a week by the busload from all over the world—generally have one question: how did the government get people to accept not being able to park their cars in front of their homes? The answer is that the question is wrong. In reality, citizens petitioned the government to redesign the district in order to get cars off the streets.

---

[15] Morris, Craig. "Citizens own half of German renewable energy." Energy Transition. The German Energiewende. 29 October 2013. Accessed February 4, 2016. http://energytransition.de/2013/10/citizens-own-half-of-german-renewables/.

[16] Dezeen Magazine. "Architects "are never taught the right thing" says 2016 Pritzker laureate Alejandro Aravena." Interview with Alejandro Aravena. 13 January 2016. Accessed February 5, 2016. http://www.dezeen.com/2016/01/13/alejandro-aravena-interview-pritzker-prize-laureate-2016-social-incremental-housing-chilean-architect/#comment-2455523622.

In 1992, French soldiers who had occupied that region of Germany since the end of World War II left their barracks in Vauban. It was the end of the Cold War, and the Allies agreed that reunited Germany no longer needed occupying. The city of Freiburg wanted to turn the area into a new residential development.

And it wanted to make the neighborhood special, so Vauban was to be developed largely by *Baugruppen* (literally, construction groups). In these groups, citizens meet with an architect to divvy up the three-dimensional space inside a complex to be built. No housing developers are involved, so costs are lower because the middleman is cut out. In return, the new tenants meet regularly (often weekly) with each other and the architect to discuss what everyone wants. What you save in money you spend in time—but you are also developing a relationship with your future neighbors in the process. In other words, you are building your community while you build your new home.

The Vauban *Baugruppen* had a special request: they wanted cars off of their streets so that the streets could become public spaces where children can play. The city initially said no. It had no choice; a federal German law (still) stipulates that at least one parking space must be built for every new apartment constructed. So the citizens sat down with city officials to come up with a compromise: car-free side streets, with parking only on the main road alongside the new tram line, and central parking garages on three corners of the neighborhood. People could then reach their cars within a two-minute walk and drive up to their front doors for quick loading and unloading.

Each household then agreed to purchase a single parking space at a cost of 12,000 euros at the time, a price unremarkable for the city. But the citizens wanted more: if they did not have a car, they could opt out, but they would have to produce an affidavit each year attesting that they had not purchased a car. In a handful of cases, residents have tried to cheat the system by having a friend or relative register the car, but compliance is above 99 percent across the roughly 4500 residents of Vauban—and even that figure understates the success. Only two of the three garages originally planned have been built, and neither of them are fully booked. The lot for the third garage that was never built is now used as a wild playground by local children, and it will remain an informal playground because the option of the third garage must be retained just in case everyone switches to cars in the future. It is unlikely they will do so. The tram stop is closer to most people's homes than the parking garages are, and there are also two buses. Residents of Vauban experience how inconvenient and expensive cars are once cities are built for people.

Several *Baugruppen* in Vauban also chose to build Passive House complexes. This type of architecture does without an active heating system, using instead a combination of high-tech and low-tech to provide passive heat: advanced insulation and large windows on the southern façade to let in sunlight in the winter. Broad-leaf trees in the front yard provide shading to protect from overheating in the summer, as do overhanging balconies facing the south. The balconies are extended far enough so that the glazed façades are completely shaded on the day of the summer equinox, June 20. Imagine that: buildings that stay cozy all year without air-conditioning or central heating—even in chilly Germany.

The citizens got a lot more from their city government in the process of developing Vauban. Officials took the opportunity to start asking citizens what they want more often. For instance, the spaces between complexes were to be public parks. Each household on the ground floor has its own small yard overlooking a giant common area the size of a football field. Instead of designing these areas according to some urban planner's liking, the city turned over the budget for each of these public parks to adjacent residents, who then decided at their *Baugruppen* meetings what should be done with the money. Each park looks totally different. One has a giant climbing rock along with an adult-sized swing (it alone is worth visiting Freiburg for). Another looks more like a modern playground for kids (the kind of thing an urban planner is likely to build). A third was essentially left wild, and local kids have carved their own paths into the woods.

Baugruppen exist all over Germany, and their success is hard to overstate, as Vauban illustrates. It's enough to make US journalist David Roberts jealous. He writes in his article "I want to live in a baugruppe":[17]

I want to live in a dense urban area where groceries, parks, schools, and restaurants are all within walking distance—where I can live comfortably

---

[17] Roberts, David. "I want to live in a baugruppe." Grist. 8 August 2013. Accessed February 5, 2016. http://grist.org/cities/i-want-to-live-in-a-baugruppe/.

without a car. I'd like for the district/neighborhood to be structured in such a way as to encourage casual encounters with neighbors. I'd like it to have a robust sense of community.

Vauban is a true community, not a collection of isolated adults in condos who, without a common area, only see each other as they drive by in their cars. Vauban residents meet each other on the street, in trams and buses, in common areas, and in the shops that line the main road.

It was citizens, not the government, who called for this progress. But the City of Freiburg can be praised for reacting properly. In Vauban, the city built the tram line immediately, even before construction on the homes was finished, to avoid the chicken-and-egg problem: who would move into a car-free neighborhood if the tram was still a few years away? The city also opened up more to citizen input; for the past decade, the government has put up the city's annual budget online for citizens to comment line by line. And they do.

Communities—Where Conservatives and Conservationists Agree

Most of all, Germany's energy transition is so successful because it is supported across party lines. As a result, the baby is not thrown out with the bathwater after every election. As Angel Gurría, Secretary-General of the OECD, put it in 2014:

> What is noteworthy about Germany's approach is that it has been more sustained and consistent than that of many other countries, which have been forced into complete U-turns in their energy transformation by pursuing needlessly costly policies.[18]

Germans themselves are aware of how important this consistency has been, and they, too, despair on seeing the back-and-forth of US politics

---

[18] OECD. "Roundtable: Climate is Everyone's Business." Opening Remarks by Angel Gurría, OECD Secretary-General. 14 May 2014, Berlin, Germany. Accessed February 5, 2016. http:// www.oecd.org/germany/roundtable-climate-is-everyones-business.htm.

in particular. Kai Schlegelmilch, an official at Germany's Ministry for the Environment, put it this way:

> The consensus and great continuity in German governmental work has a very high value for the country. After all, consistency provides planning reliability and investment certainty for the economy. It's hard for me to understand the United States in this respect because short-term policy changes are not in their own economic interest. And yet, there is agreement across party lines in the US that more energy independence is needed.[19]

As the examples of Larrieden and Wyhl show, German conservatives support the Energiewende specifically because it strengthens communities. Is there a lesson to learn here for other countries, where support for renewable energy divides more clearly along party lines?

The area where there is the greatest overlapping between different political camps—and where the most progress can be made—is communities. "Conservatives, Progressives, and everyone else likes farmers' markets, local food, mom-and-pop stores and other qualities of a thriving community," writes Jay Walljasper,[20] a US expert on the Commons.

However, a distinction has to be made between the group commonly referred to as "conservatives" in the US and the school of thought that we call "classical conservativism" below. Daniel McCarthy, editor of *The American Conservative*, makes a distinction between low-church conservatives, high-church conservatives, and no-church conservatives. For our purposes, his high-church conservatives (which he would describe himself as) are the closest equivalent to German conservatives—and hence, the most likely to want to strengthen communities, including with renewables. As he puts it, "The high church conservative's objective is to preserve the fabric of society and, so far as possible, elevate its culture."[21]

---

[19] Personal interview.

[20] See Walljasper, Jay. "The Conservative Case for a Commons Way of Life." Resilience. 16 July 2014. Accessed February 5, 2016. http://www.resilience.org/stories/2014-07-16/the-conservative-case-for-a-commons-way-of-life.

[21] McCarthy, Daniel. "What Would Burke Do?" In *The Essence of Conservatism: 10 Classic Reads from The American Conservative.* Edited by Daniel McCarthy. Langhorne, PA: The American Conservative, 2014.

When it comes to the energy sector, the focus of classical conservatism is thus not on the national level (energy independence) or the individual level (your right to have a solar roof), but on communities. Whenever Americans have come together across party lines to strengthen their communities with renewables, the progress has been tremendous.

Look closely at the classical conservatism espoused by McCarthy, and words like "society" and "community" abound. One classic of American conservative literature is Robert Nisbet's *The Quest for Community*. In it, he explains the primacy of the community for classical conservatives:

> Community is the product of people working together on problems, of autonomous and collective fulfillment of internal objectives, and of the experience of living under codes of authority which have been set to a large degree by the persons involved.

Classical conservatives like Nisbet and McCarthy oppose ideology, which they see as an imposition of a rigid utopian system (such as communism) by a powerful force from above (such as the state). They prefer local government to national government because the latter is simply too far away, as Nisbet explains:

> In our culture, with its cherished values of individual self-reliance and self-sufficiency, surrounded by relationships which become ever more impersonal and by authorities which become ever more remote, there is a rising tendency…towards increased feelings of aloneness and insecurity.

While non-classical conservatives tend to focus on things measurable and expressible in numbers (economic growth, income levels, etc.), classical conservatives are concerned about the human soul: happiness, aloneness, insecurity, and so on. And there is overlapping here with conservationists who call for a focus on communities. Here is Bill McKibben:

> Perhaps the very act of acquiring so much stuff has turned us ever more into individuals and ever less into members of a community, isolating us in a way that runs contrary to our most basic instincts.[22]

---

[22] McKibben, Bill. *Deep economy: The wealth of communities and the durable future*. New York: St. Martins Press, 2007, l. 627.

Germany has more than 2000 churches with solar roofs on them. Christian Union politician Josef Göppel is arguably the biggest supporter of community energy projects in the Bundestag today. His party colleague and former Environmental Minister Klaus Töpfer later went on to head the UN's Environmental Program, and he is now on the board at Germany's Institute for Advanced Sustainability Studies. Töpfer also wrote a forward to TV journalist Franz Alt's book entitled *The Ecological Jesus*. There is no perceived contradiction between being a conservative and being a conservationist in Germany. There shouldn't be in English either; after all, the stem of both words is "conserve."

Conservatives Are Not Libertarians

At present, libertarians—who stress individual freedom—set the tone within the US Republican Party (and to some extent among Democrats). An argument could be made that libertarians increasingly run the country though the Libertarian Party itself remains marginal.

But as powerful as libertarians are in the US, they are almost absent from the German political landscape. Up to the 2013 elections, Germany had a libertarian party in Parliament: the FDP (Free Democratic Party). Tellingly for our discussion here, the FDP was the biggest opponent of renewable energy in Germany. Equally as revealing, German voters threw them out of Parliament in the first national elections after Fukushima.

The terms "liberal" and "conservative" mean something much different today in America than they did in previous generations—and indeed, than they do in other countries. "Liberal" traditionally stood for something closer to the libertarian focus on individual freedoms than on the conservative emphasis on tradition and communities. The Liberal Party of Canada is thus more centrist than the American use of "liberal" might indicate. Likewise, the Free Democrats in Germany also called themselves *die Liberalen* (the liberals).

It is not hard to find a clear dividing line between liberals and classical conservatives. Liberals would view gay marriage, for instance, as a personal choice to bear responsibility for a loved one. Conservatives tend to oppose gay marriage as something unnatural. (Here, we see that progressives and classical conservatives will still have plenty to disagree on once they have come together on the energy transition.)

In the energy sector, delineations between classical conservatives and liberals are sometimes less clear. If someone wanted to put solar on their roof, classical liberals (modern-day libertarians) would emphasize the homeowner's individual liberty to do so. Some classical conservatives might also like the idea, but in an area of historic buildings—think: thatched roofs in England—some of them would likely complain that a solar roof changes the landscape too much. "It is hard to learn to love the new gas station that stands where the wild honeysuckle grew," Walter Lippmann once famously expressed a conservative sentiment.[23] Such views were behind the founding of the World Wildlife Foundation in England by conservatives, for instance.

Classical conservatives will thus not automatically support renewable energy projects. Some conservationists do not necessarily welcome changes to natural landscapes in the name of progress. They are likely to oppose wind projects, for instance. On the other hand, conservative farmers tend to see the land not as something to be maintained in its pristine state, but as something to live off of. They are more likely to welcome wind projects, especially those they benefit from—and in Germany, build themselves.[24] What both of these groups would agree on—and the authors of this book concur—is that the decision about a local wind farm must include the opinion of locals.

On the occasion of the 1992 Earth Summit, President George H.W. Bush famously stated that "the American way of life is not up for negotiations." In recent years, however, classical conservatives in the US—traditionally more concerned about conservation than progress—have come out against this uncompromising stance. Classical US conservative Andrew Bacevich wrote in 2011 that "the belief that

---

[23] Quoted in Kirk, Russell. *The conservative mind: From Burke to Eliot.* 7th rev. ed. Chicago, Washington, D.C: Regnery Pub, 1986, p. 80.

[24] See Franke, Alexander. "How winning over rural constituents changed the political discussions on renewables in Germany." Energy Transition. The German Energiewende. 18 November 2014. Accessed February 5, 2016. http://energytransition.de/2014/11/german-fit-helped-making-energiewende-non-partisan/.

America is privileged place in the international order [without] any obligation to live within its means" is a "shattered illusion."[25]

By "the American way of life," Bush Sr. meant specifically consumption of natural resources for greater material prosperity—something measurable in numbers, not just abstract (like happiness). In this respect, the modern US conservative movement has little in common with classical conservatism. Indeed, though the Whig Party in the US dissolved just before the Civil War, its core belief in progress is ubiquitous in America today. Perhaps the United States does not need either a Whig Party or a Libertarian Party because the two schools of thought are already so well-represented among both Democrats and Republicans today.

Across the US, there is a utility backlash against solar rooftops on homes. In Georgia, the Tea Party has come out in favor of the right of individual homeowners to make their own electricity, earning itself the moniker "Green Tea Party" in the process. Shouldn't environmentalists and proponents of renewables celebrate the support from such conservatives?

To an extent, yes. "Indiana Republicans should be championing free-market choice—not government-created utility monopolies," one Tea Party activist from the Hoosier state convincingly argued.[26] From the German perspective, one might nonetheless wish to improve on her wording: she should be calling for the right for citizens to make their own energy, not just for freedom of choice as consumers. She might also want to focus on communities, not just individuals. Paul Monaghan, director of the UK-based consultancy Up The Ethics, agrees: "It would be a shame if 'citizen-involvement' meant little more than households bearing solar-photovoltaics panels."[27] Germany's energy transition shows that citizens can have solar roofs, but also community wind farms and biomass facilities.

---

[25] See Bacevich, Andrew J. "An end to empire: After Cold War and Long War, America returns to reality." In *The Essence of Conservatism: 10 Classic Reads from The American Conservative*. Edited by Daniel McCarthy. Langhorne, PA: The American Conservative, 2014.

[26] See Pyper, Julia. "Indiana and West Virginia Look to Slash Support for Renewable Energy." Greentechmedia. 26 January 2015. Accessed February 5, 2016. http://www.greentechmedia.com/articles/read/indiana-and-west-virginia-launch-attacks-on-solar-and-wind.

[27] Monaghan, Paul. "The highs and lows of community energy across Europe." The News Cooperative. 14 January 2016. Accessed February 5, 2016. http://www.thenews.coop/100890/news/co-operatives/highs-lows-community-energy-across-europe/.

Furthermore, efficiency projects often only get started at the community level, such as when district heat networks are built.

A focus on the individual is not enough to ensure the energy transition anyway. The future will not consist merely of households (much less individuals) becoming energy independent. We will continue to rely on each other, and the grid—which will be smarter—will still connect us to each other. To that end, we need a focus on communities—and that means progressives engaging more with classical conservatives who want to strengthen communities, not just libertarians who stress individual rights.

Classical conservatives have long been skeptical of libertarians, and their criticism reveals even more overlapping with proponents of the energy transition. Russell Kirk, the author of *The Conservative Mind*, wrote a blistering criticism of libertarians from the classical conservative point of view in the late 1980s. "The libertarian asserts that the state is the great oppressor. But the conservative finds that the state is natural and necessary for the fulfillment of human nature and the growth of civilization," Kirk argued in his essay entitled "A dispassionate assessment of libertarians."

The energy transition is a political goal; as such, it requires policy intervention. It thus helps when policy support for renewables can be included as something necessary for civilization (in light of scarce resources, climate change, and pollution). Classical conservatives are open to such ideas.

Americans who admire Germany for its energy policy often frustratedly state that they wish the US government would support renewables more. "Distrust of government-directed enterprises is embedded in American political culture," laments experienced carbon taxation campaigner Charles Komanoff.[28] But Komanoff may have a friend in Roger Scruton when it comes to environmental taxation. In *The Essence of Conservatism*, Scruton says "the habit we all have of externalizing our costs" is "the real cause of the environmental problems we face."[29]

---

[28] See Komanoff, Charles. "Global warming solution staring us right in the face." Renewables International. The Magazine. 17 September 2014. Accessed February 5, 2016. http://www.renewablesinternational.net/global-warming-solution-staring-us-right-in-the-face/150/537/81797/.

[29] Scruton, Roger. "Righter Shade of Green: Conservation requires not globalist environmentalism but local stewardship." In *The Essence of Conservatism: 10 Classic Reads from The American*

Kirk did not like the libertarian focus on individuals because we live for each other, not for ourselves: "We are made for cooperation, like the hands, like the feet," he argued, quoting Marcus Aurelius in rejecting the "loveless realm" of the libertarian mantra: "I am, and none else besides me." Kirk charges that "libertarianism, properly understood, is as alien to real American conservatives as is communism" and adds that "conservatives and libertarians can conclude no friendly pact." He was writing at the end of the Ronald Reagan era, when libertarians were just beginning to take over the Republican Party. Perhaps Kirk saw the onslaught coming and hoped to fend it off.

If so, he failed. Today, the libertarians are firmly entrenched in the Republican ranks. Marco Rubio, Paul Ryan, and Ted Cruz are more libertarian than classically conservative. And today, the Koch brothers have a great influence on the Republicans. They started off as libertarians; David Koch ran against Ronald Reagan in 1980 for the Libertarian Party.[30]

For Kirk, the libertarian focus on money and individual freedom was anathema to his core belief: "the conservatives declare that society is a community of souls, joining the dead, the living, and those yet unborn; and that it coheres through what Aristotle called friendship and Christians call love of neighbor." Here, Kirk not only distances himself from US libertarians, but obliquely from then British Prime Minister Margaret Thatcher, who had claimed recently beforehand, "There is no such thing as society."[31]

German Green Conservativism

Both Germany and the US have church groups concerned about protecting God's creation. In Germany, some of them came together under the *Klima-Allianz* (Climate Alliance), primarily to fight coal projects. In the US, churches are divesting from fossil

Conservative. Edited by Daniel McCarthy. Langhorne, PA: The American Conservative., 2014.

[30] See Dickinson, Tim. "Inside the Koch Brothers' Toxic Empire." Rollingstone Magazine. 24 September 2014. Accessed February 5, 2016. http://www.rollingstone.com/politics/news/inside-the-koch-brothers-toxic-empire-20140924#ixzz3EyjxTvFK.

[31] See http://briandeer.com/social/thatcher-society.htm.

fuels.[32] Worldwatch's paper from 2002 entitled "Invoking the spirit" still provides a good overview of the role of religion in the sustainability debate.

Two things make the German political debate different from the one in the US: the complete lack of outright climate deniers among prominent political leaders in any party and the prominence of classical conservative thought not only in the Christian Union Party, but also in the Greens.

The German Green Party was founded at a meeting on 12 to 13 January 1980 in Karlsruhe. Outside Germany, Green parties are considered to be left of center, but the German Green Party started out with people from throughout the political spectrum—from conservative Christians to communists. About the only thing this group had in common was that it did not want to be in government—the German Greens were founded as an anti-establishment party. But in 1990, the German Greens suffered a major setback that caused them to sharpen their profile.

After the Fall of the Wall, the Greens from West Germany joined forces with *Bündnis 90* (Alliance 90), a group of civil rights activists from the former East Germany. To this day, the official name of the German Green Party is "Alliance 90/the Greens." By incorporating protesters who had helped topple the communist government in East Germany, the Greens probably thought their popularity was ensured. If so, this assumption turned out to be a grave mistake. The Greens remained tepid on the issue that interested voters the most in 1990: reunification of East and West. Chancellor Helmut Kohl stole the show by promising "landscapes in bloom" if he was reelected on his reunification promise. The Greens fell below the 5 percent threshold of the vote required for a party to be represented in Parliament. They faced extinction.

During the assessment of the damage, influence tilted toward the *Realos* (realists) within the party and away from the *Fundis* (fundamentalists). Radical ideas were de-emphasized as a result. The influx of party members from the former communist Germany brought in a number of people skeptical of far left ideas. Because the Protestant Church had

---

[32] See Galbraith, Kate. "Churches Go Green by Shedding Fossil Fuel Holdings." *The New York Times.* 15 October 2014. Accessed February 5, 2016. http://www.nytimes.com/2014/10/16/business/international/churches-go-green-by-shedding-holdings-of-carbon-emitters.html?referrer=&_r=5.

played a crucial role in the protests against the East German regime, a number of the Alliance 90 members were conservative Christians. In 1998, the Greens joined a governing coalition under the leadership of Joschka Fischer, putting an end to the anti-establishment party. Fischer and his *Realos* now set the tone.

Today, the largest German city with a Green mayor is Stuttgart (population: 600,000), where the Greens are in a coalition with the Social Democrats. In nearby Freiburg (population: 230,000), the Greens have formed a coalition with the Christian Union. Both cities are located in the state of Baden-Württemberg, the only German state with a Green governor: Winfried Kretschmann. "Kretsch" (as his supporters fondly call him) is a churchgoing man—and one of the few top politicians in Germany who speaks openly about his faith.[33] (In general, German politicians are subdued about their religious beliefs in public—even when they are, like Chancellor Merkel, the children of pastors.) Religion is considered a private matter in Germany, whereas the realm of politics is something for open debate among reasonable people willing to compromise.

Classical Conservatives Are Reasonable

Ideology is anathema to classical conservativism. Unfortunately, fundamentalist positions—even opposition to scientific facts—are no longer uncommon in the US debate. "To political leaders [in the US], a scientist's view is just another opinion," complains US philosophy professor Robert Crease.[34] But he immediately adds that experts are not to be followed blindly either: "I am glad that we do not live in a technocracy, ruled by experts who decide our social goals rather than advancing the goals that society establishes." In other words, experts need to state their

---

[33] See Schwarz, Patrik. ""Ich habe zu eng geglaubt"." Die Zeit Online. 23 March 2015. Accessed February 5, 2016. http://www.zeit.de/2015/12/winfried-kretschmann-die-gruenen-glaube-christentum/komplettansicht.

[34] See Crease, Robert P. "A Requiem for Technocracy." Project Syndicate. 12 June 2014. Accessed February 5, 2016. http://www.project-syndicate.org/commentary/robert-pDOUBLEHYPHEN-crease-laments-the-declining-influence-of-science-in-public-policy?barrier=true#ZWIWo8EWM-dP1cSoR.99.

case, not dictate policy. An informed public must be consulted and have input, because energy projects affect people's happiness and feeling of security. There are cases—such as the many times when the Germans were told their lights would go out without nuclear power—when the public is right and many experts are wrong.

In return, policymakers need to rely on experts when there is a consensus among them. There isn't always, which is all the more reason for decisions to be made by society in open debates—exactly what is happening with the Energiewende. After all, the energy transition is a political response to societal demands, not something the energy market naturally brought about on its own—and certainly not something brought about by incumbent energy experts (who are not interested in the disruption the transition entails anyway).

One major difference between Germany and the US is the relative lack of fundamentally ideological political leaders in Germany. The gradual shift within the Republican Party in the US over the past half-century toward a more radical, uncompromising stance has been widely noted by Americans.[35] The Energiewende would not have happened without a political consensus, and an unwillingness to compromise is anathema toward consensus-building. The good news is that classical conservatives in the US are also upset about the drifting of the Republican Party toward radicalism.

In 2011, David Hoeveler lamented, "Much recent conservative writing possesses a dogmatic quality—a metaphysics of the marketplace and the absolutism of 'American exceptionalism.'"[36] Another classical conservative wrote in 2005, "Among the worst aspects of the collapse of traditional conservatism is that my children will grow up in a world in which vulgar and belligerent nationalism will be presented to them as the alternative to leftism."[37] These classical conservatives do not reduce people to

[35] Perlstein, Rick. *Nixonland: The rise of a president and the fracturing of America.* 1. Scribner trade paperback ed. New York NY: Scribner, 2009.

[36] Hoeveler, J. D. "American Burke: Irving Babbitt formulated conservatism for a world in whirl." In *The Essence of Conservatism: 10 Classic Reads from The American Conservative.* Edited by Daniel McCarthy. Langhorne, PA: The American Conservative., 2014.

[37] Woods Jr., Thomas E. "Community or Conquest? Why militarism means bigger government and weaker families." In *The Essence of Conservatism: 10 Classic Reads from The American Conservative.* Edited by Daniel McCarthy. Langhorne, PA: The American Conservative., 2014.

individual market actors. Rather, they stress that man has a soul. "Our deepest need is not for things but for each other," writes Dermot Quinn.[38]

In stressing Edmund Burke's social contract "between the living, the unborn, and the dead," classical conservatives cut at the root of a Whigish emphasis on progress. Weyrich and Lind call for "caution in innovation" in *The Essence of Conservativism*.[39] This idea underlies the Energiewende, which eschews high-tech fixes to problems brought about by high-tech. Why try complex solutions (like nuclear or carbon capture and storage) if something simple (wind and solar power in combination with efficiency) will work? Originally German, this "caution in innovation" has been imported to the US as the "precautionary principle," and it is at the heart of the Energiewende.

Originally, the idea of the Energiewende was imported from the US. In 1976, energy researcher Amory Lovins essentially called for "caution in innovation" when he delineated the "soft energy path"[40] from the "hard energy path" that the world was going down. The path that Germany is taking with the Energiewende is the soft energy path as described by Lovins; leading German energy thinkers have explicitly stated that his writings inspired them.[41]

Classical conservatives like to talk morals, and climate change is a moral issue. Debate with fringe science-denying Republicans, and you may draw US journalist David Roberts's conclusion: "I just don't think there's any way to make the facts of climate change congenial to the contemporary US conservative perspective." But in drawing that conclusion,

---

[38] Quinn, Dermot. "Too Small to Fail: Wilhelm Roepke teaches that small is beautiful - and conservative." In *The Essence of Conservatism: 10 Classic Reads from The American Conservative*. Edited by Daniel McCarthy. Langhorne, PA: The American Conservative., 2014.

[39] Weyrich, Paul, and William S. Lind. "The Next Conservatism: By rejecting ideology and embracing "retroculture", the Right can reverse America's decline." In *The Essence of Conservatism: 10 Classic Reads from The American Conservative*. Edited by Daniel McCarthy. Langhorne, PA: The American Conservative., 2014.

[40] Lovins, Amory B. "Energy Strategy: The Road Not Taken?" Foreign Affairs. October 1976 Issue. Accessed February 5, 2016. https://www.foreignaffairs.com/articles/united-states/1976-10-01/energy-strategy-road-not-taken.

[41] See Morris, Craig. "The Energiewende—made in the United States." Energy Transition. The German Energiewende. 13 April 2015. Accessed February 5, 2016. http://energytransition.de/2015/04/german-energiewende-made-in-the-usa/.

Roberts focuses on the low-church group that US classical conservatives are also concerned about.

The Climate March held in New York in October 2014 revealed that partisanship and inclusiveness are central issues for the US environmental community. A debate flared up after the march at EnergyCollective. com about whether the organizers had truly brought everyone together— or whether conservatives had been left out (or left themselves out). No such debate occurs in Germany after climate change protests; there is simply no perceived partisanship. US progressives and classical conservatives could reach out to each other to create the same common ground that the energy transition enjoys in Germany. In "The next conservationism," conservatives Weyrich and Lind call for "the return of trains and streetcars as alternatives to dependence on automobiles."[42] These classical conservative voices need to be in the foreground for the energy transition to succeed.

---

[42] McCarthy, Daniel, ed. *The Essence of Conservatism: 10 Classic Reads from The American Conservative*. Langhorne, PA: The American Conservative, 2014; Kindle Edition.

7

The 1990s: Laying the Foundations for the Energiewende

In the fall of 1990, Christian Union politician Matthias Engelsberger was about to retire after 21 years in German Parliament. Aside from party loyalty, he had little to show for his two decades in the Bundestag. No major legislation bore his name, and he had not spearheaded any major policy initiatives. The last session of Parliament was to be held on 5th October. It was to be Engelsberger's last, and he aimed to go out with a bang. He had come up with a two-page bill that would allow citizens to sell electricity to the grid at a price not set by the utility.

A Bavarian, Engelsberger represented a region with numerous owners of small hydropower plants, with which Germany had originally electrified many rural areas. But as in Schönau, large utilities increasingly insisted that these competitors close. In previous negotiations, he had only managed to get utilities to pay a few pennies per kilowatt-hour for this electricity—not enough for most small hydro to be profitable. His bill specified that 75 percent of the retail rate would be paid for electricity from hydropower, waste gas, and biowaste—along with 90 percent of the retail rate for solar and wind power. With retail rates coming in at around 17 pfennigs, the bill would increase support for these renewable energy sources several fold.

© The Editor(s) (if applicable) and The Author(s) 2016
C. Morris, A. Jungjohann, *Energy Democracy*,
DOI 10.1007/978-3-319-31891-2_7

With large stakes in a central energy supply with large power plants, utilities were hostile to the idea of promoting distributed renewable energy. The power firms' political allies in Parliament and government—at the forefront, the Ministry of Economics—shared this skepticism. In particular, the libertarian FDP—the small coalition partner for Chancellor Kohl—denounced the allegedly market-distorting effects of feed-in tariffs. Instead, they argued for a European Union (EU)-wide quota system, which would leave it up to utilities to pick and choose between various renewable energy sources (and also prevent new players from entering the market, a fact they did not explicitly mention). Against this backdrop, the introduction of a nationwide feed-in tariff "seemed next to impossible," as professor Christoph Stefes from the University of Colorado puts it.[1]

An Unlikely Coalition of Parliamentary Backbenchers

It was not easy getting the bill on the agenda. East and West Germany were reuniting, and there was no dearth of issues to address. At an internal preparatory meeting of the Christian Union just days before the final parliamentary session, Engelsberger had a hard time getting his party colleagues to discuss his bill at all. Chancellor Helmut Kohl wanted to officially accept the Oder-Neisse Line—the current border between Germany and Poland—as a friendly gesture to neighboring countries, who were worried about a reunified Germany possibly wishing to expand again. But Kohl faced tremendous pushback from the conservative wing of his own party, as the decision would mean officially giving up for good areas of Eastern Europe that used to be German.

Heated debates went on into the evening, and the Chancellor left in a fluster as the criticism turned into personal attacks. As midnight approached, the man directing the gathering reached for the bell. Engelsberger rushed to get a word with him before the meeting—the last

---

[1] Stefes, Christoph H. "Energiewende. Critical Junctures and Path Dependencies Since 1990." In "Rapide Politikwechsel in der Bundesrepublik. Theoretischer Rahmen und empirische Befunde." Edited by Friedbert W. Rüb. Special Issue, *Zeitschrift für Politik*, Sonderband 6 (2014): 47–70, p. 57.

opportunity for his bill—was adjourned. He had only one card to play, but it was a good one: 21 years of party loyalty in Parliament.

"Wolfgang," he told his party friend, "you can't do that—what about my bill?" "Matthias," his colleague responded, "your bill is garbage. But because it's yours, I'll put it to a vote." In its last decision that day, the Christian Union approved the bill for a vote in Parliament the following week.

Just days later, the last West German Parliament held its final session on 5 October 1990. By the time Engelsberger's *Stromeinspeisegesetz* (Electricity Feed-In Act) came to a vote, so many parliamentarians had already gone home and there was no longer a quorum of members present. But no one challenged the legislation based on the quorum, and the bill passed without opposition. Social Democrat Hermann Scheer, who would later revise these feed-in tariffs in the Renewable Energy Act of 2000, made sure his party did not oppose the bill. For the Greens, Wolfgang Daniels performed a similar function, ensuring that his party accepted the legislation from their political opposition. Indeed, Daniels played an even more crucial role; he actually coauthored the legislation with Engelsberger, but the Christian Union party whip in Parliament insisted that the Green politician be removed from the bill, lest other Union politicians reject the proposal for political reasons.[2] At the time, the Christian Union and the Greens were archenemies at the federal level.

An unlikely coalition of parliamentary backbenchers across party lines thus laid the foundations of the Energiewende. These politicians realized that ministries would not advance their cause, so they drafted the bill on their own. They made use of a short time window within which the anti-renewable camp led by utilities was preoccupied with taking over the East German energy sector. As one executive at RWE later admitted: "The [feed-in tariff] was an accident. We simply did not see it coming."[3]

The Feed-in Act went into effect in 1991 and would trigger significant wind energy growth, which helped open the market for biomass and photovoltaics later. This experience with the 1991 feed-in tariff proved crucial in getting

---

[2] See Berchem, Andreas. "Das unterschätzte Gesetz." Die Zeit Online. 22 September 2006. Accessed February 5, 2016. http://www.zeit.de/online/2006/39/EEG.

[3] Stefes, p. 58.

the 2000 law on paper and through the Bundestag, again against opposition from the Ministry of Economics. Without the law from 1991, there would have been no law in 2000. And the 1991 law almost did not happen. "It hung by a thread," a Christian Union supporter of the bill described the situation.[4]

Engelsberger had not invented the idea. A US law from 1978, Public Utility Regulatory Policies Act (PURPA), also guaranteed grid access for non-utility electricity generators. PURPA is some 31 pages long and deals with various policies, with guaranteed grid access tucked away in the middle across just a few pages. In both the German and the US laws, incumbent utilities skeptical of renewables failed to recognize the radical change that these short legal passages entailed.

PURPA specified compensation for non-utility generators (third parties) based on "avoided cost" to the utility, which generally meant the cost of the fuel that the utility would otherwise have had to consume. In contrast, the German law pegged compensation to retail rates, which meant a better price for renewable energy. The reasoning that won over conservative Christian Unionists and libertarian Free Democrats in the 1990 German coalition was that fossil fuel prices did not include "external costs," such as pollution. This argument convinced the pro-market coalition because the internalization of external costs is a crucial tenant of German social-market economics.

Like PURPA, the 1991 Feed-in Act targeted non-utilities, but the EEG of 2000 changed the situation after liberalization of power markets brought competition to the previously monopolized power sector for the first time, so that utilities themselves could benefit from this compensation if they developed renewables. They hardly did, however. As of 2009, the Big Four German utilities still only accounted for 1 percent of non-hydro renewable electricity, although the country had 13 percent non-hydro green power at the time.

Instead, the big utilities launched a campaign at the beginning of the 1990s to belittle non-hydro renewables. In ads across the country, they feigned openness to new renewable energy sources even as they assured the public as late as 1994 that wind, solar, and biomass would never

---

[4] Oelker, Jan and Christian Hinsch, eds. *Windgesichter: Aufbruch der Windenergie in Deutschland.* Dresden: Sonnenbuch-Verlag, 2005, p. 326-327.

make up more than 1 percent of German power supply. By 2000, when the Renewable Energy Act would go into effect, the Feed-in Act had already increased the share of new renewables from around 0 percent to 3 percent.

Each year, German utility lobby group VDEW (the predecessor of the current BDEW) presented annual power sector figures. In 1995, the VDEW forecast a 40 percent greater share of new renewables by 2005 relative to 1994 "if we work hard" and criticized a forecast published by Öko-Institut, the organization that came out of the anti-nuclear protests in Wyhl. The pro-renewables researchers believed new renewables would grow severalfold by 2005, a stance the VDEW called "wishful thinking." Once again, the Öko-Institut forecast was quite accurate. By 2005, non-hydro renewables had grown by some 1200 percent, not just by 40 percent.[5]

Not all top German politicians trusted the utilities, with the biggest example being Social Democrat Hermann Scheer. A former military officer and an expert in peace policy, Scheer knew that German heavy industry saw its core business in building big machines in large numbers. He was looking for a way for them to make money peacefully. Renewable energy seemed obvious. He called the VDEW's skepticism "empty and mendacious" and charged that the power sector was paying lip service to renewable energy even as they blocked every policy proposed.

Scheer was the exception in German Parliament. In general, the utility warnings influenced top German politicians. In 1994, the country got a new Environmental Minister, who promptly repeated the message from the utility ad campaign: "Solar, hydropower, and wind power will not be able to make up more than four percent of our power supply even in the long term."[6] Her name was Angela Merkel.

---

[5] See Morris, Craig. "Halftime for the Energiewende." Renewables International. The Magazine. 25 March 2015. Accessed February 5, 2016. http://www.renewablesinternational.net/halftime-for-the-energiewende/150/537/86407/.

[6] See Solarenergie-Förderverein Deutschland e.V. "Sonne, Wasser oder Wind können auch langfristig nicht mehr als 4 % unseres Strombedarfs decken." Nachdruck aus Solarbrief 1/97 Seite 35 vom 24.03.1997. Accessed February 5, 2016. http://www.sfv.de/briefe/brief97_1/sob97135.htm.

Demanding Something That Utilities Had All Along

In the late 1980s, a number of policies inspired Engelsberger to pro-vide greater compensation in the Feed-in Act than was provided in PURPA. In 1989, the Swiss town of Burgdorf began offering a Swiss franc for a kilowatt-hour of solar power—several times more than even the Feed-in Act would later offer. Solar power simply cost that much at the time, so if you wanted it, that was the price to pay.[7] The idea was in the air. In 1987, a solar group in Hamburg called for full-cost com-pensation to get photovoltaics out of the vicious circle; there is no market because solar is so expensive, and solar is so expensive because there is no market.[8]

In Germany's Ministry of Education and Research, a few Christian Union officials also grew weary of subsidizing pilot projects. They real-ized that what wind and solar power in particular needed was not some breakthrough, but economies of scale. Solar and wind had reached the "Valley of Death"—the point where the R&D push became ineffective and a market pull was needed.

Citizens also started pulling. In Aachen, a military officer named Wolf von Fabeck became concerned about the effects of acid rain in the 1980s. He also realized that the German Armed Forces would not be able to protect nuclear plants, which were therefore not a low-pollution option for him. So after speaking with his local pastor, he formed a small group of solar campaigners at his church. They pooled enough money to buy 12 solar panels, which they assembled on a stand. The price was still astro-nomical; for 12 solar panels with an output of only 600 watts in total, you could have afforded a Volkswagen Golf. "It took some doing to convince my wife that this would be a good investment," Fabeck remembers.[9]

---

[7] See Thönen, Simon. "Ein früher Pionier der Energiewende." Der Bund. 22 August 2011. Accessed February 5, 2016. http://www.derbund.ch/bern/Ein-frueher-Pionier-der-Energiewende/story/11332092.

[8] Janzing, Bernward. *Solare Zeiten: Die Karriere der Sonnenenergie; eine Geschichte von Menschen mit Visionen und Fortschritten der Technik*. Freiburg: Picea-Verl., 2011, p. 94.

[9] Conversation with the author Craig Morris.

His group then visited local farmers' markets with the solar stand to demonstrate that you could run normal household appliances on solar power. Fabeck and his friends brought along a mixer, which seemed to incredulous passersby to be running on thin air. There was no power connection, and the campaigners did without a battery because people would have thought the battery might have been fully charged beforehand. When TV journalist Franz Alt covered the solar campaigners on his show, demand for Fabeck's solar demonstration surged nationwide. Visitors learned that the technology worked but was still too expensive—and that a market, not more R&D, was needed to bring the cost down.

At home in Aachen, the campaign met with resistance from the local utility and from the state's Economics Minister, who declared, "I am not willing to finance a workshop for solar friends via the retail rate." But one official from the state's Economics Ministry, Dieter Schulte-Janson, never gave in to his boss's arguments. He repeatedly sided with the campaigners, who had also convinced the City Council of Aachen that 2 deutsche marks—12 times the retail rate—for photovoltaics was not so much more than the utility already paid for peak power. Fabeck also argued that he was only asking for something that utilities had had all along: rates that cover the full cost of an investment with an additional reasonable profit added on.

The City Council of Aachen adopted a resolution requiring the local utility to provide this compensation for solar power. Under pressure from utility giant RWE, the local utility ignored the request. City officials then passed another four resolutions over two years before the town's utility finally played along. By that time, some 40 German municipalities had adopted full-cost compensation for solar power.[10] Aachen was not the first to implement the policy, but its influence led to its widespread adoption across the country.

One reason why utilities did not fight harder to stop these policies was their limited scope. In the smaller Bavarian town of Hammelburg, a similar effort was spearheaded by local schoolteacher Hans-Josef Fell, who became a coauthor of the 2000 Renewable Energy Act as a Member

---

[10] See Solarenergie-Förderverein Deutschland e.V. "Historisches zur kostendeckenden Vergütung bis zu ihrer Aufnahme in das EEG vom 1. Aug. 2004 mit Folien zum Aachener Modell." Accessed February 5, 2016. http://www.sfv.de/lokal/mails/wvf/kostendeckende_Verguetung_bis_hin_zum_EEG_2004.htm.

of Parliament. The total project was budgeted for a mere 15 kilowatts, no more than five home rooftops.[11] Another town that implemented full-cost compensation before Aachen did was Bonn. Both these towns limited this policy support to 1000 kilowatts of photovoltaics.[12] In the heyday of German solar installations from 2010 to 2012, Germany would install nearly that much per hour around-the-clock on average. On the other hand, the average panel had an output of around 50 W in the early 1990s (compared with 200 today), so such a budget was enough to cover the installation of 20,000 panels in a single city. It was a start.

Importantly, the budget impact of the policy would remain minor because the project size was so limited. In Aachen, retail rates were expected to rise by only a few percentage points, if even that.[13] The small scale reinforced what utilities already thought about renewables: they are too small to matter anyway. And regardless, the utilities had bigger fish to fry, starting with the opening of markets in the former communist East Germany.

The Wild East: Reunification Opens the Former GDR to Western Utilities

The power sector in the East was especially important because East German power supply was running mostly on old, inefficient, and air-polluting coal power plants. A wave of modernization was to take place and Western utilities saw a once-in-a-lifetime chance to expand their operation for decades. The German Democratic Republic (GDR) lacked the money to import natural gas. Instead of using gas for heat, as was already common in West Germany, the East Germans largely used lignite briquettes. The result was terrible pollution. It was in the East German heating and power sectors that Germany would clean up the air significantly during the first few years after Reunification. The German gov-

---

[11] See Fell, Hans-Josef. "Erfolgreiche 20 Jahre für die Hammelburger Solarstromgesellschaft." Accessed February 5, 2016. http://www.hans-josef-fell.de/content/index.php/presse-mainmenu-49/regional-mainmenu-71/783-erfolgreiche-20-jahre-fuer-die-hammelburger-solarstromgesellschaft.

[12] Janzing, p. 101.

[13] Solarenergie-Förderverein Deutschland e.V., *Historisches zur kostendeckenden Vergütung*.

ernment has estimated that 10 percentage points of its carbon emission reductions since 1990 (24.7 percent as of 2012) came from shutting down old factories and the cleanup of the energy sector in East Germany alone.

For German power providers, who had dealt with stagnant energy demand during the 1970s and 1980s, Reunification represented their first real opportunity for expansion in decades. Only months after the fall of the Berlin Wall in November 1989, representatives of West Germany's three largest power companies held a secret meeting with the Industry Minister of East Germany. They succeeded in convincing him that they should be allowed to take over power supply. The main argument: give us the entire power sector, and we will keep the lights on.

Contracts were signed in August 1990, but when the news became public, the Parliament of East Germany rebelled against plans in the final months of its existence. So did numerous municipalities, along with five other large West German power providers, who felt left out of the negotiations. To pacify these five, each was given its own share of the East German pie. But still, eastern municipalities were left out.

Around 140 of them eventually came together to be represented by West German legal expert Peter Becker. He and his team had been crucial in creating the network. "We went around and visited all of the mayors to make sure they understood what was at stake and to create a mood favorable for municipal utilities," he says.[14]

Becker got the case heard by the German Constitutional Court based in the southwest town of Karlsruhe, but in this case, court was held in East Germany—as a friendly gesture toward the East. Everyone met in a conference hall owned by the East German Railway Company in the small town of Stendal. "A podium had hastily been put together, and you could still smell the stale cigarette smoke in the curtains," Becker remembers the improvised venue with a laugh.[15]

Contracts in hand, the eight West German power providers offered 49 percent ownership of regional utilities to the municipalities as a com-

---

[14] Opitz, Florian. *TV Documentary: Die Macht der Stromkonzerne.* taglicht media für WDR and BR. With the assistance of Beate Schlanstein and Thomas Kamp. Akte D. 2015. Accessed February 5, 2016. https://www.youtube.com/watch?v=qsmw_UFXdos.

[15] WDR. "Die Macht der Stromkonzerne." Accessed February 5, 2016. http://www1.wdr.de/fernsehen/dokumentation_reportage/wdr-dok/sendungen/die-macht-der-stromkonzerne-100.html.

promise. Had city governments accepted the offer, they would have been minority shareholders for the foreseeable future.[16] They refused.

The power firms reacted by announcing the indefinite suspension of further investments in the East. In 1991, a total of 60 billion marks over the next few years was at stake.[17] The Western firms also continued to mention the specter of blackouts during the cold winter. The strategy proved unconvincing in the winter of 1990–1991—the first after Reunification—because peak power consumption fell from 26 to 16 gigawatt as the East German economy collapsed.

The Social Democrats and the Greens had been calling for municipal utilities all along, but they were in the opposition at the time. Industry disagreed, calling for the quick privatization of the entire East German power sector because they feared a lengthy court case would slow down their own investments. Eventually, Chancellor Kohl became convinced that municipal utilities would be an excellent way to strengthen struggling East German cities.

But the decision was the court's to make, not the Chancellor's. In December 1992, the judges proposed a compromise, which the last municipality signed in July 1993. By the time the court case ended, 164 East German municipalities were part of the case.

Becker and the East German cities had won. "When we finally had the agreement signed, we all had tears in our eyes," Becker remembers.[18] Nonetheless, West German firms still had large parts of the East German power sector to take over. The handover process went on for years in any case. It was mid-1995 before the first 100 East German municipalities had permits to run their own utilities. Nonetheless, the East German cities managed to get ownership of their power supply faster than the campaigners in Schönau did, and their struggle had begun years earlier.

The battle for community ownership of utilities in the former East Germany was crucial for the same reason it was in Schönau: decisions about energy supply could then (at least in theory) focus on what was

---

[16] Becker, Peter. *Aufstieg und Krise der deutschen Stromkonzerne: Zugleich ein Beitrag zur Entwicklung des Energierechts.* Bochum: Ponte Press, 2011, p. 73.

[17] See Leuschner, Udo "Energiechronik: Stromstreit beigelegt." July 1993. Accessed February 5, 2016. http://www.energie-chronik.de/930702.htm.

[18] Opitz, *Macht der Stromkonzerne.*

best for the community, not for shareholder value. As an indirect benefit for renewables, the court case turned utility attention away from the Feed-in Act for the moment. Two other discussions also tied up the attention of utilities: liberalization of energy markets, and the ongoing discussions about a possible nuclear phaseout.

Liberalizing European Energy Markets

In Europe, gas and electricity markets were liberalized starting in the late 1990s. Customers were to be able to choose their gas and power providers; the days of monopolies were to come to an end. The push for liberalization came from the European Commission, however, not from EU member states. Various countries implemented it more or less reluctantly and at various depths.

Germany had a long history of protecting its monopoly utilities. "Germany would never have been able to roll out competition on its own," Peter Becker writes in his history of the German power sector.[19] A law from 1935 codified the status quo of utility monopolies at the time. The goal was to prevent the "detrimental effects of competition," as the law itself states.

The power grid itself is what economists call a "natural monopoly," meaning that redundant grid lines are not built in the name of competition. Even today, numerous neighboring countries—such as France, Denmark, Poland, the Netherlands, and the Czech Republic—have largely state-owned high-voltage networks. The German transmission grid is not state-owned, however; it consists of four parts, each run by its own monopoly that faces no competition.

But below the grid level, there are good reasons to break up ownership between power generation, transmission, and distribution. The process is called "unbundling." In Europe, the idea is to ensure that a company that owns, say, power plants does not also own high-voltage transmission lines. One could also go further and require the unbundling of fuel resources (coal mines) and (coal) power plants, which Europe did not. By

---

[19] Becker, p. 86.

requiring separate ownership of power plants and transmission grid lines, Germany ensures that no company can make grid access unnecessarily expensive or difficult for competing power generators in order to protect its other power generation assets. For instance, a company that owned both power plants and power lines could charge competitors high transmission fees, making their power plants less competitive—and thereby protecting its own power generation assets.

Likewise, a company that owns cheap lignite resources along with the power plants to burn the lignite in could run at cutthroat profit margins in order to put price pressure on more expensive competitors (which was basically RWE's business strategy in the beginning of the twentieth century). When the competing company folds, the firm with lignite resources and power plants could take over the other's customers and raise its rates again now that there is no competition. Germany has not completely unbundled its utilities to prevent this from happening. For instance, RWE and Vattenfall (a subsidiary of the Swedish government) both mine coal and operate coal power plants in Germany.

For German (and European) utilities, liberalization marked the end of an era. Gone were the days when profits were guaranteed. For roughly a century, profit margins were negotiated with the governmental regulator. The goal was to prevent price gouging of customers while simultaneously ensuring a reasonable profit for the companies, so they could stay in business.

This setup had a perverse outcome: because the profit margin was fixed, additional profits had to come from greater investments. Your investments could not go bad once you had convinced the regulators to let you go forward with them. Utilities thus tried to convince the regulator that more power plants needed to be built. The result, when power consumption stagnated in the 1980s, was a surplus of generation capacity, with captive ratepayers left holding the bill.

Regulators should have prevented this outcome, but they simply lacked the expertise. As the story of the struggle in Schönau reveals, German utilities were never even properly audited. It was common practice for one firm to be audited and its cost structures then applied to other utilities, without any further checks being performed.[20]

---

[20] Becker, p. 55.

Liberalization was not rolled out until 1998 in Germany, two years after the EU legislation was finalized. In the years before, the debate raged about what liberalization would and should look like. The discussion started in the early 1990s. It was the end of the Cold War, and capitalism had defeated communism. Now, publicly owned services—railways, postal services, energy supply, and so on—were to be privatized. In 1992, the European Commission's proposal to liberalize the power and gas markets met with a mixed reaction among member states at a meeting in April in Portugal.

Most EU member states wanted regulated "third-party access" (TPA). Here, a regulator would ensure fair treatment of power producers that did not own grid lines. Germany supported a slightly different concept called "negotiated third-party access" (negotiated TPA), meaning that the country would do without a regulator and hope that the firms themselves could reach agreements. In cases of dispute, the matter would be decided by antitrust authorities.[21] At the time, France still defended its concept of a single buyer, which would have protected state-owned power provider EDF—and hence the country's nuclear sector.[22] The German Greens wanted liberalization in order to break up power monopolies because they felt that the process would hurt the nuclear sector.[23] The Greens may have drawn that conclusion from monitoring the UK. In 1989, the British Electricity Act started the privatization of the power sector in that country, but it included the Non-Fossil Fuel Obligation, which required power suppliers to get a certain share of non-fossil electricity. This clause was added after calculations revealed that nuclear power would not be profitable on a privatized market. Thanks to a British civil servant and

---

[21] Brunekreeft, Gert. "Negotiated Third-Party Access in the German Electricity Supply Industry." Accessed February 5, 2016. http://www.vwl.uni-freiburg.de/fakultaet/vw/publikationen/brunekreeft/MAILAND_Sept01.pdf.

[22] Leuschner, Udo. "Energiechronik: EU weiterhin ohne klares Konzept zur Liberalisierung des Strommarktes." November 1994. Accessed February 5, 2016. http://www.energie-chronik.de/941101.htm.

[23] Leuschner, Udo. "Energiechronik: Grüne beharren auf Ausstieg und fordern "Entmonopolisierung" der Stromwirtschaft." January 1995. Accessed February 5, 1990. http://www.energie-chronik.de/950103.htm.

intervention from the European Commission, in its final version the clause also applied to renewable electricity.[24]

By the end of 1992, no EU agreement had been reached, with the Netherlands, Belgium, Greece, Spain, France, Italy, and Luxembourg having the most objections against TPA.[25] As skeptical as energy lawyer Becker is of his own government's willingness to open its energy sector, the Germans themselves saw French resistance as the biggest obstacle. In 1994, the feeling was that German utilities would face unfair competition if they opened their markets and France continued to protect its own.[26] That December, European energy ministers agreed in principle to open gas and electricity markets with the signing of the European Energy Charter. It was a memorandum of understanding. They did not agree on the specifics yet.

Meanwhile, France continued to protect its single-buyer concept, which slowed down negotiations from the German perspective.[27] At the end of 1995, German utility lobby group VDEW warned the government that the French proposal was incompatible with negotiated TPA. The government then promised to continue to work for better terms but not sign the current proposal.[28]

When the French government still did not budge, German Economics Minister Rexrodt added a clause to the German proposal in March 1996. It imposed restrictions for competition on a liberalized German market from countries that had not opened theirs. Finally, in June 1996 a European agreement was signed in Luxembourg based on Rexrodt's proposal.

---

[24] Toke, David, and Volkmar Lauber. "Anglo-Saxon and German approaches to neoliberalism and environmental policy: the case of financing renewable energy." *Geoforum*, no. 38 (2007): 677–87.

[25] Leuschner, Udo. "Energiechronik: Vorerst kein EG-Binnenmarkt für Energie." December 1992. Accessed February 5, 2016. http://www.energie-chronik.de/921202.htm.

[26] Leuschner, Udo. "Energiechronik: Monopolkommission plädiert erneut für Liberalisierung des Energiemarktes." July 1994. Accessed February 5, 2016. http://www.energie-chronik.de/940702. htm.

[27] Leuschner, Udo. "Energiechronik: Frankreichs "Single-Buyer-Konzept" blockiert Liberalisierung des Strommarkts." March 1995. Accessed February 5, 2016. http://www.energie-chronik. de/950308.htm.

[28] Leuschner, Udo. "Energiechronik: EU-Richtlinie für Liberalisierung des Strommarkts vorerst gescheitert." December 1995. Accessed February 5, 2016. http://www.energie-chronik.de/951201. htm.

German utilities could at least live with the outcome. They continued to complain about details, but the German press was largely unsympathetic to their concerns. Left-leaning *Frankfurter Rundschau* expressed a concern that is now commonplace in Germany: small power consumers would increasingly cross-subsidize big consumers on the liberalized market planned. The center-right *Frankfurter Allgemeine* thought the fears German utilities had of a one-sided aggressive expansion strategy by France's EDF were overblown. But conservative daily *Die Welt* was not so sure, arguing that the German government may have made too many concessions to the French in an effort to be an "example to follow in European integration."[29]

Based on the European treaties, member states then had to implement the EU regulation into national law. After more than half a century, Germany would finally have to modernize its Energy Management Act from 1935. The VDEW and the *Bundesrat*, the chamber of Parliament representing states' rights, pleaded for a long transitional period, primarily because eastern Germany allegedly needed more time.[30]

In the end, the agreement between the *Bundestag* and *Bundesrat* included a compromise: not only municipal utilities, but also all retail power providers were given single-buyer status up to 2005. Otherwise, grid operators had to open up their power lines to other providers. They were not allowed to deny access unless renewable energy or power from cogeneration units needed to be curtailed due to technical constraints. Germany's new Energy Management Act went into effect on 29 April 1998, just a few months before the parliamentary elections in September 1998.

Germany's big utilities had stressed all along the risks involved in liberalization, and the press reports from back then mainly focus on those concerns. But as usual, these companies downplayed their own opportunities—and they were huge. A study published in August 1997 by the

---

[29] Leuschner, Udo. "Energiechronik: Die Richtung stimmt - aber noch keine gleichwertige Öffnung der Strommärkte." June 1996. Accessed February 5, 2016. http://www.energie-chronik.de/960603.htm.

[30] Leuschner, Udo. "Energiechronik: VDEW fordert längere Übergangsfristen für den europäischen Strommarkt." September 1996. Accessed February 5, 2016. http://www.energie-chronik.de/960902.htm.

Mercer Consulting Group underscored how beneficial liberalization was likely to be for the big players. The corporate consultants found that the process would lead to consolidation, with up to 30 percent of power providers disappearing. The public can be excused—especially in light of the focus on the alleged risks to utilities—for thinking that a move from monopolies to greater competition would necessarily serve them well, for instance by lowering prices. That *could* have been the outcome. But then there was that old adage of German social market economics, which stated that a free market leads back to monopolies. True market competition requires governmental regulation. And Germany had just adopted negotiated, not regulated third-party access. Large German utilities thus not only prepared for unfair competition from France, but also for tremendous opportunities to expand outside Germany themselves in most other countries. If East Germany looked like a big new market, the EU was much vaster.

The final issue that helped keep the attention of German utilities off of renewable energy was the possibility of a nuclear phaseout.

The Energy Consensus Talks Mark the Beginnings of the Nuclear Phaseout

In 1986, during the aftermath of the nuclear accident in Chernobyl, the Social Democrats officially adopted a call for a nuclear phaseout in their party platform, which the Greens had always called for. The SPD wanted to phase out nuclear within 10 years. But these two parties were still in the opposition, and in 1990 Helmut Kohl was reelected based on his success in reunifying East and West Germany; thanks to his excellent diplomatic connections with the Allies and other European neighbors, he had managed to placate concerns about Germany once again becoming an uncontrollable force in Europe—not a trivial accomplishment just two generations after World War II. It was clear that the Christian Democrats would remain in power for a while—and their position, along with coalition partner the Free Democrats (FDP, libertarians) was that nuclear was still needed. Furthermore, the power supply in East

Germany required revamping, so the government began working on an energy concept.

Lignite made up 90 percent of East German power supply in 1989, and the goal was to bring that level down to 50 percent. East Germany already had small nuclear reactors in Greifswald and Stendal. The construction of two new, modern reactors with a total capacity of 2.6 gigawatts at those locations was discussed.

Then Environmental Minister Klaus Töpfer argued in June 1991 that Germany "cannot do without nuclear power at present." At the time, he was also open to the option of building new nuclear plants. That October, Economic Minister Möllemann (FDP) presented the government's Energy Concept, which included an option of new nuclear plants.[31]

The federal government's plans were controversial. Elsewhere, there was movement in the opposite direction. In 1992, Hamburg-based utility HEW implemented a policy handed down by the city council to fulfill a citizen petition. The utility resolved to forgo nuclear power "as quickly as legally possible and economically feasible for the Company."[32]

In 1993, the SPD's Gerhard Schröder handled negotiations with Chancellor Kohl's coalition with the goal of producing an energy consensus. The SPD was in the opposition, but Kohl wished to achieve a consensus that would survive elections; after all, energy infrastructure lasts for decades. At the time, Schröder was still Minister-President of the state of Lower Saxony; he would later succeed Kohl as Chancellor. Schröder had produced his own energy concept, including a ban on new nuclear power construction added to the German constitution. The federal coalition wanted an exception for research purposes in order to build the European Pressurized Reactor (EPR). In return, the government was willing to limit the service lives of existing nuclear plants to 30 or 40 years. Schröder did not compromise, however, arguing that the

---

[31] Leuschner, Udo. "Energiechronik: Möllemann legt Energiekonzept vor mit Option für neue Kernkraftwerke." October 1991. Accessed February 5, 2016. http://www.energie-chronik.de/911001.htm.

[32] Leuschner, Udo. "Energiechronik: Ausstieg aus Kernenergie in die Satzung der HEW aufgenommen." June 1992. Accessed February 5, 2016. http://www.energie-chronik.de/920608.htm.

EPR would never be competitive on price.[33] The Greens were also invited to these first negotiations, which would come to be known as the Energy Consensus Talks.

By the time the next parliamentary elections were held in 1994, the Greens were calling for a complete phaseout in two years, which would have been the date of the SPD's 10-year deadline from 1986.[34] Informal party leader Joschka Fischer—always adept at *Realpolitik*—supported the call for a phaseout but felt that a specific deadline would be a trap that could backfire on the Greens if it could not be upheld. He therefore wanted to leave the exact timeline open for the time being. For their part, the SPD had already backed away from their strict 10-year demand in 1991.[35] In 1994, they nonetheless reiterated their call for a phaseout (without a deadline) along with environmental taxation and greater support for renewables.[36]

Meanwhile, the nuclear sector itself had begun making preparations for a nuclear phaseout. After the first Energy Consensus Talks failed in 1993, a utility spokesperson expressed his dismay, saying that the firms themselves would have to close down their nuclear reactors in 15 years if nothing changed.[37] By late 1994, the nuclear utilities had started to bargain. In return for more flexibility in the rules for nuclear waste storage, nuclear plant operators would be willing to forgo the final repository proposed in Gorleben, a site that remain contentious. (The issue of a final repository remains unsolved today.) Specifically, the nuclear plant operators said they would be willing to store nuclear waste in intermediate facilities, such as on the grounds of the plants themselves, while

---

[33] Leuschner, Udo. "Energiechronik: Suche nach neuem Energiekonsens am Referenz-Reaktor gescheitert." October 1993. Accessed February 5, 2016. http://www.energie-chronik.de/931001.htm.

[34] Leuschner, Udo. "Energiechronik: Grüne wollen binnen zwei Jahren aus der Kernenergie aussteigen." February 1994. Accessed February 5, 2016. http://www.energie-chronik.de/940220.htm.

[35] Leuschner, Udo. "Energiechronik: SPD will über Nutzungsdauer der laufenden KKW mit sich reden lassen." November 1991. Accessed February 5, 2016. http://www.energie-chronik.de/911105.htm.

[36] Leuschner, Udo. "Energiechronik: SPD hält an Ausstieg aus Kernenergie fest." August 1994. Accessed February 5, 2016. http://www.energie-chronik.de/940802.htm.

[37] Leuschner, Udo. "Energiechronik: 'Tiefer Dissens' hinsichtlich Kernenergie beim ersten Gespräch über Energiekonsens." March 1993. Accessed February 5, 2016. http://www.energie-chronik.de/930301.htm.

policymakers and the public tried to agree on a final site. In addition, plant operators offered to close one or two reactors immediately.[38]

Overcapacity on the power market made that offer easier for the power firms. The utilities were coming to realize that no new power plants of any kind would be needed after the investment wave in Eastern Germany. By 1995, Veba and PreussenElektra—two of the eight biggest utilities—had announced that no new power plant capacity would be necessary for the foreseeable future.[39] Another one of the Big Eight, VEAG, said it expected lower power sales for the following two years.[40]

In 1994, the nuclear reactor at Würgassen—the oldest commercial reactor in the country at the time—was closed for economic reasons when fissures were found in the reactor hull.[41] The plant thus closed after less than 30 years of operation. The nuclear plant at Stade would later also be closed officially for economic reasons.[42]

When the same coalition of the Christian Union and the Free Democrats were reelected in 1994, nuclear protests reached record levels. They also increasingly became destructive. From 1995 onward, protestors sabotaged railway lines on which nuclear waste was to be transported. The violent acts occurred all over the country, from Stuttgart in the southwest to Dessau in the east. In the northern German city of Hamburg, the offices of HEW, Deutsche Bank, and railway company Deutsche Bahn were also vandalized. And near the proposed final repository in Gorleben, protesters repeatedly sawed apart railway lines, covered them with gravel, and set up barricades.

---

[38] Leuschner, Udo. "Energiechronik: "Spiegel" berichtet über angebliches EVU-Papier zur Aufgabe von Gorleben." November 1994. Accessed February 5, 2016. http://www.energie-chronik. de/941105.htm.

[39] Leuschner, Udo. "Energiechronik: PreussenElektra sieht gegenwärtig keinen Bedarf für neue Kernkraftwerke." June 1995. Accessed February 5, 2016. http://www.energie-chronik.de/950604. htm; and Leuschner, Udo. "Energiechronik: "Kraftwerkskapazitäten nicht ausgelastet"." May 1995. Accessed February 5, 2016. http://www.energie-chronik.de/950506.htm.

[40] Leuschner, Udo. "Energiechronik: Stromabsatz der VEAG sinkt weiter." May 1995. Accessed February 5, 2016. http://www.energie-chronik.de/950507.htm.

[41] Leuschner, Udo. "Energiechronik: Kernkraftwerk Würgassen wird aus wirtschaftlichen Gründen stillgelegt." June 1995. Accessed February 5, 2016. http://www.energie-chronik.de/950603.htm.

[42] Leuschner, Udo. "Energiechronik: Dow verzichtet auf eigenes Kraftwerk." September 1997. Accessed February 5, 2016. http://www.energie-chronik.de/970911.htm.

Though a large part of the German public supported the idea of a nuclear phaseout in some form, these radical acts met with great criticism. Left-of-center newspaper *Frankfurter Rundschau* wrote that these vandals were not so much sabotaging the transport of nuclear waste as they were sabotaging the nuclear protest movement.[43]

As the demonstrations intensified, the government proposed changes to its Nuclear Power Act in 1997. The amendments were to make it easier for power companies to override objections to nuclear disposal sites and loosen requirements for the latest safety technology. Reviews were to be conducted without public input. The German government was reacting to public resistance by shutting out the public.

The federal government also wanted to restrict input from state governments. In formulating the law, it tried to make sure that the *Bundesrat* would not have a veto right. In that chamber, the SPD and the Greens would be able to muster a majority to block the law. When the government went ahead and adopted the amendments in the summer of 1998, lawsuits were immediately filed challenging the constitutionality of the process. But before those legal challenges could truly move forward, there would be elections in September 1998, which made the issue moot.

The Forgotten (and Missed) National Climate Target

In addition to nuclear, reducing emissions had been a major policy concern at least since Chancellor Kohl spoke of the "grave threat of climate change" in Parliament in 1987. In preparation for the first Climate Summit in Rio de Janeiro in 1992, Chancellor Kohl began focusing more on carbon. He wanted Germany to be seen as an international leader on fighting climate change. In 1990, the cabinet passed the target of a 25 to 30 percent reduction in carbon dioxide emissions by 2005 relative to 1987 levels for West Germany. Based on the recommenda-

---

[43] Leuschner, Udo. "Energiechronik: Serie von Anschlägen auf Bahnstrecken erreicht neuen Höhepunkt." February 1997. Accessed February 5, 2016. http://www.energie-chronik.de/970202.htm.

tion of the Parliament's Enquete Commission, Environmental Minister Klaus Töpfer had originally pushed for a 30 percent reduction, but the Ministry of Economics prevented that. After unification, the government reaffirmed a target of minus 25 percent for the newly unified Germany with the base year of 1990, the commonly used baseline international by then.[44]

Including East Germany in the calculation lowered ambition: the shutdown of the dirtiest and inefficient factories and coal plants in the former East Germany, which would have been closed anyway, guaranteed a big emissions drop over the next few years. These "wall fall profits" should have allowed Germany to reach its target more likely than without unification.

However, Germany would miss this national climate target by a wide margin. While emissions dropped year by year in the first half of the 1990s, mainly thanks to the cleanup in eastern Germany, progress slowed down considerably in the early 2000s. The German government itself now estimates that reunification reduced carbon emissions by around 10 percent.[45] It wasn't until 2012 that Germany reached a 25 percent reduction in carbon emissions, surpassing its official target for the Kyoto Protocol of 21 percent. Interestingly, Germany was neither praised widely for overshooting its Kyoto target, nor criticized for missing the self-imposed target for 2005. It wasn't until the nuclear phaseout of 2011 that the world began closely—and critically—following German energy policy.

The setting of that 2005 target nonetheless had positive benefits. First, although the failure was not widely reported, it encouraged German policymakers to take future target attainment seriously. Second, it led to a serious review of targets among the German energy experts. For instance, in 1995 the German Environmental Agency calculated that German emissions might even rise by 2005 even if targets for better gas mileage

---

[44] Werland, Stefan. "Deutsche Klimapolitik unter Schwarz-Gelb und Rot-Grün." Magisterarbeit, Internationale Beziehungen / Außenpolitik, Universität Trier, Juli 2005. Accessed February 5, 2016. http://www.deutsche-aussenpolitik.de/resources/monographies/werland.pdf, p. 53.

[45] Fischer, Severin. "Zeit für eine Reform des deutschen Klimaziels." Phasenprüfer. 27 November 2014. Accessed February 5, 2016. http://phasenpruefer.info/zeit-fuer-reform-deutschen-klimaziels/.

were met.[46] Likewise, analysts at Prognos said in 1995 that they expected only a 10.5 percent reduction in carbon emissions by 2005—essentially, the level Germany had already reached.[47]

In the mid-1990s, this process started when politicians came together in the Energy Consensus Talks, partly to see how the goal could be reached. They involved some old-school wheeling and dealing. In return for a loosening of their opposition to nuclear, the SPD—a party traditionally representing unionized coal miners—received continued support for subsidized domestic hard coal mining in 1995. The deal did not sit well with the Green Party, the SPD's potential coalition partner. Prominent Greens openly criticized the swapping of "coal financing for nuclear waste disposal."[48]

The talks failed to produce a consensus about nuclear, but gradually progress was being made in other fields, such as efficiency and renewables. For instance, in 1994 environmental protection was added to the Constitution as a goal after the talks revealed a clear consensus.[49] But nuclear power was being sidelined in the process, as German daily *Süddeutsche Zeitung* commented ominously in 1995: "Because nuclear power remains controversial, agreements will be presented on other energy policy aspects. Everyone will act as though nuclear simply no longer plays a role in energy policy."[50]

---

[46] "Es trifft die Industrie." Focus. 10 April 1995. Accessed February 5, 2016. http://www.focus.de/politik/deutschland/co2-reduktion-es-trifft-die-industrie_aid:151370.html.

[47] Leuschner, Udo. "Energiechronik: "Die Bundesregierung wird ihr CO2-Minderungsziel nicht erreichen"." December 1995. Accessed February 5, 2016. http://www.energie-chronik.de/951207.htm.

[48] Leuschner, Udo. "Energiechronik: Geheimgespräche über Energiekonsens verärgern FDP und alarmieren Grüne." January 1997. Accessed February 5, 2016. http://www.energie-chronik.de/970104.htm.

[49] Deutscher Bundestag. "Wie Umwelt- und Tierschutz ins Grundgesetz kamen." Deutscher Bundestag. Accessed February 5, 2016. https://www.bundestag.de/dokumente/textarchiv/2013/47447610_kw49_grundgesetz_20a/213840.

[50] Leuschner, Udo. "Energiechronik: "Tiefer Dissens" hinsichtlich Kernenergie beim ersten Gespräch über Energiekonsens." March 1993. Accessed February 5, 2016. http://www.energie-chronik.de/930301.htm.

The Big Players Still Doubt Solar

So what was happening with renewables? photovoltaics—the solar cells that generate electricity directly from sunlight—still had astronomical prices. In 1990, a kilowatt of installed photovoltaics would have cost around 10 times as much as in 2013.[51] It was still mainly considered a technology for outer space.

To understand where photovoltaics was in the 1990s, we need to briefly go back a bit further. The history of solar has some remarkable parallels with the history of wind power. In both cases, big firms dabbled in the technology, primarily to show it would not work. In 1962, Chancellor Adenauer gave the Atomic Minister a mandate for solar research when the ministry changed its name to the Research Ministry.[52] Little progress was made; tellingly, renewable energy was collectively referred to as "additive energy" at the time. Nuclear—that was the real deal.

In addition, what was needed for solar power was not further research, but deployment. And as with wind power, however, it would not be until new players entered the scene that true progress would be made with deployment. Finally, projects pursued by large companies were generally subsidized generously, whereas the do-it-yourselfers went ahead with their projects at great entrepreneurial risk.

The year 1975 was arguably the year when the two solar camps— incumbents and new players—both entered the scene. That year, AEG, BBC, Dornier, Phillips, and RWE founded the Solar Energy Workgroup at RWE's headquarters. Today, none of these firms (or their successors) is a major producer of solar equipment. But other firms that later joined the Solar Workgroup included some who were still in the sector much later or even today.[53] For instance, Bosch sold its solar division in 2013 during the consolidation of the sector. Likewise, Wacker was the largest maker of solar silicon at the time and is currently one of the three biggest producers internationally, with its main solar silicon facilities still in Germany.

---

[51] Fraunhofer. "Photovoltaics Report." Powerpoint presented in Freiburg on 17 November 2015. Accessed February 5, 2016. https://www.ise.fraunhofer.de/de/downloads/pdf-files/aktuelles/photovoltaics-report-in-englischer-sprache.pdf.

[52] Janzing, p. 12.

[53] Janzing, p. 20.

Back then, solar arguably meant thermal (heat) as much as electricity. The largest solar thermal array built in Europe at the time (1976) was the size of the swimming pool it was built to heat. Funding from the German Research Ministry covered half of the cost. Solar homes were also built with subsidies in Freiburg and Aachen. Phillips and RWE were behind the one in Aachen. There, the researchers did not believe the final product was even suitable as a home (for instance, the windows were kept small to prevent heat losses). They used computers (made by Phillips) to simulate residents taking showers, turning the stove on and off, and so on. RWE was mainly interested in the project as a part of its "all-electric" campaign. The firm was not a major supplier of natural gas, so a house that ran completely on electricity would consume what RWE sold (coal power). To this day, large German utilities like RWE and Eon have notable renewable energy investments outside Germany in markets where these new assets will not conflict with any stranded assets—as they do in Germany.

AEG's "commitment" to solar in the early days also reveals how big engineering firms worked to make photovoltaics fail. On the island of Pellworm, an astonishingly large solar array with 17,568 panels was installed in a field near Germany's first wind farm. The 300-kilowatt array was completed in 1983. Sheep grazed around and underneath the panels, which were propped up on wooden struts. But the firm's project manager stated at the time that solar power would never be big in Germany; the project's purpose was to research off-grid projects in the Third World. To do so, the firm had picked the northernmost location in Germany with some of the worst solar conditions for the country—in order to study applications in areas with solar conditions far better than even the best in Germany.[54]

Few experts saw grid-connected photovoltaics on the horizon. After Chernobyl, the European Commission gave Freiburg's Fraunhofer Institute for Solar Energy Systems (ISE) funding for an off-grid project. Originally, a building in the countryside outside Stuttgart three miles from the nearest power lines was to be equipped with solar panels and batteries in a pilot project. But when the local utility, which had previously demanded a prohibitive price for the grid connection, heard of

---

[54] Janzing, p. 60.

the proposal, the building was quickly connected to the grid before the project could even get going.

Fraunhofer ISE then looked elsewhere for a suitable structure. It found a hikers cottage nearby in the Black Forest. Built in the seventeenth century, the cottage had a diesel generator for electricity. The researchers covered the roof with solar panels and hooked them up to batteries. The diesel generator was still on hand if more energy was needed. The project cost 140,000 marks, compared with 380,000 for a grid connection.[55] Later, a small wind generator was added.

The project was a success and drew a lot of attention, so Fraunhofer ISE followed up in the early 1990s with a solar home. Located directly next to other residential buildings in Freiburg, the home could have had a grid connection at the usual price. But the researchers wanted to find out whether a normal family in Germany could get all its household energy needs—both heat and electricity—from solar without any grid connection. To do so, they installed photovoltaic panels for electricity and solar thermal collectors for space heat and hot water. Batteries were also hooked up, and so was a fuel cell; the researchers wanted to test the seasonal storage of excess solar energy as hydrogen from the summer for the winter. Furthermore, the building used Passive House architecture; the southern façade had a large share of highly insulating glazing, whereas the other sides were closed off and also heavily insulated. Sunlight entering the southern façade then would heat up the building in the winter. Instead of having small windows, like the RWE/Phillips project in Aachen, this building had practically all glass on one side.

And unlike the previous project in Aachen, this one included a real family, who got to live in the home for a year without paying rent. The project was more or less successful, with the weak point being the fuel cell (which remains the weak point in many projects today). But by 1991, when the off-grid solar house project was conducted, politicians were already working on incentives for grid-connected photovoltaics, which would eventually make going off-grid unattractive. Connect your solar array to the grid, and you use it as both a battery and a backup generator (so you don't have to buy either). Seasonal storage—the crux of the off-grid solar home

---

[55] Janzing, p. 80.

project in Freiburg—is not an issue. Subsequent projects—like Freiburg-based architect Rolf Disch's 1994 Heliotrop, a rotating building that tracks the sun—pick up on many of the approaches implemented in the off-grid solar home: photovoltaics, solar thermal, insulated glazing on the southern façade, Passive House architecture, and so on. But the Heliotrop is connected to the grid and therefore has no storage issues.

In retrospect, the Fraunhofer ISE project can therefore be seen as a dead-end with its focus on off-grid applications, though it was inspiring in every other respect. After that project, the focus shifted entirely to grid connections thanks to policy support, which the solar researchers did not see coming. "Politicians did more to promote photovoltaics than we did as researchers," Adolf Goetzberger, the founder and still director of Fraunhofer ISE during the solar home project, stated in 2009 with reference to feed-in tariffs.[56] Indeed, in 1996 the once off-grid solar house got a grid connection—not because it needed more electricity, but because the institute decided to avail itself of the option that German law then provided of selling solar power to the grid.[57]

Solar Startups Get Going

Clearly, solar energy needed more support for deployment, not more R&D—as was the case with wind power. And as with wind power, the main progress came from new players.

The German solar startup scene arguably also got going the very same year in which the industry Solar Energy Workgroup was created. The German Solar Society (DGS) was founded in 1975; it brought together citizens of all walks. Near the nuclear protests in Wyhl, the town of Sasbach (population: 5500) began to hold Solar Days in 1976—a gathering of early solar tinkerers. By the end of the 1970s, Sasbach probably had the greatest density of solar hot water collectors in the world in the midst of its orchards and vineyards.[58]

---

[56] Personal contact.

[57] Janzing, *p.* 106.

[58] Janzing, *p.* 45.

The products these garage tinkerers produced were not bad. In the late 1970s, the German government wanted to donate some solar thermal systems to Egypt. A survey of eight products on the market was therefore conducted so that the best one could be picked. An electrician from Sasbach, Werner Mildebrath, won, even beating the big companies like engineering giant BBC.[59] Nonetheless, 10 years later Mildebrath would go out of business. The bigger firms had begun copying his design, and they also had the capital to scale up quickly, which made them more competitive. Mildebrath's failure to grow as fast as the big boys would be repeated numerous times in future years, not only with solar thermal, but especially with wind power and later with photovoltaics.

As the Sasbach Solar Days continued to grow, the local Chamber of Commerce intervened to stop what had become a regional trade show because none of the firms were members. In the third year of the event, 1978, more than 25,000 people visited—five times more than lived in the village.[60] The trade show therefore moved to the nearby town of Freiburg in 1980, where it became the Eco-Trade Show with a larger focus. But still, the event could not cover the growing interest. So in 1991, a competitor tradeshow dedicated solely to solar thermal and photovoltaics opened up some 90 minutes north in Pforzheim. In 2000, the two tradeshows would merge to become Intersolar in Freiburg before moving to Munich in 2008—and expanding to North America and Asia in later years. Today, Intersolar is a gigantic industry trade show, but it retained its down-home character for quite a few years. In 2003, the author Craig Morris wore a business suit to the event and was asked by a woman at a stand, "Why are you wearing that?"

Another crucial event was founded in 1986, just weeks before the accident at Chernobyl. That April, a symposium on photovoltaics was held in Staffelstein for the first time, attracting 72 people, 30 of whom were speakers. The next year, the event was repeated, and 120 people attended. By 2000, 1000 people attended, and today the event is arguably the main conference without a tradeshow for the German solar sector held in the country. The organizer was the founder of International

---

[59] Janzing, p. 43.
[60] Janzing, *p.* 45.

Battery Consulting (IBC), which is now called IBC Solar. In 1982, the founder began importing panels from Japan because Siemens and AEG—the two German manufacturers—did not want to talk to him. "They were not interested. They were big, and I was small," he remembers.[61]

Another newcomer to the solar sector was Alfred Ritter, owner of the eponymous chocolate firm. After Chernobyl, some of his plantations in Turkey were contaminated, so he began looking not only for new suppliers—but also new sources of energy. He started at home. But when he could not find anyone who could install an environmentally friendly heating system in his house, he founded his own efficiency and solar heating firm called Paradigma in 1988. Later, he would bump into Georg Salvamoser, a heating system installer, at a music festival in Freiburg, where a solar campaigner had set up a photovoltaic panel on a small table. The two began talking and would go on to found both solar planning firm SAG and Solar-Fabrik, a panel producer. In 1998, Solar-Fabrik opened up a new production line, which also became Europe's first zero-emissions production plant. It got all its energy from solar and a cogeneration unit running on biomass.

Phoenix Solar is another midsize German solar firm that has its beginnings in a small citizen's campaign. In 1989, Aribert Peters founded the Energy Consumers Association, a consumer advocacy group. In April 1994, it launched a campaign called Phoenix. The goal was to buy residential solar hot water systems in bulk—several collectors along with a 300-liter tank—in order to bring the price down from 12,000 marks to 5,250 marks. The campaign was a grand success and soon covered 20 percent of the market. But one of the cost-cutting proposals drew criticism: people were encouraged to install the systems themselves. Gerhard Stryi-Hipp, now head of energy policy at Fraunhofer ISE, was head of solar lobby group BSW at the time. A study he had conducted found that 90 percent of the problems detected in Phoenix systems were caused by do-it-yourselfers. Still, the campaign was a popular success. Phoenix would later become a solar company and split into two separate firms, one for solar thermal and one for photovoltaics. During this discussion, the campaign *Solar—na, klar* was also founded. For its efforts to bring consumers together with professional solar installers, it received an

[61] Janzing, *p.* 83.

award for the best national European renewable energy campaign from the European Commission in 2001.[62]

By the time the Phoenix campaign started, there was clear evidence that deployment would bring prices down dramatically for photovoltaics as well; no technological breakthroughs were needed. In 1995, a study published by an independent engineering group found that a 20-megawatt production line would cut the price of a 2-kilowatt array nearly in half. The engineers found economies of scale all along the production chain. Furthermore, "No photovoltaic production facility in the world uses the best manufacturing processes available today at industrial scale with modern machines." For instance, when Solar-Fabrik opened its first module production line in 1997, a lot of the work was still manual. Essentially, bigger and better manufacturing processes, not more research in the lab, would drive down costs of photovoltaics panels.

Thus, even at the end of the 1990s any large engineering company could have entered into the sector, built the largest automated photovoltaics production line to date, and brought prices down dramatically on a growing future market. The money needed for the most modern 20-megawatt production line was estimated at 26 million deutsche marks, not an astronomical sum for a large corporation.[63] But Germany's RWE and Siemens doubted that there was a market for photovoltaics. In 1994, Siemens closed its photovoltaics production line in Germany. ASE, a solar subsidiary of RWE and Daimler Benz, followed suit with one of its three photovoltaics production plants in Germany in 1995. One of the two that remained open specialized in dedicated photovoltaics for outer space, not the crystalline silicon modules that would conquer the market on earth. That year, Germany installed 4.5 megawatts of photovoltaics, up from 3 megawatts annually from 1992 to 1994. Germany thus only made up just over 1 percent of the global market, which came in at around 350 megawatts in the mid-1990s. But, from 2009 to 2012 the Germans would install a quarter of global photovoltaics capacity. Around 2004 to 2006, it installed nearly half.

---

[62] Deutsche Bundesstiftung Umwelt. "DBU-Projekt "Solar—na klar!" als europaweit beste nationale Kampagne für Erneuerbare Energien ausgezeichnet." 27 November 2001. Accessed February 5, 2016. https://www.dbu.de/123artikel24792_.html.

[63] Janzing, p. 124.

By then, policy support drove the market, but in the 1990s numerous grassroots campaigns cropped up in Germany even without policy support, demonstrating that there was market demand. Under the leadership of a young campaigner named Sven Teske, who would later become the author of the annual *Energy (R)evolution* report, Greenpeace launched a bulk purchasing campaign of its own. Called Cyrus, the project aimed to sell photovoltaic arrays with an output of 2 kilowatts for less than 30,000 deutsche marks, putting the kilowatt price under 15,000 deutsche marks. Like the Phoenix campaign, this one met with criticism; installers spoke of unfair competition and accused Greenpeace of ruining the market. "Their protest completely surprised and shocked us," Teske remembers.[64]

But when 4400 Germans pledged to make an order at that price, Greenpeace at least demonstrated that market demand was greater than Siemens and RWE believed. When the campaign then took bids from suppliers, five of the 45 bidders complied with the strict criteria, and the lowest price was a revolutionary 13,500 marks per kilowatt, a full 10 percent below the target many thought unrealistic.

Other campaigns held at the time without any governmental subsidies include the solar roof on the stadium of Freiburg's football club. In 1995, a group of local activists and sports fans came together to produce a special offer in cooperation with team management. Two years earlier, the team had entered the Bundesliga, Germany's premier league. In its second season there (1994–1995) the team finished third. The small stadium built for a second-league team would no longer suffice. Far more locals wanted to see their scrappy team beat professional Bayern München at home than the old stadium had seats. When additional seats were added, those who had invested in the new solar array on the expanded roof were pushed up in the waiting list. That year, 340 kW of photovoltaics was installed in Freiburg alone, 220 of which came from Solar-Fabrik, while the big players continued to doubt there was demand.

But the most interesting example of a solar newcomer is SMA. It is not a firm that laypeople are likely to have heard of. The company does not make solar panels, solar cells, or even solar silicon. It makes inverters— devices used to connect solar arrays to the grid. The firm was founded

---

[64] Janzing, p. 124.

in 1981, 10 years before Germany would mandate that grid connections must be provided for solar arrays. The company originally provided electrical equipment for early wind generators, but inverters for photovoltaics were part of the company's research from the outset. The firm's founder, Günther Cramer, set up a business to serve a market—grid-connected photovoltaics—that did not yet even exist and that others did not even see on the horizon. Cramer's motto was revealing: "Be realistic and attempt the impossible." By 2015, more than half of the inverters used to hook up solar arrays to power lines came from SMA—worldwide.

Cramer is no longer with us, nor is George Salvamoser (founder of Solar-Fabrik), Hermann Scheer (the main renewable energy proponent in German Parliament), or Reiner Lemoine (cofounder of Solon and Q-Cells). None of these early proponents of renewable energy lived long enough to retire. In the unfair battle between newcomer startups and incumbents, they put in long, stressful hours—and paid a high personal price to move the energy transition forward.

By the time parliamentary elections were held in September 1998, Germany thus had three completely new players on the solar market: Phoenix, Solar-Fabrik, and Solon. The latter supplied most of the solar panels to the new roof for the Bundestag in Berlin. SolarWorld then went public in 1999, when Q-Cells was also founded. Numerous other new firms could also be mentioned, but the point is that all of them were newcomers. Not a single one came from an existing energy corporation.

It was these startups that would benefit from the policy changes to come. But from 1991 to 1998, the 90 percent share of the retail rate offered under the Feed-in Act was not enough of an incentive. photovoltaics still needed more support. It would have to wait until the next decade.

Wind power, on the other hand, had begun growing quite nicely. And that started to bother incumbent utilities.

Utilities Fighting the Rise of Wind Power in Court

Starting in 1991, anyone in Germany who wanted to build a turbine had the right to connect it to the grid and receive 90 percent of the retail rate for the power exported to the grid. The rate was the same all over

Germany, regardless of the turbine's location. The greatest profit could therefore be posted in areas with the best wind conditions, generally along the coastline.

The boom started immediately once the law was passed. In 1993, for instance, the market grew by slightly more than 50 percent when 664 turbines were completed, bringing the total up to 1797. The turbines were not only growing in number, but also in size and power. Towers were becoming taller, rotor blades longer, and generators bigger. In 1992, the average turbine installed in Germany had an output of 170 kilowatts. By 1998, the average size had grown nearly fivefold to 785 kilowatts.[65]

By 1995, the amount of wind power in the north was beginning to concern utilities and grid operators, although wind power still only made up around 1 percent of total power supply in Germany. The uneven spread of wind farms across the country meant that wind power affected only certain parts of the grid. It also affected only certain grid operators financially. The Feed-in Act stipulated that wind power had to be paid for, but there was no mechanism for the sharing of this cost burden across the country.

One of the first utilities to complain was EWE, the municipal utility of Oldenburg. In early 1994, its CEO complained that the wind power he was forced to purchase from third parties was 75 percent more expensive than the electricity he would otherwise buy from Preussenelektra, one of the big utilities.[66]

Other utilities, such as HEW—the municipal in Hamburg that wanted to step away from nuclear power—took a much different approach, offering a 10-pfennig bonus on top of the 16.53 pfennigs required by law. The move was one of the few cases aimed at giving greater local support to wind power than provided by the federal Feed-in Act.

In May 1995, three regional utilities joined forces with the VDEW (the utility lobby group), German industry lobby group BDI, and the

[65] Bundesverband WindEnergie e.V. "BWE-Info II: Zahlen und Fakten zur Windenergie." 28 February 2002. Accessed February 5, 2016. http://www.bund-lemgo.de/download/zahlen-fakten-windenergie-bwe-info.pdf.

[66] Leuschner, Udo. "Energiechronik: Keine Genehmigung für kostendeckende Subventionierung von Solarstrom." January 1994. Accessed February 5, 2016. http://www.energie-chronik.de/940115.htm.

lobby group for power plant engineers VIK to challenge the constitutionality of the Feed-in Act.[67] A ruling from late 1994 in a separate case encouraged them; a court found that German subsidies of domestic hard coal mining were an inadmissible levy. The "coal penny," a surcharge tacked onto power prices, provided funding to protect mining jobs in Germany; German hard coal had long been more expensive than hard coal on international markets. Germany could have gone the way of the UK under Prime Minister Margaret Thatcher and switched to cheaper coal imports, but Chancellor Kohl's government was more conservative than libertarian and did not wish to endanger so many jobs within the country.

Subsidies for hard coal miners continued nonetheless—indeed, they will not be phased out completely until 2018. But the court's ruling meant that funding had to be shifted away from power rates and into the governmental budget. The plaintiffs in the case against the Feed-in Act now hoped that the court would rule along the same lines for wind power because, like the coal penny, it was tacked onto power prices. If the utilities could get these costs moved into the federal budget, they would disappear from utility balance sheets.

To produce a legal challenge, the utility plaintiffs began paying a lower rate than the one required by law in a single case to produce a legal challenge. Politicians across the country expressed their dismay at the entire affair. Spokespeople from various parties called on the utilities to stop breaking the law. The press was no less critical, accusing the utilities of wanting to keep their monopolies for themselves even at the expense of climate protection.[68] In addition, German antitrust authorities stated that they could not see how the Feed-in Act damaged competition.

Frustrated with the public response, the VDEW said the dispute was being depicted as "big, bad monopolists beating up on the little guys trying to save mankind." The head of the lobby group complained that almost all the small hydropower generators had already been in existence when the law was passed, so not a single additional kilowatt-hour of elec-

---

[67] Leuschner, Udo. "Energiechronik: Stromversorger wollen Musterprozeß um Stromeinspeisungsgesetz erreichen." May 1995. Accessed February 5, 2016. http://www.energie-chronik.de/950501.htm.
[68] Ibid.

tricity was being generated from these units because of the law. To drive the point home, he coined a new term—instead of windfall profits, these generators were now receiving waterfall profits.

In September 1995 a first-instance court in Karlsruhe ruled in favor of Badenwerk, a southern German utility that had challenged the feed-in tariffs for small hydro. The court specifically linked the justification for the unconstitutionality of the coal subsidy to the feed-in tariffs paid for renewable electricity. But that same month, a first-instance court in Freiburg came to the exact opposite ruling for another utility, again concerning hydropower. Higher courts would have to resolve these different readings. And that indirectly meant legal limbo for the wind sector because this challenge also affected their own feed-in tariffs.

In January 1996, the German Constitutional Court refused to hear an appeal of the first-instance court's ruling that found the Act was unconstitutional based on a technicality. The constitutional judges complained that the lower court had not properly explained its ruling. The whole process was then thrown back to the lower court instead of being settled once and for all.

Then, a big player entered the scene in March 1996, and the court cases took on a different angle. Preussenelektra charged that individual utilities had to carry too much of the burden. This complaint had little to do with the constitutionality of the levy itself, and hence the ruling against the coal levy was unrelated. Politicians also pointed out that the largest section in the two-page Feed-in Act pertained specifically to such concerns. If a grid operator had more than 5 percent of its power supply from third-party renewable energy generators, it could pass on the additional costs to the next grid operator in the supply chain. Some utilities would reach 5 percent renewable power before others—and have to buy this power above the wholesale rate. The situation was clearly unfair.

The effect on the wind market was palpable. In 1995, more than 1000 turbines were installed for the first time in a single year, bringing the total up to around 3500—equivalent to roughly 40 percent annual growth. But in 1996, the market slumped to around 800 turbines. Firms that had struggled to expand with the market now had to slam on the brakes. The first ones went bankrupt, such as Tacke in 1997. Originally taken over by Enron, Tacke later became GE Wind Energy in 2002 after Enron's bankruptcy. Today, GE is one of the first companies we think of in the wind power sector, but the company was a latecomer. It was completely absent in the crucial years of innovation.

The same can also be said of Germany's Siemens, which did not exist in the wind sector until 2004, when it took over Denmark's Bonus Energy. The Danish firm folded in the wake of stagnation on the Danish wind market after the elections of 2002. Today, when news media write about the German wind market, Siemens is often mentioned—but the company makes up less than 10 percent of the German market. Enercon, a startup in the 1980s, makes up nearly 40 percent of the German market, but receives much less international press coverage. Outside Germany, too few people realize that the country's biggest firms have been of little help in the energy transition. Newcomers made the difference.

Fortunately, the wind sector itself had become big enough to start defending itself by the mid-1990s. In the early 1990s, Johannes Lackmann, a solar and wind pioneer from Paderborn, paid for two staff members out of his own pocket Germany's first renewable energy lobby group: the *Bundesverband Erneuerbare Energien* (BEE). The organization had worked on a shoestring budget with volunteers up to then. "Some people buy a fancy car," he explains, "I created a lobby organization for renewables."

Roughly 10,000 people were employed in the industry as early as 1995. The German Association of Equipment Producers VDMA began expressing its concerns about the impact of this legal uncertainty on the wind sector. When German industry lobby group BDI published a position paper calling for the cost of feed-in tariffs to be charged to taxpayers rather than ratepayers (in line with the coal decision), the VDMA distanced itself from the paper.[69] Other groups, such as the metalworkers' union and farming associations (because farmers were installing so many wind turbines), also began openly expressing their support for wind power during these years.[70]

---

[69] Leuschner, Udo. "Energiechronik: Differenzen um das Stromeinspeisungsgesetz zwischen BDI und Anlagenbauern." January 1997. Accessed February 5, 2016. http://www.energie-chronik.de/970118.htm.

[70] Toke and Lauber, 2007.

By November 1997, the urgent need to bring back certainty with better legislation had become clear. The utility Schleswag already had 10 percent wind power in its lines. It could pass on costs above 5 percent to PreussenElektra, but the road stopped there for the second utility.[71] Either PreussenElektra would have to absorb the additional costs, or additional wind power simply would not be paid for.

Or the law could be amended. Politicians and energy experts were already discussing ways to spread the cost equally across the entire country as a surcharge on top of the kilowatt-hour price. But as if to demonstrate its own paralysis, the government announced at the end of 1997 that it did not wish to change the Feed-in Act in the near future.[72]

Paralysis Before 1998 Elections

A train wreck was clearly approaching, but the Kohl government would not act. Indeed, there was growing impatience with the coalition in many areas. In the early 1990s, a new word—*Reformstau* (reform backlog)—was coined. Necessary reforms were stuck in the pipeline. Each year, German linguists choose a new Word of the Year. In 1997, the year before the parliamentary elections that Helmut Kohl would finally lose, the Word of the Year was *Reformstau.*

The "landscapes in bloom" that the Chancellor had promised during Reunification were few and far between in eastern Germany. Unemployment remained high. But Kohl *klebte am Sessel*—he was "stuck to his chair," as the Germans say when a leader refuses to step down at the right time. When Kohl turned 65 in 1995, there was talk about his crown prince, Wolfgang Schäuble, taking over the throne. But Kohl had become too important to himself; retention of his political power was more crucial to him than the strength of his own party. When he did not step down

---

[71] Leuschner, Udo. "Energiechronik: "Härteklausel" im Stromeinspeisungsgesetz nennt jetzt klare Belastungsgrenze." November 1997. Accessed February 5, 2016. http://www.energie-chronik. de/971102.htm.

[72] Leuschner, Udo. "Energiechronik: Regierung will an Einspeisevergütung für Windstrom vorläufig nichts ändern." October 1997. Accessed February 5, 2016. http://www.energie-chronik. de/971018.htm.

and make way for Schäuble, he claimed that the upcoming rollout of the euro required his continued presence in the chancellery. To Kohl's credit, Schäuble's later stance on the euro revealed that he put monetary stability above political concerns, whereas Kohl himself saw the euro as part of EU peace policy.[73] But at the time, Kohl's statement was an insult to Schäuble—and an indication to the German public that Old King Kohl was not going to leave voluntarily. He would need to be ousted.

At the end of September 1998, Helmut Kohl was voted out of office after 16 years as Chancellor—the longest term of any German Chancellor since Otto von Bismarck in the nineteenth century. German voters elected the Social Democrats and the Greens into office. It was a coalition the country had never had. Furthermore, it was the first complete shift in a coalition; up to then, at least one party from the previous coalition had always remained in power with a new partner. German politics had thus maintained a high level of continuity, but radical change might now be in the offing.

Gerhard Schröder of the SPD became Chancellor, with Joschka Fisher of the Greens becoming Foreign Secretary—and hence Vice-Chancellor. The two already knew each other from more than a decade of opposition to nuclear power.

---

[73] Schuler, Katharina. "Kohl hat das Wort." Die Zeit Online. 19 March 2015. Accessed February 5, 2016. http://www.zeit.de/politik/deutschland/2015-03/helmut-kohl-interview-altkanzler/komplettansicht.

8

Green Capitalism Made in Germany

Volkmar Lauber is a soft-spoken, retired professor of political science from Austria. He spent a good part of his career defending German-style feed-in tariffs against what he calls "the neo-liberals at the European Commission in Brussels." But when he started off in the early 1970s, he didn't have Germany on his radar. "Germany seemed to be a conservative country, not one with an innovative movement," he says today. The US attracted him more, so he went to Harvard, where he got a second law degree in 1970 (after graduating in law at the University of Vienna).

At the University of North Carolina at Chapel Hill, he followed up with a political science PhD on the economic growth controversy in France. But in 1974, nuclear protests in Germany drew his attention. "What was happening in Wyhl didn't fit the narrative of 'obedient' Germans," he explains.

In his own home country, nuclear protests were also flaring up, but the US press hardly covered the events. As a German speaker, however, he could read the European media reports. Construction had begun on a nuclear plant at Zwentendorf—the first in Austria—in 1972. Four years later, the Austrian government announced that three others would be built when Zwentendorf was completed.

© The Editor(s) (if applicable) and The Author(s) 2016
C. Morris, A. Jungjohann, *Energy Democracy*,
DOI 10.1007/978-3-319-31891-2_8

In those years, Lauber was trying to publish his way into a permanent position in higher education in the US. It wasn't easy, especially as a single parent. And US journals were not always interested in his unusual stance. A Midwestern political science journal rejected a paper of his questioning public support for nuclear power policies in Germany and France in 1978. "What angered me the most was that I actually knew what was going on," while the reviewers still adhered to the image projected by governments.

Eventually, the protests over the Zwentendorf nuclear plant had escalated so much that a national referendum was held. Construction had essentially been completed in early 1978, but protests (inspired in part by the Danish anti-nuclear movement) had become extreme. On 5 November 1978, the referendum came to an incredibly tight outcome: 50.47 percent of Austrians rejected the plant, with 49.33 percent voting in favor. All this happened before the Three Mile Island accident brought the risk of radioactivity into the foreground in the United States in 1979. Like the protesters in Germany, the Austrians had more on their mind than just radioactivity.

As he investigated the situation, Lauber found a political struggle, not primarily concern about radioactivity. The unions wanted to build more hydropower plants, and the conservative government wanted to build nuclear partly to weaken the unions. The public, including many conservatives, were then not pleased with the government's heavy hand; for instance, it attempted to bring the fuel rods in secretly. Turnout for the referendum was thus relatively low at around 61 percent of voters because, Lauber remembers, "a lot of conservatives felt that the whole nuclear plant issue simply not been handled right."

In 1982, Lauber accepted a professorship at the University of Salzburg in Austria. There, he gradually became something of an academic activist back home in Austria. He was even asked to apply for the position of CEO at an Austrian utility. "I had no chance whatsoever, but I made it on TV," he laughs today. Most of all, the experience taught him what his role would be. "I wasn't trained as a public debater, so that wasn't for me." In his research, he came to focus more and more on environmental policy.

He also started writing about European policies when Austria finally became a member state in the mid-1990s—"I felt someone had to do that at my university department"—which eventually made him an expert on German feed-in tariffs when the European Commission became increasingly hostile to them in the late 1990s. Germany was clearly creating a vast market with this new policy. Feed-in tariffs allowed newcomers to rapidly deploy renewables while electricity incumbents remained opposed. Through this deployment, an equipment industry developed, innovation skyrocketed, and big incumbent players were forced to treat startups fairly. So why were European officials critical of this policy?

"The EU started saying no to feed-in tariffs at the latest in 1997," Lauber says. And he thinks the stance in Brussels was partly ideological. "The single internal market was the big thing, and the Commission saw that big business was their best ally towards that goal." Back in 1983, the CEO of Volvo had created the European Round Table of Industrialists, which initially brought together 17 leading executives. "The CEOs didn't send lobbyists; they went themselves," Lauber remarks.

He had a hard time understanding the language that these businessmen (and they were all men) were using. They spoke of "free markets," but what they wanted was an oligopoly. "They wanted the appearance of competition—the way incumbents want competition."

Since the late 1990s, Lauber has thus supported the argument that feed-in tariffs lead to freer markets. To put a finer point on his dilemma, he took it upon himself to prove that fixed prices actually increase competition. After all, feed-in tariffs do primarily two things: ensure grid access for renewable energy and set a minimum price.

The Chicago School and the Freiburg School

In the US, one school of economic thought—neoliberalism—became mainstream in the second half of the twentieth century. Because of its impact on policymaking in Washington DC, it has—a bit inaccurately—become synonymous with the Washington Consensus, but it is also named after the Chicago School of Economics, where many of its most prominent proponents were employed.

For neoliberals, a free market is by definition one in which price-setting is left up to the market—or, to quote Adam Smith, to the "invisible hand" of the market. Smith wrote in the eighteenth century, but his concept of the "invisible hand" may be the most common term in economics today. The Chicago School was crucial in making Smith's concept a buzzword in America. Today, students learn in every Economics 101 course that the interplay of supply and demand determines prices—or at least, the more this happens, the freer a market is said to be. As one US college economics textbook explains:[1]

> ...when the government prevents prices from adjusting naturally to supply and demand, it impedes the invisible hand's ability to coordinate the millions of households and firms that make up the economy. This corollary explains why taxes adversely affect the allocation of resources: taxes distort prices and thus the decisions of households and firms. It also explains the even greater harm caused by policies that directly control prices...And it also explains the failure of communism. In communist countries, prices were not determined in the marketplace but were dictated by central planners.

In this view, price-setting should be left up to the market's invisible hand. The most opposite extreme to free markets is planned economies, such as former communist countries.

Yet, Adam Smith was not talking about price-setting when he spoke of the invisible hand. Indeed, the term itself is a passing metaphor, not a central theme, used only once in each of his two major tomes. In the first book, he speaks of "an invisible hand" redistributing wealth; in the second, of consumers preferring to buy domestic products even when there is free trade—"an invisible hand" would thus lead Americans to "buy American" today. (Since Americans clearly buy from Asia today, it seems that Smith may have overestimated people's taste for patriotic purchasing. We misinterpret his "free hand," but even a correct understanding suggests that the notion is wrong.) A more central idea in Smith's writings was that of the "impartial spectator," that voice inside each of us that tells us how our peers would judge our actions. Smith saw people not as individuals, but as members of communities.

One could therefore argue that Smith never intended his "invisible hand" to become his main idea, much less to be understood as a pricing

---

[1] Mankiw, Nicholas G., and Mark P. Taylor. *Microeconomics*. 3rd. ed. Andover: Cengage Learning, 2014, p. 10.

mechanism or a call for lower taxation, as it is often portrayed by a group of economists from the Chicago School. Germany's Freiburg School has a different (we argue: correct) reading of Adam Smith. Also known as ordo-liberalism, the German school of economic thought has the same roots as neoliberalism. But the two have grown in different directions. The Freiburg School stresses "order"—the "ordo" in ordo-liberalism.

Another Economics 101 textbook mentions Smith's invisible hand in a way that allows us to clearly highlight the difference between the two schools of economics:

"Adam Smith made the most famous observation in all of economics: Households and firms interacting in markets act as if they are guided by an 'invisible hand' that leads them to desirable market outcomes."[2]

Both the Chicago School and the Freiburg School would agree that the market should be left alone in order to come up with desirable outcomes. But as the quote above indicates, mainstream economics in the US believes that outcomes are per se better the more the market is left alone. Adam Smith himself put things differently: "By pursuing his own interest [an individual] *frequently* [emphasis added] promotes that of the society more effectually than when he really intends to promote it."[3]

Ordo-liberals agree with Smith that the market *frequently* leads to desirable outcomes. When it does not, the state has to intervene in order to prevent an undesirable result. State intervention is therefore seen as necessary toward guaranteeing fair competition on a market in the Freiburg School. The Chicago School, in contrast, calls for deregulation out of principle. In doing so, it misreads Adam Smith.

The Freiburg School contends that competition tends to restrict the number of firms, creating oligopolies and monopolies, not more competition—a stance highlighted by the authors of the second Energiewende book of 1985.[4] In this view, regulators and antitrust authorities are called upon to constantly monitor and occasionally intervene in the market, not leave it to its own devices. As the editor of German economics daily

---

[2] Mankiw, Nicholas G. *Principles of economics.* 2. ed. Fort Worth Tex.: Harcourt College Publ, 2001, p. 7.

[3] Smith, Adam. *The Wealth of Nations.* Middletown, DE: Shine Classics, 2014, p. 168.

[4] Hennicke, Peter. *Die Energiewende ist möglich: Für eine neue Energiepolitik der Kommunen; Strategien für eine Rekommunalisierung.* 3. Aufl. Frankfurt a.M.: S. Fischer, 1985.

Handelsblatt wrote in 2011 during anticapitalism protests pertaining to the euro crisis, ordo-liberalism saves capitalism from itself:[5]

> Wilhelm Röpke [a German economist considered a forefather of ordo-liberalism] understood that a market economy cannot fulfill the requirements it needs to stay alive. Rather, it needs constant protection, care, and reforms. It is an iterative process, not a final outcome. Because everything is incomplete—including information about markets and the products sold on them—market imperfections are found everywhere...Market economics is more modest than capitalism because it knows its limits. It sees itself as a man-made principle of order in which state organizations repeatedly have to intervene.

Ordo-liberalism was implemented in Germany as the official economic policy in the 1950s by Ludwig Erhard of the Christian Union. As German Economics Minister from 1949 to 1963, Erhard implemented the same shock therapy in West Germany that was later applied to developing countries and the former Soviet Union. The deutsche mark was introduced in 1948, when Germany did not yet have a government (it came in 1949). The country was still a collection of zones occupied by Allied troops. The deutsche mark was implemented under the purest neoliberal terms without any means of state regulation. But immediately afterward, Erhard would begin to add on control mechanisms wherever he saw this unregulated market leading to undesirable outcomes. Regulators perform this role everywhere, such as with the Volcker Rule in the US for stock trading, saving the market from itself. Those who propose carbon taxation and other ways of internalizing external environmental costs call for these things because the market fails to price in all its impacts.

Since Erhard's first forays in the late 1940s, this pattern has repeated itself, as the German proposal for negotiated third-party access during the liberalization of energy markets shows (see the previous chapter). Germany eventually established a grid regulator when the power market proved it could not monitor itself. The Germans repeatedly start with

[5] Steingart, Gabor. "Angriff auf die Marktwirtschaft." Handelsblatt. 17 December 2011. Accessed February 6, 2016. http://www.handelsblatt.com/politik/international/kapitalismuskritik-wo-geld-klingelt-da-herrscht-die-hure/5808584-2.html.

neoliberal deregulation but end with ordo-liberal state intervention and oversight when the unregulated market leads to undesired outcomes.

In contrast, neoliberalism is more clearly the dominating force in the English-speaking world today since at least the Thatcher/Reagan era of the early 1980s, though some claim (or perhaps hope) that its dominance ended with the financial crisis of 2008. Its influence, as Lauber and others have documented,[6] also extends to the EU level in Brussels. In common parlance, neoliberalism is synonymous with small government, low taxation, and deregulation, but like any sophisticated school of thought it is richer than these simple labels. For instance, we find economist Ronald Coase among the early members of the Chicago School. He became famous for his work on property rights and taxation to cover "external" costs—impacts not covered by the price of a product or service. Pollution is a classic example of external costs; when we purchase oil, the price does not include the expenses needed to clean up the environmental mess caused by its extraction, transport, and use—at least, not without proper environmental taxation.

Green Budget Germany is a nonprofit that lobbies for environmental taxation; in 1999, it managed to get the country's Ecological Tax Reform implemented. Each year, it also awards a prize to someone who has been especially helpful in promoting environmental taxation. The award is called the Adam Smith Prize. Germany uses Mr. Invisible Hand to promote environmental taxation, not call for less taxation. Clearly, the German reading of Adam Smith is different from the mainstream interpretation of his work in the English-speaking world. And that different reading opens up a few policy options.

How Feed-in Tariffs Bring Costs Down

To quote the former publisher of center-right German daily *Frankfurter Allgemeine*, ordo-liberalism is the "visible hand of economic prosperity" and a "third way" between Keynesian state intervention and laissez-faire

---

[6] Jacobs, David. *Renewable Energy Policy Convergence in the EU: The Evolution of Feed-in Tariffs in Germany, Spain and France.* Global Environmental Governance. Farnham: Ashgate Publishing Ltd, 2012.

neoliberalism.[7] Markets cannot always be trusted, but the Freiburg School also holds that governmental bureaucrats are not necessarily smarter. Ordo-liberalism is not micromanagement. If the market is a football game, then ordo-liberals want referees. The refs might sometimes make bad calls, but the game is fairer with them than without. Neoliberals, in contrast, want competing teams to decide whether a foul was committed without refs. The neoliberal stance essentially prefers market failures to governmental failures. Ordo-liberals argue that complete laissez-faire capitalism produces more failures overall.

Eventually, however, players who make do without referees end up squabbling over an alleged foul, and the game is held up. On markets, companies that have to work things out among each other also spend a lot of time litigating instead of focusing on their core business. This situation favors large companies with big legal departments; smaller firms prefer clear rules so they can get back to business. The real choice we face is therefore not between allegedly free, unregulated markets, but between rules set forth by the government before the game starts and rules set forth by courts while the game is underway.

In the energy sector, we can see how this thinking led to feed-in tariffs. First, governmental bureaucrats are agnostic about what technology should be supported; markets know more about that. In auctions and quota systems, markets determine technology prices, while the government sets the volume to be auctioned. But different technologies need different levels of support; the idea behind feed-in tariffs has always been to produce the proper mixture of renewable energy sources, not provide the most support to the cheapest source. The first feed-in tariffs from 1991 therefore provided two different levels of compensation so that a wider range of renewable energy sources could be supported.

Feed-in tariffs do not pick technology winners. The policy does not decide, say, whether wind turbines with or without gearboxes should be built or whether thin-film or crystalline photovoltaics should be installed. That decision is left up to investors: to the market. If you think you can

---

[7] Jeske, Jürgen. "The Visible Hand of Economic Prosperity." Project Syndicate. 25 February 2015. Accessed February 6, 2016. http://www.project-syndicate.org/commentary/germany-economic-progress-policymaking-by-j-rgen-jeske-2015-02?barrier=true#qI1OuhtEAdqKrPVR.99.

generate more electricity for the buck with monocrystalline instead of polycrystalline photovoltaics go ahead—you run the entrepreneurial risk. In contrast, in R&D programs the government bureaucrats run the risk; companies get the money whether their products work or not, provided they receive funding (a decision made by bureaucrats). So with feed-in tariffs, Germany decided it wanted more green electricity, but it left the how-to up to the market.

Feed-in tariffs represent an example of the government stepping into an oligopolistic (and occasionally even monopolistic) market and requiring the main players to allow access to smaller startups. The purpose is to introduce more competition on the marketplace by opening the door for new players and investors. Feed-in tariffs represent a kind of state interventionism to ensure greater competition by adding new entrants on an oligopoly market. The same goal was pursued in the US with the PURPA legislation from the late 1970s, which inspired Germans later to implement feed-in tariffs. Early Danish legislation prior to 1990 also served as a role model for German legislators.[8] As German politician Hermann Scheer, coauthor of the feed-in tariffs from 2000, understood from the outset, incumbent utilities were not going to drive a transition; they had a status quo to defend.[9] If Germany wanted an energy transition, new players would be needed, and they would have to be able to compete on terms not dictated by the more powerful incumbents.

Germany first implemented feed-in tariffs in its 1991 Feed-in Act, but the policy was fundamentally revised in the 2000 Renewable Energy Act (EEG). Though the Germans did not invent the idea, the way feed-in tariffs were redesigned in the EEG has served as a model for up to 100 countries and countless additional states and municipalities. The EEG is one of the most emulated pieces of legislation in the world in any field.

As described in the previous chapter, the 1991 Feed-in Act linked compensation to the retail rate. There were two tiers: the lower tariff was initially 75 percent of the retail rate for hydropower, bioenergy, and energy from waste, with 90 percent of the retail rate for solar and wind. By the end of the decade, it had become clear that these price levels needed tweaking.

---

[8] Krawinkel, Holger. *Für eine neue Energiepolitik: Was die Bundesrepublik Deutschland von Dänemark lernen kann.* Frankfurt am Main: Fischer, 1991.

[9] Scheer, Hermann. *Sonnen-Strategie: Politik ohne Alternative.* München: Piper, 1998, p. 41.

In 1997, the law was amended to provide 80 percent of the retail rate for hydropower and bioenergy, producing three different tariffs. But although photovoltaics also clearly needed greater support, the level for photovoltaics was left unchanged. Solar power needed a lot more money, not just a little more. And Chancellor Kohl's government refused to give solar what it needed because the actual price seemed exorbitant at several times the retail rate. All the other feed-in tariffs offered were still lower than the retail rate.

In 2000, the new government of Social Democrats and Greens made fundamental changes to German feed-in tariffs in the EEG. First, feed-in tariffs were unlinked from the retail rate. It had fluctuated throughout the 1990s already, leading to different levels of support over the years—regardless of what the technologies needed. Furthermore, liberalization was expected to lead to lower retail rates (which did happen for the first few years), and the result would have been a decrease in compensation for renewables. The fledgling market would have collapsed. The EEG therefore linked compensation to what was actually needed for each technology. For biomass and photovoltaics, the EEG makes a distinction between different system sizes, with large biomass units assumed to produce energy less expensively than small ones. To prevent windfall profits, large units therefore receive smaller feed-in tariffs. Again, the target return is the same for biomass, wind, and solar—and for systems small and large. For wind power, a distinction is made not on the size of the turbine, but on wind conditions. If a turbine is installed in a windy area, it generates more electricity than the same turbine installed in a location with less wind. To prevent windfall profits, the feed-in tariff for wind power can be reduced from around 9 cents to 5 cents after at least five years of production.

Government officials used established formulas to calculate feed-in tariffs for the different renewable energy technologies. To provide investors with greater security, the rates are fixed on the date of a system's grid connection for a certain time frame (generally, 20 years in Germany). The law also puts pressure on industry to bring costs down by reducing those 20-year guaranteed rates for new systems on a scheduled basis, such as annually. In other words, connect to the grid today, and you get one rate for 20 years, but connect next year, and the rate will be slightly lower for 20 years. This policy thus rewards builders who move quickly.

It is easy to calculate feed-in tariffs. They are the most complicated for bioenergy. Biomass has fuel costs, which would be added to the purchase price and operations & maintenance to produce the total cost of energy from such a system. It is then assumed that the system would run for 20 years (or at least be amortized within that time). The amount of electricity a system of a given size could realistically generate in 20 years is then estimated in kilowatt-hours. Divide the number of kilowatt-hours by the total cost, and you get the cost per kilowatt-hour.

When banks finance coal plants and gas turbines, they perform a more sophisticated version of the basic calculation above. The challenge for generators with fuel costs is knowing what fuel will cost in the future. For biomass, the fuel cost is somewhat stable, though relatively high. The calculation of kilowatt-hours from a biomass unit is also relatively easy in Germany because grid operators have to buy the electricity. These variables therefore do not fluctuate much, so risk is lower, and banks reward low risk with low interest rates. Feed-in tariffs therefore lead to low interest rates—at least, when banks understand that the risks are low.[10]

Banks are also more willing to lend to new players when grid access is guaranteed and it is known in advance that profitable rates will be paid for power. "We were a company without any history, but we got a loan covering 1.3 million euros just on the strength of the project and a mere 0.6 million euros—five percent—as equity," one German community project planner explains. "Without FITs, it would not have happened."[11]

Fossil fuel prices fluctuate wildly, in contrast, making it hard for banks to calculate whether a coal plant or a gas plant will be a good investment five, 15, or more years from now. This risk leads to higher interest rates, which in turn raise the cost of capital. However, bank loans are usually taken out only for the upfront purchase price, not the fuel, which is generally financed with normal operating income. Gas turbines are relatively cheap to build; most of the electricity price is related to the cost of the

---

[10] Grau, Thilo, Karsten Neuhoff, and Matthew Tisdale. "Verpflichtende Direktvermarktung von Windenergie erhöht Finanzierungskosten." Publication of the German Institute for Economic Research Berlin. *DIW Wochenbericht*, no. 21 (2015): 503–8.

[11] Conversation with Josef Pesch of Fesa.

fuel in that case. So, banks that finance gas turbines factor in the risk of fluctuating fuel costs when designing a loan for construction of the plant. After all, if fuel costs rise, the gas turbine generates fewer kilowatt-hours and therefore makes less money.

For renewable energy sources that do not have a fuel price, the risks are even lower than for fossil energy and bioenergy. The wind and the sun are available for free. Wind and solar therefore need schedules of guaranteed prices more than fossil fuel plants because the former are so capital-intensive; far more of the money for the project is due during construction. Furthermore, fossil fuel plants can ramp up and down to react to market prices, whereas wind and solar react to the weather. You cannot "turn on" a wind turbine or solar array to sell more power when market prices are higher. Some other form of compensation than fluctuating spot market prices therefore seems sensible if you want wind and solar at all.

Other countries have tweaked feed-in tariffs to include things Germany does not have. For instance, France also adjusts feed-in tariffs by location for solar (which Germany does for wind), with a lower rate paid in the sunny south to prevent windfall profits. France and other countries also adjust feed-in tariffs for inflation, which Germany does not. Furthermore, German feed-in tariffs are guaranteed for 20 years, whereas other countries have different time frames for the policy. But in general, feed-in tariffs do three things:

1. Ensure a grid connection for generators of renewable energy (priority dispatch).
2. Specify that renewable electricity cannot be curtailed unless power supply would otherwise be jeopardized; in particular, power firms cannot refuse to take up renewable electricity from third parties simply if their own power production would be offset.
3. Set a minimum price for renewable energy that provides a reasonable return.

If these three requirements are met, feed-in tariffs not only ramp up renewables, but they also do so by opening up the market to new players. In 2007, former US Vice President Al Gore made a convincing argument to a Senate committee that the US should adopt feed-in tariffs, though he used the word only in its abbreviated form:[12]

> We ought to have a law that allows homeowners and small business people to put up photovoltaics generators and small windmills and any other new sources of widely distributed generation that they can come up with and allow them to sell that electricity into the grid without any artificial caps. At a rate that is determined not by a monopsony—of course, as you know, that's the flipside of a monopoly. You can have the tyranny of a single seller, and you can also have the tyranny of a single buyer. If the utility sets the price, it'll never get off the ground. But if it's a tariff, if it's regulated according to what the market for electricity is—the same way public utility commissions do it now—then, you might never need another central station generating plant... Let people sell it.

From the German perspective, it is interesting that Gore only speaks of "small windmills," when in fact German community cooperatives now install utility-scale wind farms on their own, such as one with more than 80 MW and a budget of 127 million euros in 2015.[13] (Perhaps Gore is focusing too much on what individuals can do and not including what communities can do, which is much more.) Otherwise, Gore hits the bull's-eye. The energy transition needs a policy to facilitate market entrance for newcomers. To do so, it will have to break the market dominance of incumbents. It wouldn't be hard—as Gore points out; all we need is something that provides the same market access and reasonable profit that monopoly utilities in the US have enjoyed for the past century. Now what would that be....

---

[12] politicsTV.com. "Al Gore on coal and new power grid." March 21, 2007. Accessed February 6, 2016. https://www.youtube.com/watch?v=0X4REcq6qk0&mode=related&search.

[13] Morris, Craig. "How big can a community wind farm be?" Renewables International. The Magazine. 11 May 2015. Accessed February 5, 2016. http://www.renewablesinternational.net/how-big-can-a-community-wind-farm-be/150/435/87457/.

How Feed-in Tariffs Create New Markets

When Germany began offering a 6 percent return on solar power in its adjusted feed-in tariffs of 2004, the neoliberal school began screaming "subsidy!" In reality, the policy only offered investors in renewable energy what the conventional power sector had been getting all along. From North America to Europe, the twentieth century was a time when utilities would present their costs to regulators, who then ensured an "adequate profit margin" (to quote the German law from the 1930s that remained in effect until liberalization in 1998) in setting rates utilities could charge.[14] In much of the US, this arrangement is still the status quo. In Germany, it led to nearly 70 years of easy profitability at German utility RWE—so much so that the firm did not even hold executive board meetings for years.[15]

Although the target ROI was the same for every technology, the feed-in tariff for photovoltaics was around three times the retail rate starting in 2004. Roughly twice that amount had been offered in various towns in Germany in the previous decade (a sign that costs had already been cut in half), but those policies had quite small budgets. The new national scheme had no budget restrictions at all.

The result was an unprecedented boom.

By 2007, Germany was producing 45 percent of the world's solar electricity.[16] At the time, a lot of solar silicon was recovered from waste in the semiconductor sector. But semiconductors are measured in square millimeters; solar panels, in square meters. The result was an immediate solar silicon shortage. The German boom brought prices up in other countries, including the US. In 2006, *The Economist* wrote about the trend:[17]

---

[14] Gesetz zur Förderung der Energiewirtschaft (Energiewirtschaftsgesetz). Vom 13 Dezember 1935, Fassung von 1978. Deutscher Bundestag. Accessed February 6, 2016. http://www.energieverbraucher.de/files/download/file/0/1/0/448.pdf.

[15] Radkau, Joachim, and Lothar Hahn. *Aufstieg und Fall der deutschen Atomwirtschaft*. München: oekom Ges. für ökologische Kommunikation, 2013.

[16] Morris, Craig. "Global electricity overview for 2014." Renewables International. The Magazine. 19 June 2015. Accessed February 6, 2016. http://www.renewablesinternational.net/global-electricity-overview-for-2014/150/537/88299/.

[17] "Green Dreams: The flood of money into clean energy is better news for society than it is for investors." The Economist. 16 November 2006. Accessed February 6, 2016. http://www.economist.com/node/8173054.

Governments that try to pick winners often choose losers. Subsidies distort investment: since the German government fixed the price for solar power at munificent levels, the country has been sucking in huge numbers of solar panels that could be put to better use in sunnier climes.

History has not been kind to this interpretation. Germany's deployment of wind and solar when the technologies were expensive is now widely celebrated as the reason why significant production capacity has been set up worldwide, leading to plummeting prices for the benefit of developing countries in particular. "The Germans were not really buying power—they were buying price decline," says Hal Harvey, a US clean energy expert. This outcome was intentional; the Germans understood what they were doing all along: "we can afford this—we are a rich country. It's a gift to the world," says German energy analyst Markus Steigenberger.[18]

The passage from *The Economist* holds that feed-in tariffs are subsidies because the prices paid are higher than wholesale power prices. Wholesale prices are based on the merit order—a misnomer, since there is no "merit" to the scheme other than marginal generation prices, which are primarily fuel prices. And since wind and solar have no fuel price, it is unclear how they are supposed to compete in such a pricing scheme. In addition, the European Court of Justice ruled in 2002 and again in 2014 that feed-in tariffs are not state aid, the EU term for subsidies. The reasoning in both cases was that EU member states were required to submit binding goals for renewable energy by 2020, so they must be allowed to have policy support for renewables.[19] Calling feed-in tariffs subsidies also leads to misunderstandings. The whole idea behind feed-in tariffs was to bring prices down over a couple of decades, with a long-term strategy of building up a capital goods (machinery) sector for wind turbines, solar

---

[18] Gillis, Justin. "Sun and Wind Alter Global Landscape, Leaving Utilities Behind." The New York Times. 13 September 2014. Accessed February 6, 2016. http://www.nytimes.com/2014/09/14/science/earth/sun-and-wind-alter-german-landscape-leaving-utilities-behind.html?smid=nytcore-ipad-share&smprod=nytcore-ipad&_r=2.

[19] The reasoning in the PreusssenElektra case was that subsidies were paid by governmental budgets, and that FIT were financed by consumers and not paid into a state fund but directly to suppliers and from there to renewable power generators.

panels, and so on. Reports about Germany reducing its "subsidies"[20] for solar feed-in tariffs therefore always mistook goal attainment to be a withdrawal of support. While the Germans succeeded in slashing the cost of photovoltaics, the media erroneously reported that the Germans had fallen out of love with solar.

This Economist's thinking was very shortsighted anyway. Manufacturers were already responding to the silicon shortage with dedicated solar silicon production facilities. Deployment also brought about a number of cost-cutting innovations. In the 1990s, solar panels were screwed onto support systems with nuts and bolts, for instance. Today, they are quickly clicked into place with automatically locking antitheft systems. Most R&D focuses on things like increasing cell efficiency, not such mundane matters that dramatically speed up the work of installers crawling around on roofs—and bring costs down substantially.

Shortly after *The Economist* published that critique, engineering firms from Germany, Switzerland, and the US started offering turnkey photovoltaics production lines. Up to then, production equipment had been cobbled together from various industries. The screen printers, chemical baths, furnaces, and singling units used had different throughput rates, which made the process unnecessarily inefficient. In addition, each machine required its own expertise. When turnkey photovoltaics production lines appeared, each unit was made to match the other, and a single Chief Technology Officer with a team of normal technicians sufficed. Finally, the world had state-of-the-art automated photovoltaics production lines. Big ones.

Prices then began plummeting at the end of the 2000s. And it would not have happened without the market created with feed-in tariffs. Today, it is universally recognized that Germany's commitment to solar when it was expensive was the key factor in making solar cheap today for developing countries (*The Economist's* "sunnier climes"). As Swedish energy expert Tomas Kåberger says, "German feed-in tariffs for solar have surprisingly turned out to be a very effective policy for development aid."[21]

---

[20] "Germany set for dip in renewable energy subsidy." Yahoo News. 15 October 2014. Accessed February 6, 2016. http://news.yahoo.com/germany-set-dip-renewable-energy-112212843.html?utm_source=dlvr.it&utm_medium=twitter.

[21] Statement made at a conference in 2015, confirmed in personal correspondence.

What about *The Economist's* charge that German feed-in tariffs are "munificent"? If we judge feed-in tariffs by the target return, we get a much different picture. In the EEG of 2000 only around 9 cents was paid for a kilowatt-hour of wind power, falling to around 5 cents if a turbine in a windy location produced a lot of electricity; the goal was to prevent windfall profits. Small biomass units, in contrast, needed nearly 20 cents, while large ones could easily make do with less than 10 cents. But small rooftop photovoltaics arrays needed more than 50 cents.

Note that this return was a target, not a guarantee. The government did not promise investors that the wind would blow or the sun shine—or that the price of raw biomass would not increase. Moreover, investors ran the entrepreneurial risk of improper installations and system failures. The price paid for a kilowatt-hour was promised, not the number of kilowatt-hours that would be generated or the profitability of each plant. Not surprisingly, returns differed significantly from one project to another. So many community wind farms were built in areas with marginal wind resources that the average profit margin was found to be closer to 2 percent in 2012.[22]

Two explanations have been given for this disappointing result. Some suspect that turbine manufacturers and planners may have been overly eager to sell their machines and therefore did not properly advise community and citizen investors. Others believe that at least some of these committee projects went forward anyway because people simply wanted to make their own energy—and a 2 percent return is not worse than what the average citizen can get from a bank these days.

For large utilities, even a 6 percent return is unattractive. Corporations are clearly used to double-digit returns, as the following list illustrates:

1. Depending on the type of investment, German grid operators have a legally guaranteed return of 7 to 9 percent for investments in grid upgrades. But the firms were not happy with a single-digit return, so they sued for an 11 percent return (and lost the court case).[23]

---

[22] Morris, Craig. "Germany overpays solar relative to wind." Renewables International. The Magazine. 29 July 2014. Accessed February 6, 2016. http://www.renewablesinternational.net/germany-overpays-solar-relative-to-wind/150/537/80598/.

[23] Morris, Craig. "Citizen ownership of grids." Energy Transition. The German Energiewende. 19 July 2013. Accessed February 6, 2016. http://energytransition.de/2013/07/citizen-ownership-of-grids/.

2. In the US, the guaranteed return for grid investments is around 12 to 13 percent.[24]

3. Before the "utility death spiral," Bloomberg estimated that the profit margin for Germany's eight biggest utilities was 15 percent. (Tellingly, the "death spiral" had brought them down to "5.4 percent," roughly the target return for feed-in tariffs.)[25]

4. Saudi investors became interested in German photovoltaics auctions when a government official told them a 15 percent return might be possible.[26]

5. The US investment group Blackstone became interested in the offshore wind sector in Germany when it found that feed-in tariffs might provide returns up to 20 percent.[27]

6. An OECD survey found that corporate clean energy projects have a return in the range of 13 to 21 percent.[28]

Corporate Offshore Wind, Community Onshore Wind

A comparison of onshore and offshore wind in Germany is especially illustrative in this respect. Communities are largely behind German onshore wind farms, whereas no offshore wind farm has gone up yet as a community project in the country. Proponents of the Energiewende as a

---

[24] Bernhardt, John. "The Power Of Local Energy." Forbes Online. 28 April 2014. Accessed February 6, 2016. http://www.forbes.com/sites/realspin/2014/04/28/the-power-of-local-energy/#2c9d15fd3137.

[25] Mengewein, Julia, and Rachel Morison. "Germany's New Coal Plants Push Power Glut to 4-Year High." Bloomberg. 27 June 2014. Accessed February 6, 2016. http://www.bloomberg.com/news/articles/2014-06-26/germany-s-new-coal-plants-push-power-.

[26] Morris, Craig. "15% return on solar investments in Germany for Saudis?" Renewables International. The Magazine. 20 February 2015. Accessed February 6, 2016. http://www.renewablesinternational.net/15-return-on-solar-investments-in-germany-for-saudis/150/452/85658/.

[27] Morris, Craig. "Germany opens first commercial offshore wind farm." Renewables International. The Magazine. 16 November 2014. Accessed February 6, 2016. http://www.renewablesinternational.net/germany-opens-first-commercial-offshore-wind-farm/150/435/83313/.

[28] Fulton, Mark, and Reid Capalino. "Investing in the Clean Trillion: Closing the Clean Energy Investment Gap." Ceres Report. Accessed February 6, 2016. http://et-advisors.com/wp-content/uploads/Ceres_CleanTrillion_Report.pdf.

bottom-up movement have therefore sometimes spoken of the offshore wind sector as a gift to the large energy companies that resisted the community movement all along—an incentive for them to finally start taking part.[29] Offshore wind farms are all big, whereas German communities sometimes build a single turbine.

Offshore wind also receives more favorable treatment in German law than onshore wind does. Onshore wind farms receive a feed-in tariff of 5 to 9 cents (depending on wind conditions) for 20 years. Offshore wind farms have the same target return for 20 years, but they can opt to get it all in the first eight or 12 years (the firms can pick). They need not wait the full 20. After the first 8 or 12 years, the feed-in tariff then drops to a level that merely covers costs without providing a profit.

Onshore wind farms also suffer more when grid connections are delayed. The law specifies that a connection must be provided and stipulates who has to cover what cost.[30,31] It also requires the connection within a reasonable time frame—and that is exactly what community wind farm investors complain about. In some cases, community projects have charged that big utilities tried to drag out the connection process. The wind farms forgo income until the grid connection has been provided. In the offshore sector, however, a date for the grid connection is agreed, and if it is not provided on time, the grid operator has to pay the wind farm operator 90 percent of the revenue that would have been due for power generation (which is then estimated based on wind conditions).[32] The grid operator then passes on these expenses to ratepayers without having to soak up any of the losses. Neither the wind farm investors nor the grid operator has much of a financial incentive to work things out quickly

[29] "Windenergie Report Deutschland 2012." Fraunhofer (IWES). Editor Kurt Rohrig. Stuttgart: Fraunhofer Verlag, 2013, p. 87: "Neben all den Erwartungen bezüglich Offshore-Windenergieanlagen wird schnell vergessen, dass onshore die weitaus größere Windleistung installiert ist. In Anbetracht der großen Preisunterschiede zwischen Offshore- und Onshore-Windenergie ist dieser Trend eher eine politische als eine technische/ökonomische".

[30] See the "Infrastrukturplanungsbeschleunigungsgesetz." Up to then, the legal situation had been somewhat unclear. The government's position was essentially that it was "natural" for the grid to be expanded when central station power plants are built, but not for small distributed units.

[31] Based on conversation with Josef Pesch.

[32] Morris, Craig. "Is offshore wind the big story?" Energy Transition. The German Energiewende. 14 October 2014. Accessed February 6, 2016. http://energytransition.de/2014/10/is-offshore-wind-the-big-story/.

under this scheme; captive ratepayers end up holding the bill. This distinction in the treatment of onshore (generally community-owned) and offshore is glaring.

Finally, by all appearances, the profit margins for offshore wind are drastically higher than the target of 6 percent in practice, as Blackstone's estimate above shows. Though some projects in Germany have been held up because grid connections could not be provided on schedule, the first ones that have gone into operation clearly outperformed expectations. Alpha Ventus was the first offshore wind farm; a pilot project, it was financed partly with governmental research funding and went over budget. It also produced 15 percent more electricity than expected in 2011, the first full year of operation.[33] Baltic 1, Germany's fully commercial wind farm, then exceeded its expected performance by more than 20 percent in its first year.[34]

These achievements are fantastic and have been roundly celebrated. But far more power production also means dramatically higher returns. When solar was overpaid for a few years (see below), there was a great uproar about the impact on ratepayers, driven largely by the PR departments of industry—but there has been no dramatic wave of protests about the cost impact of offshore wind, even though a study in 2015 found that the increase in the cost of the energy transition over the next decade would almost exclusively be the result of offshore wind alone.[35] It seems that community projects, as numerous as they are in the country, lack PR departments.

The situation is similar in the UK, where communities are largely left out of the onshore sector as well. Although the country has far better wind conditions than Germany does, the British pay much higher rates for wind

---

[33] Morris, Craig. "Offshore wind farm posts 50 percent capacity factor." Renewables International. The Magazine. 29 April 2013. Accessed February 6, 2016. http://www.renewablesinternational. net/offshore-wind-farm-posts-50-percent-capacity-factor/150/505/62324/.

[34] Morris, *Offshore wind farm posts 50 percent capacity factor.*

[35] Agora Energiewende. "Energiewende: Kostenscheitel in Sicht: Der Erneuerbaren-Energien-Ausbau treibt die EEG-Umlage immer weniger in die Höhe. Ab 2023 wird sie wahrscheinlich sogar sinken. Das zeigt der „EEG-Rechner" von Agora Energiewende." Agora Energiewende. 7 May 2015. Accessed February 6, 2016. http://www.agora-energiewende.de/de/presse/agoranews/news-detail/news/energiewende-kostenscheitel-in-sicht/.

power, as UK renewable energy consultant Mike Parr has pointed out.[36] British energy expert Dave Toke sees corporate ownership of the wind power industry in the UK as the reason: "in the UK the Government has, by design or at least in effect, allowed the major electricity companies to cream off what economists call an 'economic rent' from renewable energy in order to compensate for the lost production from their power stations."[37]

Overpaying for Solar (Temporarily)

So how much did Germany overpay solar? Because there were different size categories, it's hard to say in general, but a study published by top German economists in 2014 found that the returns ranged (depending on project size) from 10 percent to nearly 40 percent in 2004 in sunny Bavaria, when full-fledged support for photovoltaics was first offered.[38] By 2009, most projects in Bavaria had excessive returns close to 40 percent; in the north, they will have been considerably smaller—but also significantly above the 6 percent target. But after 2009, returns tapered off, reaching the target return of 6 percent in 2012. The same study found that photovoltaics investments led to a net loss of up to 10 percent as late as 2000; likewise, it found that wind farms nation-wide had never averaged more than a 3 percent return—only half of the target. Citizens nonetheless forged ahead with their community wind farms. And when citizen investors in photovoltaics got hefty returns otherwise only enjoyed by corporations, there was great—and orchestrated—dismay.

With photovoltaics, German policymakers obviously had trouble chasing the price of a technology that was about to move faster than anyone expected. From 2010 to 2015 alone, the cost of an installed solar array fell worldwide by more than 50 percent, and the price is expected to drop

---

[36] Parr, Mike. "The myth of expensive offshore wind: it's already cheaper than gas-fired and nuclear." Energypost. 31 March 2015. Accessed February 6, 2016. http://www.energypost.eu/myth-expensive-offshore-wind-already-cheaper-gas-fired-nuclear/.

[37] Toke, David. "So why is renewable energy so much more expensive in the UK compared to other countries?" Dave Toke's green energy blog. 31 December 2014. Accessed February 6, 2016. http://realfeed-intariffs.blogspot.de/2014/12/so-why-is-renewable-energy-so-much-more.html.

[38] Breitschopf, Barbara, Simon Bürer, and Lucas Lürich. "Verteilungswirkungen der Marktförderung des EEG in den Bereichen Photovoltaik und Windenergie (onshore)." Accessed February 6, 2016. http://www.impres-projekt.de/impres-wAssets/docs/Verteilungswirkungen_Strom.pdf.

an additional 20 percent by the end of this decade.[39] In 2004, German policymakers hoped to one day bring down the cost of a solar panel from around three dollars per watt to below one dollar per watt. That threshold was crossed in 2011. As Volkmar Lauber has pointed out in his research, policymakers in other countries (especially the UK) have overpaid for conventional technologies whose cost structure was not moving at all. At least the Germans missed a rapidly moving target.

Most importantly—and here we see how feed-in tariffs always create a market—the policy puts downward pressure on equipment prices constantly, even when the rates paid for the electricity generated exceed the target. Initially, rates for newly installed systems were to drop by 5 percent annually, but by 2009 it had become clear that prices for photovoltaics systems were falling faster than the law expected, so legislators struggled to reduce feed-in tariffs to keep up with plummeting system prices.

Because tariffs are paid for electricity generated, not for the arrays themselves, investors (such as homeowners) still go shopping on a market of installers to compare prices. Price pressure is therefore always put on suppliers.

Feed-in tariffs thus produce competition between companies. Other policies produce competition between energy sources, with the cheapest one winning. Large energy firms speak of competition all the louder when natural gas competes with coal—so no one will notice that the companies themselves hardly compete. But when most people think of competition, they think of firms competing, not energy sources sold by an oligopoly.

Net Metering, Tax Incentives, and Other Policy Options

Aside from several pilot projects, Al Gore's call for feed-in tariffs in the US has gone largely unheeded. The policies used instead fall into three categories: net-metering (mainly for residential solar), tax

---

[39] IRENA. "Renewable Power Generation Costs in 2014." The International Renewable Energy Agency (IRENA). Bonn, January 2015. Accessed 10.20.2016. http://www.irena.org/DocumentDownloads/Publications/IRENA_RE_Power_Costs_2014_report.pdf; and Morris, Craig. "Solar keeps getting cheaper." Renewables International. The Magazine. 13 September 2014. Accessed February 10, 2016. http://www.renewablesinternational.net/solar-keeps-getting-cheaper/150/452/81702/.

credits, and auctions for Renewable Portfolio Standards. This list is not exhaustive; some interesting work has been done on green power marketing in the US,[40] and some states (such as California) have provided utilities with financial incentives to promote efficiency and conservation, not just power consumption. Furthermore, the US is clearly ahead of Germany when it comes to real-time power pricing and the smart grids. In these and numerous other respects, Germany can learn a lot of lessons from the US. But a quick review of the main policy mechanisms available in most parts of the country reveals some shortcomings that need addressing.

Net-metering is often proposed as the natural successor to feed-in tariffs for solar after grid parity.[41] Households and small businesses will simply run the meter backward. The charm of this policy is that, in a way, it isn't one at all; its proponents specifically say that the switch to net-metering will mean that no subsidies are required. But the argument is specious; those who make it fail to realize that they are calling for an end to policy support for renewables even as subsidies for coal and oil continue after two centuries of profitability in those sectors.

As we see from the current backlash against residential solar in the US, policy support will probably continue to be needed for photovoltaics (and other types of renewable energy) when it comes to guaranteed grid access, regulating who is responsible for what grid service, specifying fair fees for services, and so on. After all, those who net-meter will still use the grid—constantly, in fact. When households are empty during the day at a time of peak solar power production, solar electricity is sold to the grid. When people come home at night, they consume power from the grid as the sun goes down as the array (and whatever storage is installed) cannot cover all their needs. If grid fees are often not a flat monthly rate but instead a surcharge tacked onto the amount of power consumed, these households pay less of the grid they still use. Utilities therefore argue that

---

[40] Bird, Lori, Claire Kreycik, and Barry Friedman. "Green Power Marketing in the United States: A Status Report (2008 Data)." Technical Report NREL/TP-6A2-46581. Accessed February 6, 2016. http://www.nrel.gov/docs/fy09osti/46581.pdf.

[41] Mints, Paula. "The 12-step Solar Program: Toward an Incentive-less Future." Renewable Energy World. 24 January 2011. Accessed February 6, 2016. http://www.renewableenergyworld.com/articles/2011/01/the-12-step-solar-program-toward-an-incentive-less-future.html.

solar homeowners are mooching off the grid. This issue will never go away for grid-connected photovoltaics, so policies will always be needed.

In addition, profit margins will increase to unreasonable levels under net-metering as solar becomes increasingly cheaper than power from your wall socket, whereas feed-in tariffs can be designed to keep profit margins at a reasonable level. If the price of solar power from your array is 12 cents (the feed-in tariff for solar power from small rooftop arrays in Germany in 2015) and your retail rate is 15 cents, your profit margin is 25 percent. If the retail rate rises to 18 cents, the margin is 50 percent. As time passes, solar will get cheaper and retail rates higher, so net-metering is a recipe for increasingly unreasonable profits. The threat of widespread grid defection in Australia this decade is one example of the positive feed-back resulting from people going off-grid with photovoltaics and storage; as more households and small business do so, the grid becomes more expensive for everyone else, thereby increasing the incentive to defect from the grid. Properly designed feed-in tariffs would ensure, in contrast, that investors only get a modest return regardless of the retail rate.

Another popular policy in the US is tax credits. The Production Tax Credit (PTC) for wind power incentivizes power production, which is good; we want to encourage people to generate energy, not simply build systems that may or may not perform as planned. Other schemes in the US, such as the Investment Tax Credit (ITC), fall into this trap by providing payment relative to the system price upfront. This type of tax credit exists for solar at the federal level (30 percent) and in many states. For instance, in 2015 North Carolina offered an additional 35 percent tax credit on top of the federal ITC, whereas Louisiana provided a 50 percent credit, bringing the total tax credit up to 80 percent. In Louisiana, you therefore only pay for 20 percent of your solar array, and you get to net-meter in addition. Install your array on a roof shaded by trees, and your power production will plummet—but you still only pay for a fifth of the purchase price for the array, so who cares? Upfront bonuses that incentiv-ize system purchases more than power generation are therefore ineffective and should be avoided.

Tax credits are problematic in terms of social inclusion as well. While feed-in tariffs provide the same target return for everyone, those with high peak tax rates benefit more from tax credits. State ITCs sometimes refund you if your tax burden becomes negative, but the PTC for wind power does not. If your business's tax burden is low, the incentive to invest in renewables is therefore smaller. Entities with no taxable income—non-profits, pension funds, and so on—cannot avail themselves of the PTC at all without cooperating with a company that can write off some taxes, which greatly increases transaction costs.[42] In Germany, transaction costs are quite low with feed-in tariffs.[43]

An investor's tax burden also varies from year to year, making it impossible to calculate payback for a long-term investment in renewable energy via tax credits. A single project therefore receives different levels of policy support depending on who the investors are, not what level of support the renewable energy source needs—with those making more money getting more help. "This is dumb policy," says John Farrell of the Institute for Local Self-Reliance.[44]

Finally, we come to auctions, which the European Commission currently promotes. Utilities worldwide are increasingly using reverse auctions to provide a certain percentage of their power supply from renewables by a particular year.[45] The charm of auctions is that government officials do not have to figure out what the price should be, as they do with feed-in tariffs. The market itself knows more, so it should be better at setting the price—so the logic goes. Auctions have proven successful in producing low bids in numerous countries, but a pattern of drawbacks has emerged. Brazil required bidders to place a deposit of 10 percent of the project's cost when submitting their bids. The purpose was to solve a problem from previous bids, in which fly-by-night firms had submitted very low bids which they then could not deliver on. But

---

[42] Fulton and Capalino. "Investing in the Clean Trillion"

[43] "Open Letter on Market Premiums (and Response)." Accessed February 6, 2016. http://projects.exeter.ac.uk/igov/open-letter-on-market-premiums/.

[44] Farrell, John. "Why tax credits make lousy renewable energy policy." Institute for Local Self-Reliance. 17 November 2010. Accessed February 6, 2016. https://ilsr.org/why-tax-credits-make-lousy-renewable-energy-policy/.

[45] See Kieffer, Ghislaine, and Toby D. Couture. "Renewable Energy Target Setting." IRENA. June 2015. Accessed February 6, 2016. http://www.irena.org/DocumentDownloads/Publications/IRENA_RE_Target_Setting_2015.pdf and Kreycik, Claire, Toby D. Couture, and Karlynn S. Cory. "Procurement Options for New Renewable Electricity Supply." NREL Technical Report. December 2011. Accessed February 6, 2016. http://www.nrel.gov/docs/fy12osti/52983.pdf.

the outcome of this Brazilian requirement was that only firms with tremendous liquidity could bid at all. The auction produced an oligopoly.[46]

"Predatory bidding" is another concern worldwide; big firms might
place extremely low bids in the beginning in order to scare off competitors from competing at all. Later, they can raise prices and divide the
market up among themselves. Such concerns have been raised in India,[47]
where the government is considering moving in the direction opposite
to the one proposed by the European Commission—from auctions to
feed-in tariffs.[48] Predatory bidding may also be behind the failure of auction winners to complete projects in the Netherlands, where 98 percent
of photovoltaics projects awarded in 2011 had not been completed by
2014—along with 92 percent of biogas contracts.[49]

Worldwide, auctions tend to produce a large number of losers and
a small number of winners.[50] Eventually, small firms just barely able to
take part in the first rounds of bidding (each of which can cost tens, if
not hundreds, of thousands of dollars) can't afford to continue bidding,
so they stop. At the request of the European Commission, the German
government began conducting pilot auctions for ground-mounted
photovoltaics in 2015. Even though the Germans know about this pitfall,
the pilot auction repeated this outcome. Only 25 contracts were awarded

---

[46] Morris, Craig. "Actual outcomes of auctions in France, Brazil, and the Netherlands." Energy Transition. The German Energiewende. 25 June 2014. Accessed February 6, 2016. http://energy-transition.de/2014/06/outcome-of-renewables-auctions/.

[47] See Engelmeier, Tobias. "Why auctions for power projects don't work in India." photovoltaics Magazine. 19 May 2015. Accessed February 6, 2016. http://www.photovoltaics-magazine.com/opinion-analysis/blogdetails/beitrag/why-auctions-for-power-projects-dont-work-in-india_100019506/#axzz3aUMmW2wG and Kenning, Tom. "photovoltaics Talk: Bidding for an Indian solar legacy at the expense of quality." PVTech. 26 January 2016. Accessed February 6, 2016. http://www.photovoltaics-tech.org/interviews/pv-talk-bidding-for-an-indian-solar-legacy-at-the-expense-of-quality.

[48] Morris, Craig. "India may move from auctions to feed-in tariffs for photovoltaics." Renewables International. The Magazine. 14 October 2014. Accessed February 6, 2016. http://www.renew-ablesinternational.net/india-may-move-from-auctions-to-feed-in-tariffs-for-pv/150/537/82362/.

[49] Morris, Craig. "Actual outcomes of auctions in France, Brazil, and the Netherlands." Energy Transition. The German Energiewende. 25 June 2014. Accessed February 6, 2016. http://energy-transition.de/2014/06/outcome-of-renewables-auctions/.

[50] For the US, see this report on contract failures in auctions: "Building a "Margin of Safety" Into Renewable Energy Procurements: A Review of Experience with Contract Failure." Consultant Report. January 2006. Accessed February 6, 2016. http://www.energy.ca.gov/2006publications/CEC-300-2006-004/CEC-300-2006-004.PDF.

to the 170 bidders, so there were 145 losers.[51] Not a single community project or individual was awarded a contract. Furthermore, the prices obtained were slightly higher than the feed-in tariffs applicable at the time. Auctions in France also led to higher rates than the feed-in tariffs for comparable system sizes.[52]

In other countries, auctions have led to delays in project development. Once the price was agreed, companies could sit back and wait while the price of renewable energy equipment fell on global markets (thanks largely to countries with feed-in tariffs deploying a lot of systems). In Germany, the firms who won the first round of bidding for photovoltaics have 24 months to complete their projects. Deadlines can help, but they should be stricter; photovoltaics projects can easily go up within 12 months (as the Spanish photovoltaics market of 2008 revealed). Generous deadlines give planners an incentive to build as late as possible in the hopes that equipment prices will continue to drop.

Likewise, an auction in Brazil produced amazingly low prices for wind power in 2013. It turned out, however, that no bids at all had been placed for photovoltaics, so the country had to create a set-aside in future auctions specifically for photovoltaics, which still could not compete with onshore wind power in the country. When the price is the main criterion, auctions do not ramp up wind, biomass, and solar energy at the same time without such specifications. Of course, the price alone need not be the only criterion in an auction; environmental impacts, a diversity of winners, and so on can also be written in as requirements.

Obviously, all policies need to be properly designed. But Germany's energy transition not only has a goal of affordability, but also greater competition between large corporations and SMEs (often called the "diversity of actors" in the German literature).[53] Countries only focusing

---

[51] Morris, Craig. "German photovoltaics auctions: told you so." Renewables International. The Magazine.30April2015.AccessedFebruary6,2016.http://www.renewablesinternational.net/german-photovoltaics-auctions-told-you-so/150/452/87253/.

[52] Couture, Toby D., David Jacobs, Wilson Rickerson, and Victoria Healey. "The Next Generation of Renewable Electricity Policy: How rapid change is breaking down conventional policy categories." NREL/TP-7A40-63149. Accessed February 6, 2016. http://www.nrel.gov/docs/fy15o-sti/63149.pdf.

[53] Hauser, Eva, Andreas Weber, Alexander Zipp, and Uwe Leprich. "Bewertung von Ausschreibungsverfahren als Finanzierungsmodell für Anlagen erneuerbarer Energienutzung."

on low prices may indeed fare well with reverse auctions. But clearly, auctions are not a good way of leveling the playing field between players big and small—or of ramping up a mix of renewable energy sources simultaneously. Furthermore, energy transitions powered by feed-in tariffs are faster by design; after all, governments set limits on volume in auctions.

Proponents of auctions often claim that they produce more competition and are more "market-based." In practice, we see that feed-in tariffs facilitate newcomer entry, whereas auctions promote oligopolies. But even in theory, both policies constitute state intervention in markets—feed-in tariffs set the price, while auctions set the market volume.

There can be a mixture of feed-in tariffs and other policies, of course. FITs can be used as the mechanism to meet targets, for instance, so they could be implemented as ways of fulfilling Renewable Portfolio Standards, a point that experts in the US have also made.[54] Likewise, feed-in tariffs have been tweaked worldwide, so policymakers have a wide range of design options.

For a transition to renewables, a proper mix of energy sources will be needed, so ramping up only the cheapest one is not helpful. Furthermore, supporting only the energy source that needs the least support is no way to develop the others that will be needed. Often, there are calls for these more expensive options to be funded with R&D, but photovoltaics and wind had left the early stage where the research push alone would have been beneficial. A market pull—deployment—was needed, and that's what feed-in tariffs provide.

The focus on low prices is especially pernicious because citizens are not just consumers. By facilitating market entrance, feed-in tariffs allow people to make their own energy and open their own companies, thereby also providing an incentive to oligopolists to move in this direction themselves—or else to lose even more of this market. Not surprisingly, the Social Economy Alliance in the UK is drawing attention to the "right to invest," as opposed to merely the "right to buy"—the latter being one of

---

Accessed February 6, 2016. http://www.izes.de/cms/upload/publikationen/IZES_2014-05-20_BEE_EE-Ausschreibungen_Endbericht.pdf.

[54] Kreycik, Couture, and Cory. *Procurement Options for New Renewable Electricity Supply* (See Footnote 45).

Margaret Thatcher's slogans as Prime Minister.[55] Tellingly, they point out that Thatcher also wanted to bring about a "crusade of popular capitalism" to "enfranchise the many in the economic life of the nation." The British campaign specifically calls for community ownership of renewable energy—because customer choice is not enough:[56]

The ability to pick between different energy suppliers is a false freedom. Those who use the energy are excluded from influencing decisions on how any surplus should be invested—into fossil fuels or renewables, imported fuel or local sources—or on how to structure prices.

Adam Smith's Relevance Today

In the end, low prices are not the only goal worth pursuing. Society must debate, for instance, if mom-and-pop shops should continue to exist or whether Big Box stores are better. This example shows what else is at stake. Small entrepreneurs cannot compete on price with large retail chains. Consumers may then have to pay more for products. In return, more people might be able to found their own business instead of working at minimum wage for a giant retailer. Big Box prices tend to bring along Big Box salaries, and people are not just consumers, but also workers and citizens. As US climate change activist Bill McKibben explains concerning his quest into food locally produced, he benefited as a person: "In my role as an eater, I was part of something larger than myself that made sense to me—a community. I felt ground, connected."[57] There is more to life than low prices.

Americans have come to associate free markets with a lack of price controls—and free trade with international trading. But 100 years ago, Supreme Court justice Louis Brandeis called for "resale price mainte-

---

[55] "A Right to Buy for the 21st Century." Social Economy Alliance. Accessed February 6, 2016. http://socialenterprise.org.uk/social-economy-alliance/news/right-to-invest-report-a-21st-century-right-to-buy.
[56] Platform. "Energy beyond neoliberalism: A new energy settlement will be an important part of the transition from neoliberalism." *Soundings: A journal of politics and culture* 59, no. 1 (2015): 96–114.
[57] McKibben, Bill. *Deep economy: The wealth of communities and the durable future.* New York: St. Martins Press, 2007, l. 1520.

nance" to protect small producers.[58] Such views go all the way back to Thomas Jefferson's ideas about citizens producing their own products. The United States has clearly gone the way of large industry, and Americans talk more about the low prices than the resulting low wages. Citizens are relegated to the status of mere consumers in the process.

Germany is clearly more comfortable with minimum retail prices. Feed-in tariffs for renewable energy are one example, but books in Germany also have a minimum price in order to protect small publishers and bookshops. Other German price controls include rent, in order to ensure affordable housing for low-income families.

In the US, price-fixing is rarer. But as in Germany, numerous US states have minimum prices for cigarettes to discourage consumption. One big example in almost all countries in the West is healthcare. In both Germany and the US, insurance firms and/or the government sat down with healthcare providers to agree on prices for products and services that ensure profitability for providers even as they prevent price gouging for patients.

In power markets, we often find regulated prices. France does not pass on the full cost of electricity to retail consumers; the remainder is passed on as an item in the governmental budget. The French government also approves and rejects power price hikes for industry. Likewise, the Spanish government limits retail rates, passing on utility losses as a budget item that may end up being paid for with tax revenue.[59,60] In most parts of the US, public service commissions review requests for retail rate changes submitted by utilities. In the power sector, at least, Germany is

---

[58] Foer, Franklin. "Amazon Must be Stopped: It's too big. It's cannibalizing the economy. It's time for a radical plan." New Republic. 20 October 2014. Accessed February 6, 2016. https://newrepublic.com/article/119769/amazons-monopoly-must-be-broken-radical-plan-tech-giant.

[59] The initial deficit from before the photovoltaics boom in 2008 was originally to be paid for with future rate hikes, not with tax revenue, but the overdraft became chronic, and the government was forced to step in and assume some of the debt.

[60] Information based on conversations with Toby Couture and Hugo Lukas. See Couture, Toby D. *FITs and Stops: Spain's New Renewable Energy Plot Twist & What It All Means*. Analytical Brief. 2012 4, no. 1. Accessed February 6, 2016. http://www.e3analytics.eu/wp-content/uploads/2012/05/Analytical_Brief_Vol4_Issue1.pdf.

and Couture, Toby D. *Booms, Bust, and Retroactive Cuts: Spain's RE Odyssey*. Analytical Brief. 2011 3, no. 1. Accessed February 6, 2016. http://www.e3analytics.eu/wp-content/uploads/2012/05/Analytical_Brief_Vol3_Issue1.pdf.

more laissez-faire; all retail power providers are free to charge what they want. In return, all German retail power consumers are free to switch power providers at the end of each month. In contrast, around 85 percent of American power consumers are still captive; US retail power rates still need to be regulated because many Americans have no consumer choice.

The biggest global example of price-setting is found, ironically, in the banking sector. Both the US Federal Reserve and the German Bundesbank (and now the European Central Bank) have one main job: setting the price of money (the prime lending rate).

Different societies have therefore come up with different areas where they feel some form of price control is necessary. What the power sector, Medicaid, books in Germany, cigarettes, and rent all have in common is a social impact. Society has decided that those prices should not completely be left up to the market. In the case of the prime lending rate, society has determined that the price of money should not be left up to governments, which—like the Weimar Republic in German history— can exploit inflation to reduce debt at the cost of great public suffering. In all these cases, we are dealing with a word that Adam Smith used much more frequently than "invisible hand": justice.

Writing in the eighteenth century, Smith was an economist at a time when the field was still more moral philosophy than mathematics. He was primarily concerned with fairness and believed that merchants should be allowed to do business without being inhibited by national trade duties. It wasn't until the twentieth century that mathematics overtook the field of economics in an attempt to make the discipline more scientific. In today's terms, Smith was less a mathematician than a philosopher. He was interested in economics for its impact on society.

His world knew no multinational firms, with the possible exception of the East India Company. He mentions the firm several times in *Wealth of Nations*, generally with criticism. He was concerned about such monopolies becoming too big to fail:

"Since [the East India Company] became sovereign, with the revenue which, it is said, was originally more than 3 millions sterling, they had been obliged to beg the ordinary assistance of government, in order to avoid immediate bankruptcy."

This is not the voice of a man calling for unfettered deregulation. It is someone who believes that government in collusion with big business was detrimental to public welfare and in particular to middle-class merchants.

Today, multinationals are everywhere, while mercantilism is no longer how we describe our economy. One wonders why the popular misreading of the "invisible hand"—a passing metaphor used by a moral philosopher to describe a mercantile economy that no longer exists—has become so predominant in modern mathematics-based economics and a global economy increasingly dominated by speculative oligopolists.

An argument could therefore be made that Adam Smith has little to say to us today. Anselm Görres, the man who heads the NGO behind the Adam Smith Prize, Green Budget Germany, disagrees. "We intentionally chose Adam Smith to name the prize after," he says. "We wanted to take the philosophy of environmental taxation back to its originator—and to point out that you cannot trace trickle-down economics back to him as some extremists do in the US."[61]

A former McKinsey consultant, Görres says he learned to appreciate entrepreneurship and American pragmatism as an exchange student in Des Moines, Iowa, decades ago. Get him talking about America, and he begins to swoon. Still, he cannot understand why such an obvious misreading of Adam Smith has gone so mainstream in the Anglo world. "When you read Adam Smith, there's a certain spirit to his writing," Görres explains. "It's not the spirit of greed. It's the spirit of fairness." And there is a German angle to the story, he says: "Smith corresponded with German moral philosopher Immanuel Kant, whose categorical imperative he found inspiring: your actions should follow rules that could be universal. Smith wanted to show that self-interest in business transactions could produce such objective fairness."

Görres understands Smith in his historical context. Smith never espoused selfishness, but self-interest—which is different. Doing what's good for you includes being likeable in the eyes of your community. "Smith makes it abundantly clear that money and fame don't lead to happiness. What leads to happiness is being loved and being lovely," econo-

---

[61] All quotes from Görres take from a personal conversation with the authors.

mist Russ Roberts explains the main tenant in Smith's *Theory of Moral Sentiments*.[62] Unfortunately, today "economists think of human beings primarily as individual and not as members of a community," laments McKibben.[63]

"Smith lived in an age when the church preached that sexuality and self-interest were sins," Görres says. "If you wanted to earn money and became rich, you were a sinner. Smith came in and told everyone that a free Christian has the freedom to be successful in business." Still, Smith did not see self-interest as necessarily a good thing. "Self-interest is neutral for Smith. It is not good or bad; it is simply there. It is a tool. As such, it can be used for good purposes. And that was Smith's message to Christianity—as with sex, you can do good things and bad things with self-interest," Görres says with a smile.

Here, we see again where neoliberalism misconstrues Smith. "The Chicago School sees its self-interest only in a good light, not dialectically," Görres argues. Over the past few decades, the Chicago School of economics has stretched Smith to his limits: "Is there some society you know that doesn't run on greed?" neoliberal economist Milton Friedman once asked on US television.[64] In fact, the people who study societies (such as anthropologists) know of no society that runs on greed—not so-called "primitive" ones, and not ours either. "Everything that is important in my life is incompatible with selfishness," Germany-based American economist Dennis Snower recently despaired over the state of mainstream economics.[65] "A revolutionary suggestion to the profession: go and observe some actual markets," a former British official in the Council of Europe recently wrote upon visiting a flea market in France,

[62] Roberts, Russell D. *How Adam Smith can change your life: An unexpected guide to human nature and happiness*. New York, NY: Portfolio / Penguin, 2014.

[63] McKibben, Bill. *Deep economy: The wealth of communities and the durable future*. New York: St. Martins Press, 2007, l. 1728

[64] *Milton Friedman on Greed*. Excerpt from an interview with Phil Donahue, 1979. YouTube. Accessed February 6, 2016. https://www.youtube.com/watch?v=RWsx1X8PV_A.

[65] Kaiser, Tobias. "Top-Ökonom Snower zweifelt an seiner Wissenschaft." Die Welt Online. 29 September 2012. Accessed February 6, 2016. http://www.welt.de/wirtschaft/article109539943/Top-Oekonom-Snower-zweifelt-an-seiner-Wissenschaft.html.

where buyers and sellers seemed more interested in getting to know each other—in creating community—than in satisfying personal greed.[66]

As the classical conservatives remind us, we are here for each other. We get our sense of worth from interactions with other people. We create communities, where we volunteer in various activities and take part in the decisions that influence us all.

Germany is transitioning to renewables with a mixture of liberalism and ordo-liberalism. The country tries to let the market solve everything—until it starts to produce undesirable outcomes. Then, the state intervenes, and the game continues under fairer rules. The focus is on the public good—what outcome society wants, not what the unfettered market would produce.

German Social Market Economics as Fettered Capitalism

Perhaps the most important thing for students of Germany to understand is how much the Germans themselves think they are pro-market, even with mechanisms like environmental taxation and feed-in tariffs, which may seem anti-market to mainstream Anglo thinking.

Naomi Klein's recent book *This Changes Everything* is a good example of how left-leaning North Americans (Klein is Canadian) believe the German story fits their anticapitalist narrative. Impressively, Klein tells the German story very accurately although it is not her main topic. Nonetheless, the Germans feel she has thrown them into an anticapitalist camp they do not belong in. Görres thinks she falls into the same trap as the "moralistic left" in Germany. "Like the church centuries ago, they still cannot accept self-interest as a neutral tool that can be used for good purposes. We need enlightened entrepreneurs to come up with efficient machines. And that's what we're doing in Germany."

"Isn't the Energiewende an example of how energy supply can be changed within the capitalist economic system?" Klein was asked in

---

[66] Wimberley, James. "Flea markets." Samefacts. Accessed February 6, 2016. http://www.samefacts.com/2015/04/economics/flea-markets/.

an interview with German media.[67] Another German reviewer criticized her for claiming that capitalism itself is the problem—because the Energiewende is taking place within capitalism—a form of capitalism with rules.[68] "Feed-in tariffs are a good example of how you can steer market forces in the direction of environmental progress," Görres says. Or, as Volkmar Lauber wrote in an article coauthored with British policy expert Dave Toke, "Although the [feed-in tariff] system is not neoliberal, it developed in an institutional setting that was shaped by ordoliberal preferences and its concern for the common good. Thus it emphasises competition, opportunities for smaller market players against monopolistic practices, and the internalisation of external costs."[69]

North American commenters on the left and the right hold Germany to be socialist.[70] Germans like Görres stress the difference between socialism and social democracy: "Planned economies have produced dreadful outcomes all over the world. Indeed, we are living more in the age of Adam Smith today that of Karl Marx." Rather than do away with capitalism, as the "moralistic left" would like, Görres wants to tap its ability to bring about rapid change—but control the direction of the change. "If you want fast change—which we need to solve our current problems—nothing will bring it about more rapidly than entrepreneurial competition. We need this dynamic, but we also need to make sure it goes in the right direction. That's why I fight for environmental taxation. I don't just want to loudly criticize present-day capitalism, but propose a better alternative."

Just as ordo-liberalism rejects unregulated markets, it also—in line with the refusal of left-wing utopias in classical conservativism—rejects

---

[67] "WDR interview with Naomi Klein: Klimawandel ändert alles." 21 March 2015. Accessed February 6, 2016. www1.wdr.de/themen/kultur/naomi-klein100.html.

[68] Ekardt, Felix. "Nicht die Konzerne - wir selbst sind das Problem." Die Zeit Online. 11 March 2015. Accessed February 6, 2016. http://www.zeit.de/wirtschaft/2015-03/naomi-klein-kapitalismus-klimawandel.

[69] Toke, David, and Volkmar Lauber. "Anglo-Saxon and German approaches to neoliberalism and environmental policy: the case of financing renewable energy." *Geoforum*, no. 38 (2007): 677–87.

[70] Rich, Howard. "Germany's Green Energy Disaster: A Cautionary Tale For World Leaders." Forbes. 14 March 2013. Accessed February 6, 2016. http://www.forbes.com/sites/real-spin/2013/03/14/germanys-green-energy-disaster-a-cautionary-tale-for-world-leaders/#3362c8914a69.

"the attempt to perfect life through increasingly comprehensive state intervention, until the smallest injustices were compensated," as the former publisher of German daily *Frankfurter Allgemeine* states.[71] The German SPD is not a socialist party, but a party of Social Democrats. In 1954, the SPD officially accepted Christian Unionist Ludwig Erhard's social market economy as its own stance, leaving behind its history of Marxist thought.[72]

"It is a common misconception, though one that abounds in American politics, that laissez-faire capitalism supposes less law and less regulation," writes American legal expert Scott Horton, who adds (paraphrasing eighteenth-century French economist François Quesnay), that "an intelligent observer would not seek deregulation, which would lead to destructive chaos, but rather regulations that coincide with laissez-faire economic principles." Ordo-liberalism is thus laissez-faire within a set of rules for the common good. Or as one Christian Union politician in Germany once put it, "People will do anything for money, even good things."[73]

Most Germans believe their energy transition is creating greater market competition, not an anticapitalist socialist/green utopia. But they are aware that the outside world misunderstands them when it comes to their special flavor of capitalism. At the end of a long conversation about the previous century, German elder statesman Helmut Schmidt, who passed away in 2015, asked famous German historian Fritz Stern a question in 2010: "Fritz, at the end of these three days can you think of a particular wish you might have?". Stern answered: "yes, I wish that the difference between communism and social democracy were more clearly defined in the consciousness of the Western world—that is, that the right-wingers were not left to define the difference by saying the two are more or less the same thing and an established failure." And Schmidt concurred.[74]

---

[71] Jeske, "The Visible Hand of Economic Prosperity." See Footnote 7.

[72] See the Godesberg Program https://www.wikiwand.com/en/Godesberg_Program

[73] The CDU mayor of Landshut, quoted by Görres in conversation.

[74] Horton, Scott. "Our Century: A Dialogue with Helmut Schmidt and Fritz Stern (IV)." Harper's Magazine. 24 September 2010. Accessed February 6, 2016. http://harpers.org/blog/2010/09/our-century-a-dialogue-with-helmut-schmidt-and-fritz-stern-iv/.

9

The Red–Green Revolution (1998–2005)

On March 25, 1999, Reinhard Loske stepped before a crowd in a drab, first-floor meeting room in the *Tulpenfeld*, a parliamentary office building in the city of Bonn. At a buffet with crackers, cheese, and white wine, Loske announced the success of a milestone for Germany's Green Party. It was also his baby. The freshman Member of Parliament from the Green Party was an expert in environmental taxation. As an economist who loves nature, he pushed the Greens to integrate economic approaches into their environmental platform. For years, he had worked at a leading research institute to promote the concept of the Ecological Tax Reform, or eco-tax (*Ökosteuer*) in short. Now, the proposal had finally become law. It was his moment—or so he must have thought.

Indeed, in retrospect the first few years of the coalition between the Social Democrats and the Greens (called the Red–Green coalition based on the parties' colors) seem like a tremendous success particularly for the Greens. In addition to getting the eco-tax, the government adopted a Renewable Energy Act in 2000. That same year, it got the nuclear sector to sign a phaseout agreement, which became law in 2002.

But the crowd at Loske's speech not only consisted of eco-tax supporters, as the present author Arne Jungjohann (at the time, legislative

C. Morris, A. Jungjohann, *Energy Democracy*,
DOI 10.1007/978-3-319-31891-2_9

adviser to Loske) remembers. Members of the peace movement had come that night to throw tomatoes, not roses. During those same weeks, the German Armed Forces were planning to march into Kosovo under the new Foreign Minister Joschka Fischer—of the Green Party. Since World War II, sending German military outside the country's borders had been a taboo. And no party was more pacifist than the Greens. Now, they were to be the ones to break the taboo.

Green Failures that Weren't

Perhaps the outcome was not so surprising. Call it the Nixon Effect. Richard Nixon was the first US president ever to visit communist China. Because of his fierce anticommunist stance, it was hard to accuse Nixon of being soft on the Commies. Likewise, with a Green Foreign Minister, the German military intervention in Kosovo—coming as it did under NATO coordination—was less likely to conjure up images of resurging German military might. In politics, fundamental reversals often happen when opponents of an idea suddenly find themselves spearheading it.

The price that the Green Party paid, however, was high. It wasn't just the about-face on the German military that hurt the Greens. The public also had trouble seeing the nuclear phaseout and the eco-tax as the breakthroughs they were. Green supporters expected much higher energy taxes and a much faster nuclear phaseout. Voters and the party base were no longer sure what the Greens stood for. As a consequence, the Greens lost a slew of local and state elections. Thousands of disappointed members left the party.[1]

Voters had a hard time getting their heads around the eco-tax in general. Study it long enough, and it is a thing of beauty. First, it's not yet another tax. Rather, the idea is to redesign the current level of taxation so that existing taxes discourage environmentally detrimental consumption.

---

[1] Berg, Stefan, Jürgen Dahlkamp, Dietmar Pieper, and Andrea Stuppe. "Grüne: Gegen die Wand." *Der Spiegel*, no. 46 (1999): 24–25. Accessed February 6, 2016. http://www.spiegel.de/spiegel/print/d-15083337.html.

The bumper sticker slogan is, "Tax the bads, not the goods." Germany's main problem during globalization and increased international competition in the 1990s was the high cost of labor. Revenue from the eco-tax was therefore devoted to offsetting payroll taxes. The result was more competitive German labor costs and a price incentive to reduce consumption of dirty energy—what was there not to like?

From the very beginning, industry protested loudly against the reform, even though firms from the manufacturing sector were mostly exempt from the energy taxes, but benefitted from lower labor costs. The impact on households was also small and gradual. The cost of a liter of gasoline increased by three cents a year for five years from 1999 to 2003, equivalent to a total of around 50 cents per gallon over the entire time frame. Had such a hike been imposed in one fell swoop, the pushback would have been significant, but stretched across five years the impact was digestible.

At the beginning of the 1990s, then Chancellor Helmut Kohl increased the petroleum tax by similar increments—once by 11 cents and then again by 9 cents. The goal was to finance German reunification. The eco-tax was revenue-neutral, whereas Kohl's tax hike on oil did actually increase the overall level of taxation. Yet, the pushback against the eco-tax was much greater, thanks to a campaign orchestrated by German automotive association ADAC, anti-tax groups, industry representatives, and tabloid BILD.

Nonetheless, the eco-tax reached all its goals. Payroll taxes dropped by 1.7 percent, leading to the creation of an estimated 250,000 new jobs. Passengers taking public transportation increased by 5 percent; car sharing, by 70 percent. Sales of cars with better gas mileage increased, as did purchases of cars running on natural gas.[2]

Yet, the public had a hard time understanding why most of the revenue from an eco-tax should go to something completely unrelated (payrolls)—why shouldn't more of this money be used to fund renewable energy and environmental projects, many asked. In addition, the tax was

---

[2] Energy Transition. The German Energiewende. "D—Environmental taxation." Energy Transition. The German Energiewende. Accessed February 6, 2016. http://energytransition.de/2012/10/environmental-taxation/.

even applied to all electricity, including renewable power, the justification being that even this type of energy should be used in moderation. Not everyone agreed that it was a good idea to weigh down fledgling renewables even further right when they were learning to fly. Thus, the eco-tax suffered from lack of public support from those who opposed it and those who supported the idea in general. It didn't help that the Greens themselves weren't confident about their own baby. The first posters to promote the project read, "the eco-tax is okay."

The nuclear phaseout posed a similar PR disaster for the Greens. Though it might seem to be a major accomplishment for the Greens in hindsight, the phaseout was viewed at the time as a betrayal of green fundamentals. The Greens had run on a platform to shut down nuclear plants within five years. Many supporters from the anti-nuclear movement even asked for an immediate shutdown of all reactors. What they got was an agreement to shut down the last ones at the beginning of the 2020s. In the short term, two reactors would be closed, but even the plant operators themselves said the main reason was economics; these reactors would have been closed anyway with or without the government's phaseout.[3,4]

By the time the details were announced, it was clear that the Social Democrats had gotten the slow nuclear phaseout they—and the utilities themselves—wanted. Hard-core nuclear protesters even called the compromise "a law to promote atomic power."

The anti-nuclear movement was being harsh. Politically, the phaseout was a major accomplishment. For the first time, German law reflected popular will on the issue, which had been a conflict for more than two decades. The Germans wanted a nuclear phaseout, just not a quick one.

---

[3] Eon had already planned to close the Stade reactor prematurely at the latest by mid-2001. Likewise, Obrigheim could have run until November 2005 but was closed in May 2005.

[4] See Leuschner, Udo. "Energiechronik: Stade als erstes deutsches Kernkraftwerk stillgelegt." November 2003. Accessed February 6, 2016. http://www.energie-chronik.de/031107.htm; & Leuschner, Udo. "Energiechronik: KKW Obrigheim darf noch bis 15. November 2005 betrieben werden." December 2002. Accessed February 6, 2016. http://www.energie-chronik.de/021212.htm.

The Nuclear Phaseout of 2002

At the time of the elections in 1998, more than 90 percent of Germans wanted some kind of nuclear phaseout. Opinions mainly differed on the time frame. Fewer than 10 percent of the public supported the construction of new reactors. In return, only 13 percent wanted an immediate shutdown. But a large group—more than 40 percent—wanted to see the existing nuclear plants used as long as possible. The reasoning was that the investments had already been made, so why throw away something that works?

The present author Craig Morris admits he was in that camp at the time, as were some other leading German energy experts who are now clear supporters of the current nuclear phaseout.[5] Claudia Kemfert from one of the country's top economic research institutes is arguably Germany's most prominent climate economist and now also an ardent supporter not only of renewables, but specifically of the Energiewende as a grassroots movement.[6] Yet, as late as 2007, she was still of the following opinion:[7]

> Because CCS [carbon capture and storage] still requires research and renewables will probably not make up more than 20 percent of German power supply by 2020, the commissions of nuclear power plants should be extended by 10 to 15 years until CCS and renewable energy can be used and are competitive.

As of 2014, Germany had completely given up pursuing CCS, partly because renewables already made up 27 percent of power supply. People like Kemfert (and Morris) have gone through a learning curve. They now understand that conventional generators, including nuclear plants, must go to make room for the fast growth of renewables. Policymakers now

---

[5] Leuschner, Udo. "Energiechronik: Große Mehrheit der Bevölkerung ist für Weiterbetrieb bestehender Kernkraftwerke." October 1998. Accessed February 6, 2016. http://www.energie-chronik.de/981004.htm.

[6] Kemfert, Claudia. *Kampf um Strom: Mythen, Macht und Monopole.* Hamburg: Murmann, 2013.

[7] Kemfert, Claudia. "Ein Zehn-Punkte-Plan für eine nachhaltigeEnergiepolitik in Deutschland." *GAIA* 16, no. 1 (2007): 16–21.

agree that the energy transition will require the early retirement of conventional generators. Germany's nuclear phaseout serves that purpose.

That insight was unfortunately not mainstream in the late 1990s. To understand the challenges that the proposed phaseout faced after the 1998 elections, we have to recognize that forecasts still included photovoltaics under the category of "other"—meaning that even proponents of renewables underestimated the future growth of photovoltaics. The target in the government's *Leitstudie* (the roadmap for the Energiewende) for solar by 2050 was 29 GW, which had been built by 2012.[8] Wind power also grew faster than expected, with Germany reaching its 2010 target for wind power in 2005.[9]

When the Greens proposed a shutdown of all 19 nuclear reactors within five years, the promise seemed unreasonable because no one expected wind and solar to grow that quickly. The call for such a quick phaseout also turned out to be politically unfeasible; once in government, the Greens lacked a clear strategy for turning their campaign promise into sound legislation. For its part, the nuclear sector was busy disseminating horror studies, showing that an immediate shutdown would lead to 150,000 job losses and compensation payments to the industry of 90 billion deutsche marks.[10] Any forward-looking strategy would have included the kind of compromises the Social Democrats (SPD) were willing to make. When should what reactor be shut down?

To be fair, the SPD hardly came into office with a clear plan for a phaseout themselves. "I want a consensus and nothing else," Chancellor Gerhard Schröder merely told his party colleagues and the Greens. In other words, the phaseout required a consensual agreement with Germany's nuclear plant owners. Furthermore, no special compensation was to be provided; the firms were not to receive money in order to close their plants. Money was all these firms wanted anyway. For the

---

[8] Morris, Craig. "Conventional power providers hurting: Schadenfreude about RWE unwarranted." Renewables International. The Magazine. 6 March 2014. Accessed February 6, 2016. http://www.renewablesinternational.net/schadenfreude-about-rwe-unwarranted/150/522/77367/.

[9] Jungjohann, Arne, and Craig Morris. "The German Coal Conundrum." Heinrich Böll Foundation, North America. June 2004. Accessed February 6, 2016. https://us.boell.org/sites/default/files/german-coal-conundrum.pdf.

[10] Munsberg, Hendrik. "Abschied vom Atomstrom." *Der Spiegel*, no. 52 (1998): 22–23. Accessed February 6, 2016. http://www.spiegel.de/spiegel/print/d-8452409.html.

right price, they would gladly step away from nuclear. In insisting on an agreement, the government was trying to learn a lesson from Sweden. In 1980, the Swedes had adopted their own nuclear phaseout in the wake of the accident at Three Mile Island,[11] only to see the industry sue the government for compensation. If the German government could get their nuclear utilities to promise in writing that they would not sue for special payments, the phaseout stood a much greater chance of standing in court, should it be challenged.[12]

Having set down his stipulations, Schröder then played the role of the generous moderator in the negotiations, with strong sympathy for the interests of the nuclear sector. For the Social Democrats, Werner Müller was involved as Minister of Energy and Economics. As a former utility executive with no official party affiliation, he became the mouthpiece of the nuclear industry within the cabinet. For the Greens, Joschka Fischer took part in the talks as Vice Chancellor. As Minister for the Environment and Nuclear Safety, Jürgen Trittin naturally also played a major role. In the background, Environmental Undersecretary Rainer Baake of the Greens helped navigate bureaucratic and legal issues. Numerous legal opinions were provided, confirming that the legal basis for claims of financial damages was minimum if the plants were allowed to remain in operation for at least 30 years.[13]

The agreement eventually reached provided a theoretical service life of 32 years. On 14 June 2000, the German government signed the agreement with nuclear plant owners, which became law in 2002. The utilities knew that the compromise would constrain their investments in nuclear for the future, but it also cleared up political risks, giving them greater certainty. Before the elections in 1998, nuclear protests had become massive and violent. But as soon as the phaseout negotiations started,

---

[11] Morris, Craig. "Angst… that the Energiewende will work." Energy Transition. The German Energiewende. 24 July 2014. Accessed February 4, 2016. http://energytransition.de/2014/07/angst-that-the-energiewende-will-work/.

[12] Raschke, Joachim. "Die Zukunft der Grünen. " Frankfurt/Main: Campus, 2001, p. 173.

[13] Matthes, Felix, C. "Ausstiegsfahrpläne und Einstiegspfade: eine Einschätzung zum Atomkonsens." Öko-Mitteilungen, no. 2 (2000): 4–7; Öko-Institut e.V. Accessed February 6, 2016. http://www.oeko.de/oekodoc/67/2000-oemi-2.pdf, p. 4.

turnout at demonstrations plummeted, though protests were still held.[14] The nuclear sector could clearly see that, in return for the promise of an eventual closure, it would get a calm and predictable business environment. The president of the Atomic Forum stated when the agreement was finally signed, "We have reached our goal to continue operating our nuclear power plants under economically acceptable conditions."[15]

The head of the German chapter of Friends of the Earth agreed that the nuclear sector had reached its goal—and that was the problem for environmentalists. Practically everyone who was not involved in the agreement disliked it, however, not just critics of nuclear. Several state governments led by the Christian Union, which was friendly to nuclear, said they would take the matter all the way to the Constitutional Court. The head of German industry association BDI called the signing "a dark day for Germany."

The nuclear industry obtained a relatively long commission of 32 years per plant, but those years were counted in expected electricity generation, not actual time. Specifically, the number of kilowatt-hours was calculated for a given reactor over 32 years of operation. If a plant needed to be shut down temporarily for whatever reason—or if it generated slightly less electricity for whatever reason—those allotted kilowatt-hours remained available even after 32 years. For this reason, Chancellor Schröder's nuclear phaseout did not have a specific deadline. But it was easy to calculate that, under business as usual, the last plant would probably close around 2022.

There was one additional loophole that the power firms themselves would later exploit: they were allowed to shut down an old plant early and transfer the remaining kilowatt-hours to a newer reactor. The reasoning was that more modern nuclear plants were considered state of the art, and hence safer.[16] At least one nuclear utility later attempted to exploit this stipulation by drawing out revisions at a nuclear reactor. The goal

[14] Leuschner, Udo. "Energiechronik: Mäßige Beteiligung bei Anti-AKW-Aktionen." September 1998. Accessed February 6, 2016. http://www.energie-chronik.de/980921.htm.

[15] Otto Majewski: „Unser erklärtes Ziel, die deutschen Kernkraftwerke zu wirtschaftlich akzeptablen Bedingungen weiterhin nutzen zu können, haben wir erreicht." Quoted In: "Deutscher Atomausstieg mit dem Rechenschieber." Ingenieur.de. 23 June 2000. Accessed February 6, 2016. http://www.ingenieur.de/Politik-Wirtschaft/Energie-Umweltpolitik/Deutscher-Atomausstieg-mit-Rechenschieber.

[16] Matthes, p. 5.

was to leave some of those kilowatt-hours unused until the next elections, when—the sector hoped—a new coalition would step away from the agreement.

At the time, not everyone believed that the phaseout was permanent. One German academic openly asked whether the label "moratorium" would not be more appropriate.[17] Next door, France remained committed to nuclear, even announcing in 2004 that 19 new reactors of the next-generation EPR design would be completed in the country by 2020, along with an EPR to be completed in Finland by 2009.[18] (As of 2016, no EPR has been completed worldwide, and France was only building one as of 2015.)

Nonetheless, Germany was not alone with its nuclear phaseout at the time. In 2003, Belgium limited the commissions of its nuclear reactors to 40 years, thereby imposing a phaseout by 2025.[19] Likewise, Denmark, Austria, and Italy had each adopted its own nuclear phaseout years, if not decades, earlier.[20] In fact, Germans themselves took their famous anti-nuclear slogan ("Atomkraft—nein, danke" or "nuclear power—no thanks") with a smiling sun from the Danes, who coined "Atomkraft—nej tak" back in 1975.

With the nuclear issue settled for the moment, German utilities could focus on other fish they had to fry.

Competition Leads to Concentration

In 1998, German utilities were forced to compete with each other for customers for the first time ever. EnBW, the utility traditionally in the southwest, came up with some creative ideas. For instance, it founded a

---

[17] Brunekreeft, Gert. "Negotiated Third-Party Access in the German Electricity Supply Industry." Accessed February 5, 2016. http://www.vwl.uni-freiburg.de/fakultaet/vw/publikationen/brunekreeft/MAILAND_Sept01.pdf.

[18] Leuschner, Udo. "Energiechronik: Union und FDP wollen Laufzeiten für Kernkraftwerke verlängern." May 2005. Accessed February 6, 2016. http://www.energie-chronik.de/050502.htm.

[19] Leuschner, Udo. "Energiechronik: Belgien beschließt Ausstieg aus der Kernenergie bis 2025." January 2003. Accessed February 6, 2016. http://www.energie-chronik.de/030110.htm.

[20] Morris, *Angst… that the Energiewende will work.*

subsidiary called Yellow, which began telling the public what color electricity was. The firm also took over a retail shoe chain so it could use outlets to reach customers throughout the country. Overall, German utilities upped their expenditures on advertising by 81 percent in the first half of 1999.[21]

Nonetheless, German power consumers were slow to change power providers. One reason was the shenanigans utilities engaged in to retain customers. More than half of those who tried to switch utilities reported difficulties, such as when the old provider did not respond other than by threatening to take out the power meter entirely.[22] Some power providers sabotaged their former customers; when technicians came by to read the meter one last time, they also unscrewed the main ceramic fuse, thereby unnecessarily separating the household from the grid. By 2004, the practice was so widespread at Badenova that the Freiburg-based utility's hotline knew exactly what advice to give former customers on their last day of service when they called to report a blackout: just go downstairs and screw the fuse back in.

The combination of more marketing expenditures and customer unwillingness to switch providers was disastrous for power firms. In 2001, Eon began using Arnold Schwarzenegger in ads to get new customers. The firm spent 22.5 million euros on the campaign, which only won over 1100 customers. The price tag per customer came in at around 20,000 euros. At an average power bill of 44 euros at the time, it would have taken the firm some 40 years just to get the money back—and centuries to turn a profit.[23]

Under these circumstances, it is perhaps not surprising that the ultimate result of liberalization was not greater competition and better customer service, but greater market concentration. German utility association VBEW (predecessor of the current BDEW) spoke of "preda-

[21] Leuschner, Udo. "Energiechronik: Viel Geld für Strom-Werbung." October 1999. Accessed February 6, 2016. http://www.energie-chronik.de/991012.htm.

[22] Leuschner, Udo. "Energiechronik: "Jeder zweite Stromkunde berichtet über Probleme beim Wechsel des Lieferanten"." July 2000. Accessed February 6, 2016. http://www.energie-chronik.de/000710.htm.

[23] Janzing, Bernward, and Dieter Seifried. *Störfall mit Charme: Die Schönauer Stromrebellen im Widerstand gegen die Atomkraft; wie eine Elterninitiative, die sich nach Tschernobyl gründet, zu einem bundesweiten Stromversorger wird.* Red.-Schluß: 1. Okt. 2008. Vöhrenbach: Dold, 2008, p. 74.

tory competition" in February 2000.[24] By the time Sweden's Vattenfall entered Germany through the takeover of four smaller utilities in 2002, Germany's Big Eight utilities had become the Big Four.

German and EU antitrust authorities were worried that the oligopoly on the German power market had become a duopoly.[25] RWE and Eon alone controlled not only more than half of power generation and 43 percent of the high-voltage grid, but also held stakes in three quarters of regional and municipal utilities—up from 45.5 percent in 1997.[26] In contrast, both Vattenfall and EnBW (the other two in the Big Four) each only made up closer to 10 percent of the market.

Chancellor Schröder's coalition hardly minded; on the contrary, in a case that drew international attention, the German Cartel Office blocked Eon's takeover of Ruhrgas—only to have to sit back and watch as German Economics Minister Werner Müller overruled the Commission and gave his personal ministerial permission.[27] The event was all the more scandalous because Müller later went through the political–industrial revolving door, becoming head of coal firm RAG, while his Undersecretary became CEO of a subsidiary. It was also later revealed that Müller had still been receiving regular payment from Eon while he was Economics Minister, including during the time when his ministry approved the firm's takeover of Ruhrgas.[28]

In these years, the German Cartel Office would work to limit mergers and acquisitions. For instance, the big utilities were not allowed to have more than a 20 percent stake in municipals. As the big firms began to reach those limits, they realized that further growth could only come

---

[24] Leuschner, Udo. "Energiechronik: VBEW sieht "Verdrängungswettbewerb"." February 2000. Accessed February 6, 2016. http://www.energie-chronik.de/000212.htm.

[25] Leuschner, Udo. "Energiechronik: Kartellbehörden befürchten Duopol infolge der geplanten Großfusionen." April 2000. Accessed February 6, 2016. http://www.energie-chronik.de/000401.htm; and Becker, Peter. *Aufstieg und Krise der deutschen Stromkonzerne: Zugleich ein Beitrag zur Entwicklung des Energierechts*. Bochum: Ponte Press, 2011, p. 122.

[26] Leuschner, Udo. "Energiechronik: Liberalisierung verstärkte Konzentration auf dem deutschen Strommarkt." December 2001. Accessed February 6, 2016. http://www.energie-chronik.de/011224.htm.

[27] Becker, *Aufstieg und Krise der deutschen Stromkonzerne*, p. 138.

[28] Leuschner, Udo. "Energiechronik: Müller bezog bereits als Minister eine Pension von E.ON." January 2005. Accessed February 6, 2016. http://www.energie-chronik.de/050119.htm.

from international expansion. EnBW tried to purchase shares of a Spanish power provider, but the Spanish government blocked the attempt, arguing that France's EDF held too many shares in the German utility.[29] In 2000, EDF—still a state-owned utility—had purchased a quarter of EnBW. The German firm operated the grid in the state of Baden-Württemberg on the French border, and EDF used German power lines in the area heavily to sell power to Switzerland.[30] The EU only approved the deal because of this historic situation, but it required EDF to sell 6000 megawatts of generation capacity to competitors in order to ensure a minimum of competition.[31]

While EDF was free to enter Germany, it was having trouble expanding elsewhere. In 2001, the Italian government prevented the French utility from taking over 18 percent of Italy's second-largest utility.[32] The problem for these other countries was partly that the French market remained largely closed; though EDF was free to do business elsewhere, firms from Germany, Spain, and Italy still had a hard time setting up shop in France. The French government was a year late in complying with the EU directive on liberalization, and even then the market was only 30 percent open to newcomers, who still could only serve industrial customers—the retail market remained closed. To defend its historic monopoly position, EDF began to speak of itself as a "public service." In reality, the state-owned company could not afford to allow competitors to come in, lest its nuclear fleet become unprofitable.

In Germany, signs of surplus capacity had become obvious; plants needed to be closed. Before the days of liberalization, these firms merely had their accounting books reviewed by regulators, and the profit margin was guaranteed—after all, the government cannot force a company to do business at a loss. Surplus capacity—plants not generating enough

---

[29] Leuschner, Udo. "Energiechronik: Madrid suspendiert Stimmrechte der Hauptaktionäre von Hidrocantabrico." May 2001. Accessed February 6, 2016. http://www.energie-chronik.de/010512.htm.

[30] Although the Swiss and the French share a border, the region is mountainous, making power pylons harder to install.

[31] Leuschner, Udo. "Energiechronik: EU-Kommission genehmigt Einstieg der EDF bei EnBW mit Auflagen." February 2001. Accessed February 6, 2016. http://www.energie-chronik.de/010201.htm.

[32] Leuschner, Udo. "Energiechronik: Rom begrenzt Stimmrechte von EDF bei Montedison." May 2001. Accessed February 6, 2016. http://www.energie-chronik.de/010511.htm.

electricity—was simply a cost item passed on to consumers. After liberalization, these power plants needed to sell electricity if they wanted to stay in business.

By October 2000, Germany's two biggest utilities, Eon and RWE, had announced intentions to close a whopping 10,000 megawatts of capacity, with each firm closing roughly half of that amount. Other smaller utilities, including EnBW, confirmed the unprofitability of numerous additional power plants.[33] More than a tenth of German power plant capacity was about to be shut down.

The result of this excess capacity during the early years of competition was lower retail power prices. From 1998 to 2000, they dropped from 17 to 13 cents per kilowatt-hour. Another reason was brief competition between RWE and EnBW. The former had a subsidiary within the latter's service area. When RWE insisted that it be able to provide power to its subsidiary directly, it was the first breach of monopoly territories. The two firms then began fighting for territory.

By 2000, utilities understood that aggressive pricing would not give them more customers.[34] The biggest companies then began trying to work together again to break up Germany into the old territories—and power prices began rising steadily again.

Newcomers struggled to stay in business. The low power prices may have cut into the profits of big utilities with big pockets, but the losses drove startups straight into insolvency.[35] And there was one other problem for new market entrants: grid fees. Throughout the first years of liberalization, there were repeatedly reports of discrimination in grid charges. New power providers had to use power lines owned by the incumbents. These established firms could then overcharge for grid usage to cut into the profits of competitors. In 2001, grid fees varied by up to 130 percent from one operator to another, and no good reason could be given for the difference.[36]

---

[33] Leuschner, Udo. "Energiechronik: E.ON und RWE verkünden Stillegung von 10 000 Megawatt Kraftwerksleistung." October 2000. Accessed February 6, 2016. http://www.energie-chronik.de/001003.htm.

[34] Becker, *Aufstieg und Krise der deutschen Stromkonzerne*, p. 116.

[35] Leuschner, Udo. "Energiechronik: "Die Newcomer auf dem liberalisierten Strommarkt kämpfen ums Überleben"." August 2000. Accessed February 6, 2016. http://www.energie-chronik.de/000809.htm.

[36] Leuschner, Udo. "Energiechronik: Netznutzungsentgelte differieren um bis zu 130 Prozent." April 2001. Accessed February 6, 2016. http://www.energie-chronik.de/010405.htm.

It was mainly over this issue that the German experiment with negotiated third-party access failed. In 2001, the European Commission called on Germany—the only EU member state that still did not have a regulator on the power market—to create one.[37] That September, German antitrust authorities began looking into wildly excessive grid fees charged by 22 grid operators.[38] The next summer, the Monopoly Commission finally recommended the creation of a grid regulator.[39] All these events also took place against the backdrop of the deregulation crisis in the US, which was unfolding at the time.

By 2003, a new German minister, Social Democrat Wolfgang Clement, had replaced Werner Müller as Economics Minister and was also directing the task force on a regulatory body. Clement previously served as Minister President of North Rhine-Westphalia, homeland to Germany's strong coal lobby and the big utility RWE. His idea was not to create a regulator within the Cartel Office, which was likely to do a good job, but add grid regulation to the tasks of the agency that already oversaw postal services and telecommunications. He hoped that being able to appoint the head of the future Network Agency would allow the government to keep better control of it.

Relations between the Cartel Office and Clement were awkward. He was the only Economics Minister in Germany to skip visiting the Cartel Office when he took office.[40] In particular, Clement faced criticism from Ulf Böge, director of the Cartel Office. Under Böge's leadership, the Cartel Office prevented the Big Four from taking over numerous municipal utilities; indeed, it was the Cartel Office that set the limit for corporate holdings of municipal utilities at 20 percent in January 2002.[41] An ordo-liberal, Böge wanted to provide fair com-

---

[37] Leuschner, Udo. "Energiechronik: Keine beschleunigte Liberalisierung des Binnenmarkts für Strom und Gas." March 2001. Accessed February 6, 2016. http://www.energie-chronik.de/010301.htm.

[38] Leuschner, Udo. "Energiechronik: Bundeskartellamt ermittelt gegen 22 Netzbetreiber." September 2001. Accessed February 6, 2016. http://www.energie-chronik.de/010902.htm.

[39] Leuschner, Udo. "Energiechronik: Monopolkommission empfiehlt allgemeine Regulierungsbehörde für Netzsektoren." July 2002. Accessed February 6, 2016. http://www.energie-chronik.de/020702.htm.

[40] See https://lobbypedia.de/wiki/Wolfgang_Clement

[41] Leuschner, Udo. "Energiechronik: Kartellamt setzt Grenze von 20 Prozent für Beteiligungen an Stadtwerken." January 2002. Accessed February 6, 2016. http://www.energie-chronik.de/020107.htm.

petition between firms large and small; he specifically criticized the Schröder coalition's penchant for national champions.[42] Even more than his predecessor Müller, Clement wanted to strengthen Germany's largest power firms in competition with foreign national champions—exactly what Böge opposed.

One of Clement's first acts on taking office was to do away with his predecessor's Task Force on Grid Access, which was also too critical of the new Economics Minister's friendly position toward the largest firms.[43] In 2005, the Network Agency would finally go into business overseeing the power market in addition to telecommunications and postal services (later, gas markets were also added). That year, however, the Red–Green coalition would be voted out of office.

From 1998 until 2005, utilities remained mainly focused on the nuclear phaseout and liberalized markets. The firms were also busy lobbying domestic legislation on an emissions trading platform to start in 2005 across the EU. Leading German politicians—such as Minister Müller—spoke out against the call for lower emissions, which they felt would endanger German industry.[44] The Red–Green coalition was divided over how to implement the EU directive for emissions trading. But the European Commission had been pushing hard for the idea since the late 1990s. There was no getting around it. And the biggest emitters of carbon—industrial firms and power generators—were not exactly sure what carbon trading was. Most markets *provide* products and services, but this one *avoids* something: carbon emissions.

With the focus on the nuclear phaseout, the market liberalization and emissions trading, the big utilities paid a lot of attention to their core business, but less so on something they failed to take seriously: renewables.

---

[42] "Kartellamtschef Böge: Abschied eines Ordoliberalen." Der Spiegel Online. 3 April 2007. Accessed February 6, 2016. http://www.spiegel.de/wirtschaft/kartellamtschef-boege-abschied-eines-ordoliberalen-a-474929.html.

[43] Leuschner, Udo. "Energiechronik: Clement löst "Task Force Netzzugang" auf." Septmeber 2003. Accessed February 6, 2016. http://www.energie-chronik.de/030902.htm.

[44] Leuschner, Udo. "Energiechronik: Energiebericht warnt vor überzogenen Klimaschutz-Zielen." November 2001. Accessed February 6, 2016. http://www.energie-chronik.de/011101.htm.

& Leuschner, Udo. "Energiechronik: Emissionshandel spaltet Regierungskoalition." June 2002. Accessed February 6, 2016. http://www.energie-chronik.de/020603.htm.

The Renewable Energy Act of 2000 Starts the Renewable Energy Boom

In 2000, the EEG went into force. As Section 1 states, the goal was to "at least double" the share of renewable energy in total energy consumption and to "considerably increase" the share of renewables in power supply. While the target for electricity is a bit vague, the term "at least" for energy (which includes heat and motor fuel) is crucial. In many countries— including the US—markets come screeching to a halt when targets are approached. A simple phrase like "at least" can prevent such a market slowdown.

The law has become associated with its two principal authors: Hans-Josef Fell of the Greens and Hermann Scheer of the SPD, who passed away in 2010.[45] "My role was principally the writing," Fell remembers. "Scheer knew how to get the legislation through parliament politically." The law was an exception, in that parliamentarians wrote it; normally, laws come from ministries, where utility lobbies have long-standing contacts and can influence policy. The EEG was a bit of a rogue law in that respect, and utilities would have had to influence specific parliamentarians like Scheer and Fell.

Fell understates his political maneuvering, however. He worked hard to get German solar firms to petition the government in favor of the law. The only firm that did not play along was, tellingly, Siemens Solar, which was not a grassroots startup. Fell also got the speechwriter for Wolfgang Clement—then governor of North Rhine-Westphalia—to include a call for "feed-in tariffs" at a speech Clement gave at the opening of a new photovoltaic facility by Shell in his state.[46] At the time, neither Clement nor his speechwriter apparently understood what the term meant, for Clement would later be a major opponent of the EEG as Economics Minister.[47]

---

[45] The other two main architects of the Renewable Energy Act of 2000 were Dietmar Schütz (SPD) and Michaele Hustedt (Greens).

[46] Janzing, Bernward. *Solare Zeiten: Die Karriere der Sonnenenergie; eine Geschichte von Menschen mit Visionen und Fortschritten der Technik.* Freiburg: Picea-Verl., 2011, p. 137.

[47] Reimer, Nick. "TAZ Interview mit Hermann Scheer: "Clement hat sich isoliert"." 5 September 2003. Accessed February 6, 2016. http://www.taz.de/1/archiv/?id=archivseite&dig=2003/09/05/a0081.

Fell's accomplishment is all the more impressive when we remember that Clement was a Social Democrat, not a Green like Fell.

Political analyst Christoph Stefes argues that the introduction of nationwide feed-in tariffs "seemed next to impossible,"[48] given the considerable skepticism of the big utilities and their political allies against renewables. So how did the law pass its manifold barriers?

In the 1990s, Fell had successfully gotten feed-in tariffs implemented in his Bavarian hometown of Hammelburg. And though Fell was not new to politics, he was new in the Bundestag. Scheer, in contrast, had been a parliamentarian since 1980. He thus brought with him two decades of political experience, including his success in keeping his own party—then in the opposition—from blocking the 1990 Feed-in Act.

For the successor legislation, Scheer once again needed to get his own party on board. The Social Democrats were traditionally the party of workers, including coal miners—not exactly a group fond of renewable energy. Scheer went into a meeting with 15 SPD politicians from the coal state of North Rhine-Westphalia especially prepared. They aimed to bring down the new legislation before it had even been adopted. But Scheer had a trick up his sleeve. When a list of the technologies supported by the previous Feed-in Act was read, Scheer raised his hand and said, "wait a minute, something is missing—mine gas."[49] The surprise on the faces of the pro-coal lobby was clear to see. Mine gas is essentially methane that escapes coal mines unused. Sometimes, it is flared off because recovery would not be economical. It can also endanger the lives of coal miners. But mine gas is clearly not one thing: renewable.

Scheer argued that making mine gas recovery economical with policy support made environmental sense. There is some logic to this reasoning, but what he was really after was a way to get this group of 15 coal proponents on his side. When someone at the meeting objected, "We cannot

---

[48] Stefes, Christoph H. "Energiewende. Critical Junctures and Path Dependencies Since 1990." In "Rapide Politikwechsel in der Bundesrepublik. Theoretischer Rahmen und empirische Befunde." Edited by Friedbert W. Rüb. Special issue, *Zeitschrift für Politik*, Sonderband 6 (2014): 47–70, p. 57.

[49] May, Hanne. "Abgeordneter, Autor, Botschafter für Erneuerbare und ein leidenschaftlicher Kämpfer für seine Überzeugungen—all das ist Hermann Scheer. Ein Besuch." Neue Energie, 04/2010. Accessed February 6, 2016. http://www.hermannscheer.de/de/index.php/presseecho-2010/720-hermanns-schlacht.

expand the policy to cover everything," Scheer knew he had reached his goal. The group of 15 immediately protested, with all of them saying: "either mine gas is included, or we will oppose the law." Scheer had broken the coal sector's opposition and forged a coalition for renewables in the SPD.

Another part of his strategy was not to talk about prices. He knew he would need to get an extremely high rate for solar in order for it to be profitable—several times the retail rate. Instead of negotiating that number with people who are unconvinced, he focused on the principle: the policy must provide the same fair chance for a reasonable return on investments for renewable energy that were offered for regulated utilities all along. Once he had enough people on his side for that idea, there was no way around discussing specific numbers. He then borrowed an old trick from retailers, proposing 99 pfennigs as the rate for solar—because, he assumed, a full deutsche mark would have seemed much more expensive.[50]

Another stumbling block was the pending ruling on whether the Feed-in Act violated EU rules on state aid. It wasn't until March 2001 that the European Court of Justice ruled that the 1990 law did not constitute illegal state aid, which is defined in the EU as funding that flows back to governmental budgets. In the case of both the Feed-in Act and the Renewable Energy Act of 2000, the funding was collected as a surcharge on electricity consumption by grid operators and paid back to renewable energy generators without ever touching a governmental budget. Ironically, the power firms behind the legal challenge started off with a proposal to switch compensation from their own books into the governmental budget; it was only later that they decided to challenge the policy as illegal state aid.[51]

The power firms behind the case may have also felt encouraged when the European Commission stated that it had not been notified of this state aid.[52] Not notifying Brussels of energy policy changes was a mistake

---

[50] Janzing, *Solare Zeiten*, p. 138.

[51] Scheer, Hermann. *Sonnen-Strategie: Politik ohne Alternative.* Serie Piper Bd. 2135. München: Piper, 1998, p. 192.

[52] Leuschner, Udo. "Energiechronik: EU-Kommission überprüft das Erneuerbare-Energien-Gesetz." April 2000. Accessed February 6, 2016. http://www.energie-chronik.de/000404.htm.

the French would later make,[53] but in this case the Commission dropped its investigation in 2002, after the European Court of Justice had stipulated that state aid requires a connection to a governmental budget, which German feed-in tariffs did not have.[54] Likewise, the German High Court also confirmed the constitutionality of the Renewable Energy Act in June 2003, finally putting an end to a slew of legal challenges.[55] Clearly, the authors of the policy from 2000 had done their homework.

When the EEG became law in 2000, retail and industry power prices were hitting rock-bottom in the third year of liberalized power markets. Lower rates were in line with expectations; the subsequent price increases resulting from the smaller oligopoly were not. So at the time the EEG was designed, the focus was on uncoupling feed-in tariffs from retail rates, lest support for renewables continue to drop. Even in the late 1990s, feed-in tariffs—which were still defined as a percentage of retail rates—had fallen. The focus was therefore on providing each technology with the specific payment it needed irrespective of the retail rate.

When the first review of the EEG was presented in July 2002, it found that the financial support provided for biomass was only sufficient for large biogas units and systems fired with waste wood (which could be had inexpensively). However, press reports confirmed that a boom was indeed underway, with hundreds of megawatts of biomass-fired systems in the pipeline.[56] As of 2013, just over 5000 megawatts of power generators running on biomass had been built.

Geothermal fared worse—no plants at all had been completed by 2002, though eight were in the pipeline. Not all those would be completed. On the other hand, Germany has little geothermal potential, so that even in 2015 the official power production statistics included geothermal electricity under "other." In 2003, technical advisors to the gov-

---

[53] Morris, Craig. "EU may rule against French feed-in tariffs." Renewables International. The Magazine. 21 August 2013. Accessed February 6, 2016. http://www.renewablesinternational.net/eu-may-rule-against-french-feed-in-tariffs/150/537/72159/.

[54] Leuschner, Udo. "Energiechronik: Brüssel stellt Beihilfeverfahren wegen EEG und KWKG ein." May 2002. Accessed February 6, 2016. http://www.energie-chronik.de/020511.htm.

[55] Leuschner, Udo. "Energiechronik: Bundesgerichtshof bestätigt Erneuerbare-Energien-Gesetz." June 2003. Accessed February 6, 2016. http://www.energie-chronik.de/030608.htm.

[56] Leuschner, Udo. "Energiechronik: Abstriche am Marktanreizprogramm für erneuerbare Energien." July 2001. Accessed February 6, 2016. http://www.energie-chronik.de/010712.htm.

ernment recommended geothermal for the long term, adding expressly that the risk of earthquakes was low.[57] Geothermal power plants are a bit like fracking facilities, in that both inject liquids into underground rock. Unfortunately, a geothermal project near Basel, Switzerland, on the German border caused minor earthquakes in 2006—and again in 2007, long after the project had been discontinued.[58] Since then, geothermal power plants have not been widely pursued in Germany.

Photovoltaics performed much better, but it was starting from a very small base. The feed-in tariff—a whopping 99 pfennigs, equivalent to just over 50 cents in euros—was incredibly high but apparently still not enough. An earlier policy—the 100,000 Roofs Program—had been implemented in 1999 and remained in force until the end of 2003. Whereas feed-in tariffs were paid for the power generated, this program provided low-interest loans to cover the upfront purchases. (At the time, a normal household array of 3 kilowatts would have cost some 23,000 euros, compared with 4000 euros in 2014.[59])

In combination with feed-in tariffs, the low-interest loans were extremely popular. In the first 3 months of 2000, 3700 applications for the loans were processed, more than in the entire previous year—and another 7000 were still piled up waiting for processing.[60] The waiting list slowed down the market. The 100,000 Roofs Program had a budget of 300 MW, which was reached at the end of 2003.

Unlike other technologies, there was also a limit on feed-in tariffs for photovoltaics: 350 MW. With that limit quickly approaching, the government resolved in June 2002 to raise the ceiling to 1000 MW.[61] Nonetheless,

[57] Paschen, Herbert, Dagmar Oertel, and Reinhard Grünwald. "Möglichkeiten geothermischer Stromerzeugung in Deutschland." TAB: Sachstandsbericht. Arbeitsbericht Nr. 84. Accessed February 6, 2016. https://www.tab-beim-bundestag.de/de/pdf/publikationen/berichte/TAB-Arbeitsbericht-ab084.pdf.

[58] Morris, Craig. "Geothermie-Experten auch vom zweiten Beben überrascht." Heise. 10 January 2007. Accessed February 6, 2016. http://www.heise.de/tp/artikel/24/24419/1.html.

[59] Leuschner, Udo. "Energiechronik: 100 000-Dächer-Programm soll schon im neuen Jahr starten." December 1998. Accessed February 6, 2016. http://www.energie-chronik.de/981231.htm.

[60] Leuschner, Udo. "Energiechronik: 100 000-Dächer-Programm vorerst auf Eis: Im Haushalt eingeplante Mittel erschöpft." April 2000. Accessed February 6, 2016. http://www.energie-chronik.de/000405.htm.

[61] Leuschner, Udo. "Energiechronik: Deckelung für Solarstrom-Förderung auf 1000 MW ange-hoben." July 2002. Accessed February 6, 2016. http://www.energie-chronik.de/020707.htm.

in January 2002, Siemens and Eon sold their remaining holdings in Siemens Solar, a solar cell manufacturer, to Shell, which was trying to put on a greener face at the time (an attempt it has since abandoned).[62] But even Shell left the solar sector in 2006, selling to SolarWorld. Likewise, RWE sold half of its shares in its solar subsidies to Schott Solar in 2002 and the other half in 2005, though the sector was still skyrocketing.[63] The conventional incumbents gave up on solar at the stupidest possible time; they still did not believe it would work.

The big winner of the 2000 EEG, however, was wind power.[64] In 2000, the 1-gigawatt threshold was crossed for new installations, a figure few imagined possible in 1991—and that volume would triple by 2002, when Germany installed 3.2 gigawatts. In 1999, a total of 7.9 TWh of electricity financed with feed-in tariffs had been generated. The EEG review of mid-2002 expected that number to reach 21 TWh by the end of the year—but it actually reached 22.4. At the time, Germany was by far the world leader in wind power, ahead of the US in a distant second place. A third of global installed wind power generation capacity was located in Germany in 2002.

Interestingly, the review states, "The government holds the growth of offshore wind turbines to be crucial." This emphasis turned out to be a gravely inaccurate assessment, and the focus on offshore drained commitment from the onshore sector. After the record 3.2 GW onshore in 2002, annual installations fell again for roughly the next decade down to 2.0 GW or lower. One reason was because so many people expected wind turbines to start being built in German waters rather than on land.

In November 2001, the first permit was issued for an offshore wind farm. Environmental Minister Trittin had already thrown his weight behind the offshore sector. As of mid-2001, permits had already been requested for a whopping 21 wind farms in the North Sea along with seven in the Baltic. At the end of 2014, only three wind farms had been completed in the German North Sea along with one in the Baltic, including two non-commercial research projects. Admittedly, a few others were

---

[62] Leuschner, Udo. "Energiechronik: Siemens und E.ON geben Solarzellen-Produktion ganz an Shell ab." January 2002. Accessed February 6, 2016. http://www.energie-chronik.de/020112.htm.

[63] Janzing, *Solare Zeiten*, p. 142.

[64] Leuschner, Udo. "Energiechronik: Erster Erfahrungsbericht zum Erneuerbare-Energien-Gesetz vorgelegt." July 2002. Accessed February 6, 2016. http://www.energie-chronik.de/020703.htm.

partly completed and already generating electricity, but the performance fell dismally short of expectations.

It wasn't just the government that was keen on offshore wind. Greenpeace also was. Community wind farm builders did not appreciate the new focus. "The criticism completely surprised us," Greenpeace energy expert Sven Teske remembers.[65] The problem was that offshore wind farms required much more upfront capital. The attempt to build a community offshore project, called *Butendiek*, eventually failed, and the required feed-in tariffs for offshore wind turned out to be twice as high as those needed for onshore turbines. During the years after 2002, the onshore wind market partly stagnated because so much time was committed to offshore projects that fizzled out.

In the end, the goal to phase out nuclear power helped proponents of renewables. They argued renewables were the only remaining energy source left to increase without Germany giving up its ambitions to reduce greenhouse gases.

A War, a Flood, and a Bumbling Bavarian Keep the Energiewende Going

All this progress could have come to a precipitous end, however, after the elections held in September 2002. It was anyone's guess who would win, and the final tally showed that the vote had been incredibly close. The Social Democrats and the Christian Union both received 38.5 percent of the vote. Only because the Greens got more votes than the FDP was the Red-Green coalition able to muster 306 seats, just four more seats than needed for a majority in Parliament.

During the election campaign, the FDP called for an end to the eco-tax, which caused a rift with the Christian Union, a potential coalition partner, whose leading politician at the time said he would not do away with the eco-tax if elected. While the eco-tax was a major topic, renewable energy was less so. Nonetheless, opposition politicians continued to call for an end to the EEG; other matters simply overshadowed that debate.

---

[65] Personal Communication with Sven Teske.

The year 2002 was the year in which the United States marched into Afghanistan and talk turned to Iraq. After 9/11, Schröder declared Germany's "unrestricted solidarity" with Americans, and the Germans were part of the coalition that sent troops to Afghanistan. But neither Chancellor Schröder nor Foreign Minister Fischer played along with plans to invade Iraq. Foreign Minister Fischer famously switched to English at a meeting on these plans, telling US Defense Secretary Donald Rumsfeld that the evidence for a link between Iraq and Al Qaeda seemed tenuous: "You have to make the case. And to make the case in a democracy, you have to convince by yourself. And excuse me, I am not convinced…And I cannot go to the public and say, well, let's go to war because there are reasons and so on, and I don't believe in that."[66]

At a time when Europeans were taking to the streets to protest US militarism by the tens of thousands, Fischer's statement made him a bit of a hero—and the German government's stance brought back loads of admiration to a coalition that had fallen out of favor with the public. A "flood of the century" around Dresden in August 2002 also gave the Chancellor an opportunity to appear in the public eye as a caring father figure just weeks before the election, also highlighting the coalition's record of a more environmentally conscious government in contrast to the opposition.

It didn't hurt that the other candidate for chancellor, Bavarian Minister President Edmund Stoiber of the Christian Union, came across as a bit awkward. Schröder racked up popularity points walking around flooded East Germany in rubber galoshes and telling flood victims he felt their pain; he promised everyone fast help without red tape. Stoiber also tried to show voters he was a likable guy, but he lacked a national platform to strut around on. So he went to his hometown to kick a soccer ball for his fans—and promptly hit in the face a woman wearing glasses. She left with a bloody cut, and Stoiber stood around embarrassed. On TV debates with Schröder, Stoiber strangely seemed out of his element, and the entire nation could see it. His favorite word seemed to be "uh," and he addressed the TV moderator as "Frau Merkel."[67]

---

[66] The Associated Press. *German Foreign Minister Joschka Fischer makes impassioned plea for peace.* YouTube. Accessed February 6, 2016. https://www.youtube.com/watch?v=CpuN-yM1sZU.

[67] Fischer, Sebastian. "Gescheiterte Kanzlerkandidaten: Als Stoibers Ende seinen Anfang nahm." Der Spiegel Online. 29 April 2007. Accessed February 6, 2016. http://www.spiegel.de/politik/deutschland/gescheiterte-kanzlerkandidaten-als-stoibers-ende-seinen-anfang-nahm-a-479966.html.

Without these events, the SPD and the Greens would probably have lost the elections, and the likely replacement coalition of the CDU and FDP might have made good on their promise to get rid of the EEG altogether. It took a natural disaster (the flood in Dresden), the US invasion of Iraq, and a bumbling Bavarian to keep the Energiewende going.

The 2004 EEG

After the elections, the old coalition began preparing planned amendments to the EEG. Like most of Germany's energy and climate laws, the EEG was always expected to be adjusted over time. The Energiewende was not implemented with the stroke of a pen; it is an iterative process of policy changes that continues today.

A large round of changes to the EEG would be implemented in 2004, but in July 2003 a special aspect was added in advance: energy-industry was exempted from the renewable energy surcharge. Essentially, generators of renewable power sold their electricity to grid operators, who paid the feed-in tariff. Under the old Feed-in Act, some grid operators had to pay more than others depending on how much renewable electricity was generated within their grid area. The 2000 EEG changed that by creating a national pay-as-you-go system; now, feed-in tariff payments were spread nationally across all power consumers, so everyone paid the same surcharge, and the grid operators balance the account among themselves each year.

The surcharge was 0.25 cents per kilowatt-hour in 2001, when renewable energy made up 6.4 percent of electricity. By 2004, the surcharge had reached 0.54 cents, with renewable energy at 9.2 percent of power consumption.[68] It wasn't a lot, but energy-intensive industry started

---

[68] Schwarz, Adrian. "Die Förderung der Stromerzeugung aus erneuerbaren Energien in Deutschland." Wissenschaftlicher Dienst des Deutschen Bundestages. WD 5—3010—109/13. Accessed February 6, 2016. https://www.bundestag.de/blob/194982/4cb1e1b813a7b5997b16adc fcbe9b4af/die_f__rderung_der_stromerzeugung_aus_erneuerbaren_energien_in_deutschland-data.pdf.

complaining vociferously. So a stipulation was added on-the-fly in 2003 specifying that, under certain conditions,[69] power-hungry firms could be largely exempt from the surcharge. The objective was to protect German firms that faced international competition. Industry exemptions would take center stage ten years later in a heated debate, but the amendment drew little attention in 2003.

Specific goals for renewable energy were also added. Instead of the 2000 EEG's vague "considerable increase" in renewable electricity, the 2004 EEG stipulated that renewables were to make up "at least 12.5 percent" of electricity by 2010 and "at least 20 percent" by 2020. Many energy experts considered both those targets to be unrealistic.[70] In fact, Germany reached 17 percent renewable electricity by 2010, overshooting the target by almost a half. By 2014, the figure had risen to 27 percent, and the target for 2020 had been raised to 35 percent in 2010—nearly double the original target from 2004. Clearly, the apparently innocuous phrase "at least" was one of the most important parts of the legislation. Had it not been included, opponents of the EEG could have called for the law to be done away with once the targets had been reached prematurely.

As the EEG entered its fifth year, its further existence was called into question based on its success. The Christian Union strongly argued that the law was only needed for the start-up phase and should be done away with. Even Hermann Scheer had stated as much in the 1990s, writing in one of his books that, after five years of such policy support, "there would be a self-supporting trend on the private market, and an avalanche would be on the move."[71] In 2004, Scheer would eat these words as he called for an indefinite extension of feed-in tariffs and priority grid access for renewables.

---

[69] To be eligible for the exemption to the renewable energy surcharge, a firm had to consume at least 100 gigawatt-hours annually; its electricity costs had to make up more than 20 percent of gross added value; and the differential cost had to be enough to "have a considerable detrimental impact on the firm's competitiveness." Note that these requirements were immediately loosened in 2004, when only 10 gigawatt-hours needed to be consumed, and power costs only needed to make up 15 percent of gross added value.

[70] Fell, Hans-Josef. "Gute Argumente für 100% Erneuerbare Energien. 100% Ökostrom bis 2030 in Deutschland sind realistisch und machbar." Positionspapier. Berlin, 19. December 2008. Accessed February 10, 2016. http://www.hans-josef-fell.de/content/index.php/dokumente/beschluss-und-positionspapiere/178-positionspapier-100-ee/file.

[71] Scheer, Sonnen-Strategie, p. 215.

Here, we have another example of the Energiewende as a steep learning curve. Fortunately, German politics is tolerant of people changing their minds. As one German saying goes, "why should I be interested in the nonsense I said yesterday." In English, the saying is "you can't teach an old dog new tricks," but the Germans seem to understand that you can't stop an old dog from learning new tricks. German Chancellor Adenauer paved the way for this approach, stating in the 1950s, "No one can stop me from getting smarter every day."

By 2004, something also clearly had to change for solar. The ceilings on feed-in tariffs and the 100,000 Roofs Program were met far faster than anyone expected, so the EEG was amended in 2004 to do away with these artificial barriers for solar. Starting in 2004, photovoltaics could grow without any limits. And it did.

Feed-in tariffs for photovoltaics were designed from the beginning to fall automatically by 5 percent annually for new systems (but not for existing ones, which locked in at the given rate for 20 years starting in the year of the grid connection). The euro had also been introduced in 2002. As a result, the magical marketing effect of 99 pfennigs had become a somewhat unsexy 45.7 cents by 2004, including the annual 5 percent reduction. To compensate for the complete loss of low-interest loans, even higher rates were then offered for arrays attached to buildings and other existing structures in the built environment, such as noise barriers along train routes and highways (the reason being that these arrays took up no additional space). The result was rates reaching up to 62.4 cents. In combination with priority grid access, these new feed-in tariffs worked like a magnet, drawing the global market to cloudy Germany.

With solar rates so high, better feed-in tariffs for biomass—which also had yet to properly get going—could also be offered without drawing too much attention. The 2000 EEG paid a maximum of around 10 cents per kilowatt-hour, but the 2004 rates went up as high as 17.5 cents (for small systems running on feedstock with certified sustainability). It would take a while for these changes to produce numbers on the market, but by 2006 Germany had its first two "bioenergy villages": Jühnde and Mauenheim. Both these hamlets (Jühnde has a population of around 1000, Mauenheim closer to 500) managed to cover their entire demand for electricity and heat from renewables. The biomass used in Jühnde is a combination of liquid manure, wood waste, and corn (the whole plant is harvested before the fruit has finished growing). These two villages set an example for scores of others, but all these success stories brought about by the 2004 EEG took a few years to come to fruition. The biomass boom therefore actually started after the 2005 elections under Chancellor Merkel.

For a brief time, it may have seemed that all would be smooth cruising for the Energiewende. Granted, the Christian Union managed to clamp down on feed-in tariffs for onshore wind power, which would fall from a maximum of 8.7 cents in 2004 to 8.03 cents by 2008 in the 2004 EEG.[72] But the wind sector continued to build at nearly 2 gigawatts annually. The next federal elections were not expected until the fall of 2006. But then, in May 2005, Chancellor Schroder unexpectedly stepped in front of the TV cameras to announce early elections in the fall of 2005.

The Early Elections of 2005

In North Rhine-Westphalia, state elections had produced a stunning result. The CDU had managed to break the SPD's 39-year hold on the governor's seat by forming a coalition with the FDP. The loss of a state governorship also meant the loss of Red–Green votes in the *Bundesrat*, essentially the upper house of Parliament. With the CDU and FDP basically being able to veto most of his legislation in the *Bundesrat*, Schröder took the election results in North Rhine-Westphalia as a sign that the public no longer supported his coalition with the Greens. He called for elections a year early.

Energy was still not in the foreground. The main issue was Schröder's Agenda 2010, his coalition's strategy for a fundamental change to the German welfare system and labor market. The policy was Schröder's attempt to do many things that his predecessor, Chancellor Kohl, had lacked the courage to address. Without making substantial concessions to

---

[72] These numbers represent the maximum level paid, which fell to 5.5 cents (2004) and 5.07 cents (2008), depending on local wind conditions. The better the location, the lower the feed-in tariffs.

the political opposition, especially the Christian Union, Schröder would have not been able to pass Agenda 2010 through both chambers.

Welfare payments were reduced slightly, and people were moved from (higher) unemployment benefits more quickly into welfare. For the general public, these changes seemed harsh, much in the way President Bill Clinton's welfare reform and British Prime Minister Tony Blair's *Third Way* were perceived in some circles as a betrayal of center-left. But the goal of *Agenda 2010*—as with the eco-tax—was increasing German competitiveness internationally, especially on labor markets. And in that respect, it worked over the long term.

But in 2005, the Germans mainly felt the pain from Agenda 2010. With his popularity dwindling against the backdrop of unpopular reforms—and with no natural disaster or American foreign-policy folly to save him—Schröder lost the election. This time, Schröder faced not a bumbling Bavarian opponent, but the calm Angela Merkel—a woman who almost never made a faux pas, but whose taciturn demeanor led power-hungry machos to underestimate her.

The Christian Union got the largest share of the vote, though only one percentage point more than the SPD. Neither the current coalition nor the opposition had a majority. A grand coalition of the CDU and the SPD was the most obvious option.

Merkel, the Christian Union's candidate for the chancellery in 2005, could have vied for the candidacy in 2002, but she wisely let Stoiber crash and burn first, knowing that her time would come. In 2005, it was Schröder's turn to put his foot in his mouth. And he did so in a round of top German politicians on a TV talk show—informally called *the Round of Elephants*—watched by 13 million Germans as the first election results were presented.

Letting his giant ego get the best of him, Schröder claimed, "I am the only one who will be able to form a coalition." The two moderators struggled to explain to him that his party had come in second, not first. Schröder became petty, accusing the media of understating his party's success and overstating Merkel's. He boasted that his decision to call for early elections had been "complicated and courageous." The more he spoke, the less likable he became.

"Mister Chancellor, I have an intellectual problem," one moderator then asked. He then explained Schröder's options, which were none. The only option was a grand coalition with Angela Merkel as Chancellor because her party had received more votes. "You don't seriously believe that my party is going to talk with Frau Merkel about such a coalition if she says she wants to be Chancellor? Let's not get carried away," Schröder stated. "She will not manage to get my party to form a grand coalition with her as Chancellor," Schröder added, turning his gaze from the moderators to Merkel on the other side of the round table.

The entire time, Merkel said nothing, merely smiling conspicuously at times as Schröder dug his own grave politically. (He later agreed with his wife that his performance had been "suboptimal.") When the head of the FDP, Guido Westerwelle, who was also at the TV round table, laid into Schröder, Merkel again sat back and watched with glee. "I don't know what you did before the show," Westerwelle said suggestively, "but the democratic tradition in this country is that the party with the most votes invites others to coalition negotiations. And that is Angela Merkel's party." After 45 televised minutes of refusing to accept he was already no longer the Chancellor, a moderator put an end to the shenanigans when he addressed Schröder one last time: "Mister Schröder—and I am calling you Mister and not Chancellor, because I don't think this kind of talk has any place on public television." At the beginning of the show, everyone still addressed him as *Bundeskanzler Schröder*. By the end, everyone simply called him *Herr Schröder*—everyone, that is, except good ole bumbling Stoiber, who called him "Bundes-Bundes, uh, Herr Bundesschröder."[73]

Overall, during the Red–Green coalition the Big Four seemed to have fared quite well. They had negotiated a nuclear phaseout which gave them long-term economic certainty. They had navigated liberalization, dividing up the market among themselves to produce an even smaller oligopoly than before. While politicians assured consumers that greater competition would bring down power prices, rates had actually risen to a new high. German consumer advocate Aribert Peters summed up the

---

[73] Kohlmann, Sebastian. ""Herr, äh, Bundesschröder…"." Der Spiegel Online. 10 July 2009. Accessed February 6, 2016. http://www.spiegel.de/einestages/elefantenrunde-2005-a-949832. html.

general sentiment quite well in August 2005, a month before federal elections: "That is not money they earned fair and square," he complained about the record profits posted by the biggest utilities. "These firms stole the money from customers by abusing their monopoly power."

Red–Green had come to an end, but it had managed to consolidate several Energiewende policies in its second term. Most importantly, the Energiewende could have taken a much different turn had the 2002 elections not been affected by a natural disaster. These seven years of policy consolidation would prove very important for the renewables sector, which was increasingly developing the political clout it would need under less favorable coalitions.

10

Healthy Democracy: Key to the Energiewende's Success

Why did the majority of German politicians support such laws as the nuclear phaseout, the eco-tax, and the EEG, which conventional utilities with lots of political clout opposed? And why isn't the baby thrown out with the bathwater after every election—what accounts for the continuity and consensus in German politics?

The answers to these questions are manifold and go beyond conventional explanations. The German public supports renewables, but not more than folks in other countries. German businesspeople and academics have strong technological and engineering capacities, but not more than their colleagues in the UK and the United States. Geographic factors don't explain Germany's success with renewables either; the country has the solar conditions of Alaska and some of the worst wind conditions in Europe. The UK is the Saudi Arabia of wind power in Europe, and the US is, well, the Saudi Arabia of wind, solar, biomass, geothermal, and ocean energy worldwide.

Nonetheless, Germany forged ahead with renewables in the hands of new market players and citizen cooperatives, pushing aside powerful energy corporations in the process. If geography and German engineering don't explain that outcome, does the country's political culture?

© The Editor(s) (if applicable) and The Author(s) 2016
C. Morris, A. Jungjohann, *Energy Democracy*,
DOI 10.1007/978-3-319-31891-2_10

British researchers Andy Stirling and Philip Johnstone recently analyzed the differing nuclear policies of Germany and the UK.[1] In their view, the countries' approaches toward nuclear power present perhaps the clearest divergence in developed world energy strategies. While Germany is seeking to entirely phase out nuclear power by 2022, the UK has for years advocated a "nuclear renaissance," promoting the most ambitious new nuclear construction program in Western Europe alongside France's. Yet, as the researchers remind us, Germany recently had one of the most successful nuclear engineering industries in the world—and a higher share of nuclear power than the UK ever did.

Patterns of public opinion have long been pretty similar in the two countries as well. "All those criteria conventionally emphasized in mainstream theory predict the opposite" of what actually happened, argue Stirling and Johnstone. So why is the UK building new reactors while Germany shuts its down?

To answer this question, the researchers looked at 30 different parameters: general market conditions, nuclear contributions to electricity mixes, strengths in nuclear engineering, costs and potential of renewables, strengths in renewable industries, scales of military nuclear interest, general political characteristics, public opinion and social movements, and contrasts in overall qualities of democracy. Most criteria suggest that the UK, rather than Germany, should be more likely to steer away from nuclear power. In fact, only two criteria clearly point in the opposite direction: the strong military nuclear interests in the UK and the "unanimous verdict" in the political science literature that Germany ranks markedly higher than the UK in terms of key "qualities of democracy."[2] It is worth noting that German nuclear experts Joachim Radkau and Lothar Hahn come to a similar conclusion in their history of the nuclear sector in

---

[1] Johnstone, Phil, and Andy Stirling. "Comparing Nuclear Power Trajectories in Germany And the UK: From 'Regimes' to 'Democracies' in Sociotechnical Transitions and Discontinuities." SPRU. Working Paper Series. SWPS 2015-18. Accessed February 8, 2016. https://www.sussex.ac.uk/webteam/gateway/file.php?name=2015-18-swps-johnston-stirling.pdf&site=25.

[2] Johnstone, Philip, and Andy Stirling. "Why Germany is dumping nuclear power – and Britain isn't." The Conversation. 8 September 2015. Accessed February 8, 2016. http://theconversation.com/why-germany-is-dumping-nuclear-power-and-britain-isnt-46359.

Germany.[3] They mainly argue that German democracy forced the nuclear sector to be open rather than secretive—and that the nuclear sector could not survive this transparency.

But how do you measure the quality of democracy? Scientists from the University of Bern and Switzerland's Center for Democracy have come up with a comprehensive toolkit for that purpose. In their Democracy Barometer,[4] they look at individual liberties, the rule of law, the quality of the legal system, freedom of speech and association, competition in and transparency of the political system, checks of powers, the capability of governments, citizen representation, and their participation in the political process. The result is a democratic quality index ranging from 0 to 100. Since 1990, the mean of democratic quality in the countries viewed has been roughly flat at around 55, right where the UK stands. Germany scores above average (60 to 64 over the last 25 years, with a slight upward trend). The US is significantly lower, between 52 and 61. Especially worrying is the steady decline for the US from 62 in 1999 to only 52 in 2011 (the last available data).

The International Crisis of Democracy

There's nothing more American than complaining about the US political system, and the present authors have nothing to add to that discussion. However, Germany's energy transition took place within the German political system, and that has made a difference. Those interested in energy democracy will want to know what those differences are. This chapter highlights the characteristics of Germany's political system by illustrating key differences with the US, largely because the authors only know these two countries well enough. Readers from other countries will hopefully find the comparison illustrative for their own situation.

An analysis that attributes policy outcomes to differences in national character is doomed to fall short. Rather, the United States and Germany

---

[3] Radkau, Joachim, and Lothar Hahn. *Aufstieg und Fall der deutschen Atomwirtschaft*. München: oekom Ges. für ökologische Kommunikation, 2013.

[4] "Democracy Barometer. Country Profiles." Accessed February 8, 2016. http://democracybarometer.org/countryprofiles_en.html.

have political systems shaped and developed over decades that allow certain conditions to prevail—and those systems continue to evolve. The good news is that, once identified, drawbacks of political systems in democracies can be remedied. One of the more disappointing things for the present author Craig Morris, an American, is seeing how often young Americans seem to have given up on their own democracy. Several times a year, he gives energy tours to groups of US students, and invariably—after learning all the great things Germany has done—a student comments, "yeah, but that won't work in the US."[5]

Of course it will—not exactly the same, but close enough. America has abolished slavery, given women the right to vote, banned alcohol, and reinstituted it. Just because improving your government is hard doesn't mean you don't have to try. The struggle for change is at the heart of democracy, which is why the present authors lose heart when they hear people young and old call the attempt to improve their own country futile. The US political system has its problems, but it is not undemocratic; giving up is. Democracy is struggle.

After World War II, Germany had a rare opportunity to redesign its democracy from scratch. The Germans not only had the much older democracies of the US, the UK, France, Canada, and elsewhere to learn from, but also its own experience from the short-lived and ill-fated Weimar Republic, which fell prey to a populist demagogue in 1933. In the wake of World War II, West Germany hit reset—under the watchful eye of the US, the UK, France, and Canada. Under such conditions, it would almost be surprising if German democracy were not an improvement on the past. Present-day German democracy also aims to prevent the Weimar failure by promoting civil discourse, sidelining antidemocratic populists, and protecting the rights of minority opinions. The winner does not take all.

Proponents of renewables in the US tend to demonize their opponents—such as the Koch Brothers—instead of analyzing the political structures that allow them to flourish. In reality, as US blogger Heather Smith put it, their success "in politics is less about their dia-

---

[5] Morris, Craig. "The laws of nature are immutable, not the laws of man." Energy Transition. The German Energiewende. 18 July 2013. Accessed February 8, 2016. http://energytransition. de/2013/07/the-laws-of-nature-are-immutable-not-the-laws-of-man/.

bolical cleverness and more about the system."[6] Germans are not better people than anyone else. In fact, lots of German businesspeople would probably love to be as influential as the Koch Brothers. German politics is designed to prevent that from happening. It therefore makes sense to compare political systems.

Some political scientists speak of the dawning age of post-democracy, in which moneyed interests govern and politicians attempt to increase prosperity, even at the expense of personal liberty. Western democracies are located at various points on a spectrum ranging from a technocratic extreme to open democracy. The former assumes that experts can determine objectively what's best for everyone, whereas the latter rejects the notion of an objective optimum for everyone, arguing instead that subjective conflicts of interest are natural—and best fought out transparently and fairly to produce compromise.

The German Lignite Association expressed the technocratic sentiment quite well in 2014:[7] "Repeated calls for the public to be included in far-reaching energy policy decisions need to take into account that the German public only has marginal expertise, if any, needed to make such decisions." Such voices have themselves become marginal in Germany, however, where drafts of energy legislation are put online for public comment. In contrast, the US public is increasingly being left out of the debate. To take one example, Florida's Public Service Commission closed hearings on power rates to the public because the issue was "too technical for public participation."[8] Often, what goes for technical expertise is really a moneyed interest—firms with stranded assets they want to protect from the democratic process.

It wasn't always this way in the US. US policy researchers Christopher McGrory Klyza and David Sousa have analyzed the development of American environmental policy over the last few decades.

---

[6] Smith, Heather. "How the Koch brothers grew their tentacles." Grist. 29 January 2016. Accessed February 8, 2016. http://grist.org/politics/how-the-koch-brothers-grew-their-tentacles/.

[7] Bundesverband Braunkohle. "Perspektiven der Kohlenutzung in Deutschland – 2014: Meinungsvielfalt trotz Polarisierung." Accessed February 8, 2016. http://www.braunkohle.de/185-0-Studie-Perspektiven-der-Kohlenutzung-in-Deutschland-2014.html.

[8] Foster, Joanna M. "Florida Utilities Move To Slash Energy Conservation Programs." ClimateProgress. 21 July 2014. Accessed February 8, 2016. http://thinkprogress.org/climate/2014/07/21/3462294/florida-utilities-to-cut-energy-conservation-programs/.

In their award-winning book *American Environmental Policy*,[9] they point out that the enthusiasm and bipartisanship for the environment during the mid-1960s and 1970s led to the enactment of numerous statutes, including the Endangered Species Act, the National Environmental Policy Act, and the Clean Water Act. The PURPA of 1978 initiated the first substantial increase of renewable electricity generation in the US. Until the early 1980s, the United States was considered a leader in environmental policy worldwide.

However, after these initial successes came a phase of pushback, driven largely by low oil prices in the 1980s. There was also a shift in the position of US conservatives on conservation. Starting in 1981, President Ronald Reagan broke with his party's tradition of conservation by becoming outright hostile to environmental issues and renewables. For instance, Reagan worked to clip the wings of the fledgling Environmental Protection Agency (EPA), which his fellow Republican Richard Nixon had created when he was president just a decade earlier. The 1990s and 2000s then saw legislative gridlock on the national level, regardless of who controlled the White House or Congress.

Comparing renewable energy policymaking in the United States, Japan, and Germany, Katrin Jordan-Korte from the German Institute for International Affairs comes to a similar finding. She explains the gridlock in US politics with the complex set of rules of legislative procedures, which makes it easier to block legislation than pass any: "The political system of checks and balances is an additional impediment for robust policy for renewables."[10] Perhaps her perspective is a bit German, however. The Americans McGrory Klyza and Sousa make a forceful argument that legislative gridlock has channeled political action and policy experimentation along other pathways. The executive can, for instance, set stricter clean-air standards via the EPA instead of introducing a carbon

[9] Klyza, Christopher M., and David J. Sousa. *American environmental policy: Beyond gridlock.* Updated and expanded edition. American and comparative environmental policy. Cambridge, Massachusetts: The MIT Press, 2013.

[10] Jordan-Korte, Katrin. *Government Promotion of Renewable Energy Technologies: Policy Approaches and Market Development in Germany, the United States, and Japan.* Gabler research. Wiesbaden: Gabler Verlag / Springer Fachmedien Wiesbaden GmbH Wiesbaden, 2011. Freie Univ., Diss. --Berlin, 2010.

price, which would require support in Congress. Environmental groups like the Natural Resource Defense Council (NRDC) and most recently the Sierra Club have also successfully challenged the operation and construction of coal plants in court.

The states have also filled the federal gap. For instance, under Governor George Bush Jr., Texas became a leader in wind power before the country as a whole got moving. More and more states implemented various regulations (Renewable Portfolio Standards, net metering, etc.), leading to a patchwork of different policy mechanisms. The downside of this fragmentation is that it opened up numerous veto points; the US policy process became vulnerable to influential—and moneyed— interest groups.

But if there is any credence to the assessment above, then German policymaking must do two things better than the US does: keep money out of politics, and prevent legislative gridlock.

Keeping Money Out of Politics and Public Debate

In Germany, donations from wealthy individuals or big corporations play a rather small role in campaign financing. In 2015, individual donations for parties summed up to a tiny €1.6 million for *all* political parties together. In contrast, the largest single individual donor in the US contributed 93 million dollars in the 2012 elections.[11] In the United States, the average cost of winning *one* seat in the House of Representatives was US$1.6 million alone in the 2012 election and US$10.4 million in the US Senate.[12] Presidential elections in the US have long counted spending in the billions.

Public funding plays a much bigger role in Germany. Introduced in 1958, it is the major income for Germany's parties. The overall annual

---

[11] "Top Individual Contributors: All Federal Contributions." Accessed February 8, 2016. https://www.opensecrets.org/bigpicture/topindivs.php.

[12] Wheaton, Sarah. "How Much Does a House Seat Cost?" The Caucus. 9 July 2013. Accessed February 8, 2016. http://thecaucus.blogs.nytimes.com/2013/07/09/how-much-does-a-house-seat-cost/?_r=0.

amount of public funding allotted to the parties is currently €157 million per year. They may use these funds for staff and administrative expenses, political work, and campaigning, which differs from year to year. The year 2009 was a "super election year" with one European, one federal, and six state elections; that year, Germany's parties in parliament spend a historical high 242 million euros altogether on campaigning[13]—3 euros per citizen—compared with US$6000 million in the 2012 US presidential election[14]—around US$19 per American. Public funding for German political parties is distributed in proportion to the latest election results plus a partial matching for private donations. For each vote that a party receives, it is granted 0.70 euros in public funding. That figure is higher—0.85 euros—for the first 4 million votes, so that small parties are made stronger. Making the parties' income depend on actual votes cast on election day gives German politicians an interest in a high voter turnout. There are no incentives for voter suppression (Box 10.1).

Box 10.1 "Voter suppression" and "gerrymandering": two terms that don't exist in German

Elections in Germany are held on Sundays, when people have the most spare time. A photo ID is required, but there is no discussion in Germany about voter suppression; no party has accused another of making it hard for citizens to cast votes. Though in a long-term declining trend, Germany still has relatively high voter turnout: just over 70 percent in the last federal election of 2013. In the 2012 US presidential election, the turnout was 55 percent in a race considered a cliffhanger between Mitt Romney and Barack Obama. Likewise, the boundaries of German constituencies reflect political realities and are not adjusted to affect election results. This practice, called "political gerrymandering" in the US, is legal there.

Like political parties around the world, German parties constantly strive for higher funding and donations. But certain campaign rules

---

[13] Niedermayer, Oskar. "Wahlkampfausgaben der Parteien." Bundeszentrale für politische Bildung. 1 September 2015. Accessed February 8, 2016. http://m.bpb.de/politik/grundfragen/parteien-in-deutschland/140330/wahlkampfausgaben.

[14] Plickert, Philip. "Kaum noch Großspenden für die Parteien." F.A.Z. 29 December 2015. Accessed February 8, 2016. http://www.faz.net/aktuell/wirtschaft/cdu-csu-co-kaum-noch-grossspenden-fuer-die-parteien-13987823.html?GEPC=s3.

serve as antidote for the parties' appetite for fund-raising. German state and local laws limit campaign billboards to a few weeks before the election. Campaign advertising on radio and television is limited to a few ads per month preceding the election. By an agreement among the states, the political parties may not purchase any advertising time on radio or television, thereby limiting them to the few officially granted campaign ads. Overall, these restrictions keep political advertising to a digestible amount for voters.

As a consequence, German politicians spend far less time on fund-raising than their American counterparts do. Back in 2010, the *New Yorker's* George Packer wrote about how Congress members no longer even move their families to Washington because they spend so much time fund-raising at home.[15] Thanks to that fund-raising, US Congresspeople basically have a three-day work week in DC (Tuesday to Thursday), much of which is also fund-raising.[16] As longtime Iowa Democrat Tom Harkin put it, "Of any free time you have, I would say fifty per cent, maybe even more," is spent on fund-raising.[17] Members of Congress devote several hours a day to "dialing for dollars." The fund-raising not only limits the time politicians can spend on policymaking and forming relationships with each other, but also increases their incentive to please corporations and big money networks. German politicians spend practically no time fund-raising, freeing them up for actual policymaking.[18]

The need for fund-raising opens up the US political system to lobby groups. The fossil fuel industry spends a lot of money to influence US politics. There is no equivalent in Germany; while Germany has no powerful oil and gas sector, it does have a mighty coal industry—but its financial influence is limited under current rules. In contrast, the fossil

---

[15] Packer, George. "The Empty Chamber: Just how broke is the Senate?" The New Yorker. 9 August 2010 Issue. Accessed February 8, 2016. http://www.newyorker.com/magazine/2010/08/09/the-empty-chamber.

[16] Packer, George. "Washington Man: He transformed himself from public servant to rich lobbyist. When the financial crisis hit, he remembered who he was." The New Yorker. 29 October and 5 November 2012 Issue. Accessed February 8, 2016. http://www.newyorker.com/magazine/2012/10/29/washington-man.

[17] Packer, *The Empty Chamber*.

[18] "Campaign Finance: Germany." Library of Congress. Accessed February 8, 2016. http://www.loc.gov/law/help/campaign-finance/germany.php.

fuel industry spent three quarters of a billion dollars lobbying the 113th Congress (2013/2014) and influencing the election cycle in 2014.[19]

Oil, gas, and coal interests in the US are also reported to have funneled an unknown amount of money through so-called dark-money groups that are not required to disclose their donors, such as organizations in the Koch network.[20] Michael Klare, a professor of peace and world security studies at Hampshire College, points out that the ability of lawmakers to get reelected depends on their success in backing legislation for the extraction of oil, gas, and coal: "It doesn't take too much imagination to calculate the consequences of this conveyor belt of financial support, both for affected communities and for the climate."

Money in the Media

Although a lot of attention is paid to money in politics, money in media deserves closer scrutiny. The media is the fourth pillar of democracy, and there are some salient differences between the US and Germany here. Germany's Channel One and Channel Two (ARD and ZDF) are publicly funded TV stations with the largest viewership alongside one commercial channel (RTL), each of which has around 12 percent of viewership. The two public stations are largely ad-free and get their money from a public budget that also funds numerous local TV channels along with radio stations; the total budget was worth 8.3 billion euros in 2014. In contrast, the US Corporation for Public Broadcasting, which funds public television and radio, makes do with a budget of less than 0.5 billion dollars annually. Both PBS and NPR then go asking for additional funding. Public TV and radio stations in Germany do no such

[19] Gold, Matea. "Koch-backed network aims to spend nearly $1 billion in run-up to 2016." The Washington Post. 26 January 2015. Accessed February 8, 2016. https://www.washingtonpost.com/politics/koch-backed-network-aims-to-spend-nearly-1-billion-on-2016-elections/2015/01/26/77a44654-a513-11e4-a06b-9df2002b86a0_story.html?postshare=2531422306006736.

[20] Moser, Claire, and Matt Lee-Ashley. "The Fossil-Fuel Industry Spent Big to Set the Anti-Environment Agenda of the Next Congress." Center for American Progress. 22 December 2014. Accessed February 8, 2016. https://www.americanprogress.org/issues/green/news/2014/12/22/103667/the-fossil-fuel-industry-spent-big-to-set-the-anti-environment-agenda-of-the-next-congress/.

thing. Per capita, Germany spends some 60 times more than the US does to provide the public with news reports not contingent on commercial advertising or other funding from businesses.

Dependency on advertisers is not only limiting in terms of the topics covered (reporters are unlikely to even investigate the wrongdoings of corporations that advertise), but also makes ratings more important—in extreme cases, the goal is not to cover news items accurately, but to create a controversy and report on it. The US media sells audiences to advertisers. To gain viewers, US news shows tend to be confrontational, whereas interrupting others is still considered impolite on the German news talk shows. Americans complain about the proliferation of pundits in the media, while German has no word at all for "pundit"—there are only journalists and intellectuals in the German media. Proponents of renewables in the US are likely to complain about the likes of Bill O'Reilly, but Keith Olbermann and Rachel Maddow would also be considered pundits in Germany, not journalists.

In the US, charges of "false balance" in the media are common, though the situation may be improving[21]; reporters often search for a dissenting opinion after speaking with one expert. In the case of climate change, where some 97 percent of climate researchers support the theory of anthropogenic causes, equal time is then given to the 3 percent to dissent. The public then believes the debate is real, not something created by the media. This way, US media contributes to an atmosphere "where 'facts' are not real—you can find an expert anywhere to deny them," argues Norm Ornstein, resident scholar at the American Enterprise Institute for Public Policy Research.[22]

Around the world, views about climate change divide up along partisan lines. In most economically advanced nations, those on the political left are significantly more likely than those on the right to view climate change as a major threat. The Global Attitudes Survey by the

---

[21] "Boykoff's Who Speaks for the Climate? Making Sense of Media Reporting on Climate Change." Yale Climate Connection. 21 October 2011. Accessed February 8, 2016. http://www.yaleclimate-connections.org/2011/10/boykoff-who-speaks-for-the-climate-book/.

[22] Ornstein, Norm. "The Eight Causes of Trumpism." The Atlantic. 4 January 2016. Accessed February 8, 2016. http://www.theatlantic.com/politics/archive/2016/01/the-eight-causes-of-trumpism/422427/.

Pew Research Center from 2015[23] found this divide to be especially true in the US, where Republican voters are much less likely than Democrats to see climate change as a problem. Globally, 31 percent of Australia's conservative voters are very concerned that climate change will harm them personally. So are 39 percent of the UK's, 45 percent of Canadian and 51 percent of German conservative voters. Only 12 percent of US Republican voters see it that way. American conservatives have moved away from conservation; German conservatives have not.

A comparison of top US and German legislators is also revealing. To the authors' knowledge, no member of the German Bundestag denies anthropogenic climate change. If one even takes part in a roundtable discussion that includes controversial figures in the climate change debate—like Marie-Luise Dött, a member of the Christian Union, did in 2010 when she spoke with climate change skeptic Fred Singer—one comes under fire from party leadership.[24] There is no debate among German lawmakers about the reality of climate change. The overwhelming consensus among politicians from left to right is that climate change is real and caused largely by humans—and that Germany has a responsibility to act. The debate in Germany revolves around what specifically should be done; critics merely argue that Germany should not do more than its fair share.

In contrast, over 56 percent of Republicans in the 114th Congress deny or question the science behind human-caused climate change. Some of those politicians hold influential positions in US Congress, even chairing science committees. The connection to the fossil fuel industry is salient; the 38 climate deniers in the Senate have taken 27.8 million dollars in donations from the coal, oil, and gas industries. (In contrast, the 62 Senators who haven't denied the science have taken "only" 11.3 million in career contributions.)[25] Some of the climate deniers, such as Ted Cruz and Marco Rubio, even run for the highest office.

[23] Stokes, Bruce, Richard Wike, and Jill Carle. "Global Concern about Climate Change, Broad Support for Limiting Emissions." Pew Research Center. 5 November 2015. Accessed February 8, 2016. http://www.pewglobal.org/2015/11/05/global-concern-about-climate-change-broad-support-for-limiting-emissions/.

[24] Böck, Hanno. "Die Klimaskeptikerfraktion der CDU." Klimaretter. 15 May 2012. Accessed February 8, 2016. http://www.klimaretter.info/politik/hintergrund/11149-die-klimaskeptikerfraktion-der-cdu.

[25] Germain, Tiffany, Kristen Ellingboe, and Kiley Kroh. "The Anti-Science Climate Denier Caucus: 114th Congress Edition." ClimateProgress. 8 January 2015. Accessed February 8, 2016. http://thinkprogress.org/climate/2015/01/08/3608427/climate-denier-caucus-114th-congress/.

Americans have long debated whether their media are too liberal or too conservative, and numerous watchdog groups have cropped up on both sides of the political aisle: Media Matters, FAIR, Accuracy in Media, and so on. There is no such debate in Germany; civil society has not perceived the need to establish its own non-governmental media watchdogs representing particular political camps—indeed, there is no equivalent of the term "media watchdog" in German. Instead, governmental officials monitor what happens with all the money spent on public media; high viewer ratings are one metric, but quality content also counts.

Granted, disenfranchised voices in the German political debate coined the term *Lügenpresse* ("lying press") in 2014. (German linguists fought back by giving the term the award *Unwort des Jahres*—literally, the Un-word of the Year.) The word is often chanted at anti-immigrant demonstrations (very much resembling events held by Donald Trump), the charge being that the mainstream media spin stories and don't allow Germans to speak their minds on certain things. The latter is certainly true. Holocaust denial, for instance, is a crime, and one TV newscaster (Eva Herman) lost her job in 2007 merely for saying that not everything during the Third Reich was bad (she specifically meant family values). The Germans had a taste of demagoguery in its purest form a century ago and remain concerned about the slippery slope back in that direction. Radical voices are marginalized in Germany to an extent that might surprise Americans. A politician like Donald Trump would face criminal charges in Germany for numerous statements; the leaders of demonstrations where *Lügenpresse* is chanted often do. Public figures like Trump are just as attractive to parts of the German public as they are in America, but German political culture—including the media—marginalizes such populists, whereas the US political and media culture gives them a bully pulpit—another word that does not have an equivalent in German.

In fact, Germany's proportional voting system and the strong role of parties as gatekeepers make it basically impossible for the super-wealthy—like Ross Perot, Donald Trump, and Michael Bloomberg—to enter the political scene in Germany in the first place. To become German Chancellor, you have to work your way up through the party (as in the UK).

With its direct presidential elections (via the Electoral College), the US political system can be praised for being more open than the German one. The German system tries to filter popular will to separate out anti-democratic forces. The US does not and is freer for it—but it remains more open to populists. So is the US media; German knows no word for "hate radio" because there are no such radio stations in Germany. The Germans had enough of it in the 1930s and 1940s.

The Energiewende has greatly benefited from the respectful debate culture within Germany. With the exception of tabloids and *Der Spiegel* (which the Germans call the Tabloid for Intellectuals),[26] most of the German media focus not on creating controversy, but on getting the facts right. Established energy experts are forced to take the general public and pro-renewable experts seriously, lest they be considered disrespectful. Without respect, it is hard to reach a consensus. And without a consensus, there can be no long-term energy policy.

The German political system is itself structured to promote consensus-building.

"Ministers Come and Go, But the Administration Stays"

In democracies, you don't want elected officials to become too powerful, so various systems of "checks and balances" have been found. But if these checks and balances are too strong, legislative change becomes too hard to push through. The German system focuses on long-term consensus across party lines, political levels (federal and state), and geographical regions, giving a large number of political actors a way of tweaking legislation to their taste without blocking it outright.

For instance, the German constitution—the Basic Law—is relatively easy to amend compared with the US Constitution. By 2009, the American constitution had been amended only 27 times during the 222

---

[26] Morris, Craig. "A guide for international Spiegel readers. Selling print by scaring readers." Renewables International. The Magazine. 17 September 2013. Accessed February 8, 2016. http://www.renewablesinternational.net/selling-print-by-scaring-readers/150/537/72925/.

years of its existence, compared with 55 changes in 61 years of Germany's Basic Law. The hurdles for constitutional amendments in the United States are much higher, requiring three quarters of votes in both houses of Congress as well as ratification in both houses of three quarters of US states. Perhaps not surprisingly, Americans tend to see their constitution as a "sacred text"—something that should not be touched; in practice, it rarely can be. In comparing the German and the US Constitution, US Law Professor Donald Kommers of Notre Dame points out that amendments to the US Constitution are tacked onto the end, with the original wording remaining in place unchanged—whereas amendments to Germany's Basic Law replace the old wording, which is not considered sacred. So whereas the Germans see their constitution in practical terms and continually tweak it, Americans consider their Constitution sacred (perhaps to compensate for the difficulty in changing it).[27]

The US Constitution has been amended in spurts. From 1792 to 1804, three amendments were passed, followed by none until the 1860s. Then, three swiftly followed in the wake of the Civil War. After that, we also find amendments passed in batches: two in 1913, two in 1933, and three from 1961 to 1971 pertaining to the civil rights movement. These clusters are indications that the US Constitution is easier to amend in certain windows of opportunity. The Energiewende has also lurched forward when specific windows of opportunity opened: the Reunification of Germany in 1991, the country's first completely new coalition in 1998, and a nuclear meltdown half a world away in 2011. In-between those events, the Energiewende movement successfully fended off pushback, whereas the US environmental movement suffered real setbacks starting in the Reagan years after significant previous progress.

Though Germany does not have a powerful presidential office like the US, the German executive nonetheless has strong capabilities to govern. The government can draft, shape, and implement policies. Coalition parties that form a government take top executive positions and are backed by a legislative majority in the *Bundestag* (the lower

---

[27] Bahners, Patrick. "What distinguishes Germany's Basic Law from the United States Constitution?" English translation of article originally published in F.A.Z. on 18 May 2009. Accessed February 8, 2016. http://law.nd.edu/news/13557-human-dignity-and-freedom-rights/.

house of parliament). In most cases, the government drafts legislation and sends bills to parliament for a vote, where it already has a majority. Furthermore, when the government submits a bill to parliament, the content reflects the political agreement from the coalition treaty, a legislative agenda that the parties negotiate at the beginning of their tenure. "For any bill the government sends to parliament, the rear cover of the coalition parliamentarians needs to be ensured," says Kai Schlegelmilch, a long-term official in the Ministry of the Environment. If the government fails to do so, the parliament groups call for fundamental changes, which delay the legislative process and open it up for lobbyists.

The situation is similar to the United States having a president with a majority in both houses of Congress from the same party. The US system has checks and balances by design—the president often faces an opposition majority on the Hill—whereas the German executive and legislative bodies are closely intertwined, as they are in the UK. In both countries, a government knows it can get legislation passed quickly, provided the chancellor or prime minister keeps her party ducks in a row. From the American perspective, one would expect radical change after every election under such circumstances; after all, the purpose of US-style "checks and balances" is to promote political negotiation.

A comparison with the UK suggests how the Germans manage to focus on consensus-building even without checks and balances. Like Germany, the UK has a parliamentary system in which the prime minister (similar to the German chancellor) is elected from the majority in Parliament. But the British are only now getting their feet wet with coalitions; until recently, the prime minister's party itself generally had a majority in parliament, so no coalition partner was needed. In contrast, no German chancellor since the Federal Republic was founded in 1949 has formed a government without a coalition partner.

Because of the proportional voting system, German political parties need to partner with another for a parliamentary majority to form a coalition. They therefore regularly work with competitors, in rare cases even with opponents. The parties agree on compromises in coalition treaties, which serve as initial policy drafts and guidelines for 4 years, thereby providing planning security. Voters seem to like continuity. In the last 30 years, Germany had only one election that brought a complete change in government—in 1998, when two center-left opposition parties won the election. Otherwise, all new coalitions have always included one party from the previous coalition with a new coalition partner. The large German political parties therefore look at each other as potential coalition partners for the future.

Polarization between the two parties in the US has increased in recent decades—and there are only two, whereas Germany has generally had five in parliament since 1990. It is also easier for new parties to come about in Germany, such as the Greens in 1980 or, more recently, the Pirate Party (focusing on the digital age) and the populist AfD (*Alternative für Deutschland*). In contrast, it is so hard for a third party to come about in the US that the Tea Party movement is essentially forced to work within the Republican Party instead of becoming its own party.

The proportional voting system (as opposed to the US-style "winner takes all" approach) results in a more diverse parliament with big and small parties. "The spectrum of political thought and party allegiance in Germany is far wider than that found in the United States," explains Greg Nees, a US consultant advising American firms in doing business in Germany.[28] The Free Democrats, a libertarian pro-business party with a profile of low taxes and civil rights, served in the 1970s and 1980s as a coalition partner for both the Social Democrats and the Christian Union. A more recent example of this diversity in Germany's political landscape is the Green Party; the German chapter is considered the most established and successful among its green peers around the world. The German Greens emerged in the late 1970s and captured a policy field the other parties did not take seriously: environmental and nuclear policy. The early electoral success of the Greens brought competition to an essentially three-party system and forced traditional parties to address environmental issues, lest voters defect to the Greens. In general, newcomer political parties start off by covering issues that existing parties overlook (such as the internet for the Pirate Party today).

Party discipline is also strong in Germany; members of parliament are expected to vote in line with their party. In contrast, the US president often has to negotiate and win votes of Congressmen and Senators from

---

[28] Nees, Greg. *Germany: Unraveling an enigma.* Yarmouth, Me: Intercultural Press, 2000, p. 4.

his own party. Congress's interests often differ from the president's. In only two of his eight years in the White House, Barack Obama enjoyed a majority of Democrats in both the House of Representatives and the Senate. Still, he lacked full support from party colleagues. The two lasting achievements of Obama's first two years in office, financial regulation and health care, required what US journalist George Packer calls "a year and a half of legislative warfare that nearly destroyed the Senate."[29] The success depended on a set of extraordinary circumstances—a large majority of Democrats, a charismatic President with an electoral mandate, and a national crisis—that did not last long and cannot simply be conjured up again. Other pressing issues like immigration, veterans' care, campaign finance, climate change, and scores of executive and judicial appointments were never passed. In contrast, Angela Merkel has been Chancellor since 2005 with a comfortable parliamentary majority in the Bundestag. Her party's legislative ideas move forward—in coordination with her coalition partner and, to a lesser extent, with the political opposition.

Ministries and Continuity

On a gloomy November day in 1994 at a *trattoria* near the Rhine River in Bonn, Germany's then capitol, a young woman in a blazer—an up-and-comer Christian Democrat named Angela Merkel—was about to meet someone important for the first time. She had just been appointed Minister for Environment, Nature Protection, and Nuclear Safety by Chancellor Helmut Kohl—and she felt like she needed to figure out who she was dealing with. She leaned forward to speak confidentially: "Mr. Schafhausen, I hear you are a member of the Free Democrats." "That is not the case, Ms. Merkel," he replied, looking up from his plate of steaming pasta. Merkel wasn't satisfied: "But people say you are with the FDP," she continued, "in which party are you then?" Schafhausen, the ministry's long-term head of the division for energy and environment, explained he wasn't a member of any political party. "For the kind of work I do, it is

---

[29] Packer, *The Empty Chamber.*

not necessary. I serve my ministers, no matter which party they belong to." Merkel smiled. "That is fine, Mr. Schafhausen. This way we can work together." In Germany, the political affiliation of government bureaucrats does matter. On the other hand, so does their loyalty toward the political leadership in government.

In Germany's political system, ministries enjoy considerable autonomy. In consultation with other ministries, one ministry drafts a bill, which the cabinet first reviews before sending it to parliament. The parliament then debates the bill, makes amendments and often adopts it, generally without major changes. For energy matters, the Ministry of Economics was in charge for a long time. Historically, the Ministry took policy positions close to the ones utilities favored. So for a long time, the Ministry was an obstacle to a progressive renewable energy agenda. To bolster their agenda after their reelection in 2002, the Green Party insisted that the responsibility for renewable energy be moved from the Ministry of Economics to the Ministry of the Environment. Renewables then had their own ministry in the government, which increased its standing at the cabinet table. (After the 2013 elections, renewables were moved back to the Ministry of Economics.)

Civil servants at ministries play a strong role in Germany's legislative process. Some top bureaucrats have been shaping Energiewende legislation for more than two decades. The aforementioned Franzjosef Schafhausen is one example. He has been with the Federal Ministry for the Environment, Nature Conservation, and Nuclear Safety since its beginnings in 1986. In his position today as the Deputy Director-General for the Environment and Energy, Schafhausen is the government coordinator of German domestic climate policies. He has been a civil servant from the beginning of the Energiewende and has worked for nine different Ministers of Environment, no matter if they were with the Conservative Party, the Social Democrats, or the Greens. "Ministers come and go, the administration stays," is no exaggeration for him. Staff retention at ministries after elections is another indication of strong continuity in German policymaking; expertise is respected. Staff continuity translates into policy continuity. In contrast, a change in the White House often brings fundamental change of senior staff between administrations, making it much harder for long-term initiatives to survive.

Of course, incoming German ministers will always have personal staff joining them, often in key positions (such as strategic planning and press relations). Civil servants have to demonstrate expertise and loyalty, not partisanship, to remain influential under new leadership. Incumbent senior career staffers are, on the other hand, important assets for incoming ministers. Over time, they have built up knowledge and experience about institutions and decision-making processes. Ministry staffers make use of their wide network of contacts for their new boss. It is not uncommon for incoming ministers from one party to retain or appoint civil servants affiliated with other parties. The most recent example was Energy Minister Sigmar Gabriel (Social Democrats) appointing Rainer Baake (Greens) as his State Secretary.

Of course, policies do sometimes change fundamentally in Germany after elections. That's often what voters want in a few areas—but not everywhere. When a course has been set and proves productive, a new government might change policy details, but the overall direction remains the same. Of course, incoming ministers want to set new priorities, but changes are often first negotiated with the same senior ministry staff that drafted the previous legislation that is now to be tweaked. Apart from last-minute items, legislation usually slows down in an election year in Germany as civil servants prepare for the time after the election and draft a working agenda for the next government. Often these agendas find their way into coalition negotiations after the election. "Often, top politicians from parties in coalition negotiations even request the agenda from us, further producing continuity," says Schlegelmilch.

In addition, Germany is a member of the European Union. As such, it is embedded in a wider policy framework with obligations and legal requirements—such as implementing a certain policy or target within a certain time frame. Take the EU's emissions trading system: in the late 1990s, regulations to implement a cap-and-trade system across Europe were discussed. After years of negotiations, the European Commission, the European Parliament, and the EU member states agreed on the final text in 2003. Over the next two years, all member states including Germany were asked to implement the directive into national law so the trading could start in 2005. In addition, the EU directive required member states to monitor and report on emissions in the future. In the years after, more EU directives and ordinances were passed down to the member states.

Being a member of the EU thus makes Germany an integrated part of a wider framework with political commitments and responsibilities across different governments, no matter which party is ruling in Berlin. No incoming government wishes to risk Germany's reputation within the EU by radically changing policy without coordination with other member states. Indeed, when Chancellor Merkel closed nearly half of the country's nuclear fleet in 2011, neighboring countries criticized her for not coordinating the plan with them—coordination has become expected. Since then, her government has focused more on communicating the Energiewende to the EU.

Cooperative Federalism

In addition to federal coalitions, a wide variety of party coalitions exist at the level of Germany's 16 states or *Länder*, three of which are city states (Bremen, Hamburg, and Berlin). Some of these coalitions are considered test labs for possible future coalitions at the federal level (such as the Christian Union and the Greens in Hessen). Germany's federalism fosters cooperation and drives consensus-building at regular conferences of the 16 state Minister-Presidents with the Chancellor. Furthermore, policy-specific working groups are held between the 16 state ministers of, say, industry with the federal industry minister.

At these *Fachministerkonferenzen*, the federal government takes account of the views and concerns of the 16 states. Over time, personal relationships and trust are developed among top decision-makers here as well. When successful, this approach saves time and reduces opposition in the two chambers of parliament. "Germany's federalism is just more coordinated and geared towards compromise" says Dale Medearis, a former EPA employee. Medearis also spent six months at the German Federal Ministry of the Environment in 1995, where he studied urban environmental development policies in both countries. Medearis observed that the US has fewer formal processes to coordinate environmental policies between the federal government and the states than in Germany. Especially in the realm of energy policy, the 50 US states operate in a universe characterized by far less accountability than their German counterparts. This situation has slowed down considerably the US's ability to implement meaningful climate and energy policy compared with Germany's.

The federal and state levels interact most publicly between the *Bundestag* and *Bundesrat*. The federal government often has to negotiate with state governments in the *Bundesrat*, Germany's second chamber. Like the US Senate, it represents the interests of the 16 *Länder* at the national level. The *Bundesrat* is also like the US Senate, in that both clearly have strong veto rights (the *Bundesrat* less so than the US Senate, however). But unlike state senators in the US, each state delegation in the *Bundesrat* directly represents a state government and reflects the political makeup of the ruling majority of each state legislature. In fact, the head of each German state is its Minister-President, a powerful hybrid between the head of the state executive (like a governor in the US) and a US Senator. German Minister-Presidents govern their states but also represent them in the *Bundesrat*.

The federal government needs to incorporate anticipated concerns of state governments and therefore proactively seeks support among the *Länder*. The political leaders of these state governments are most often leaders of their parties, so a lot of these compromises reached between regions are sometimes simultaneously made across party lines. But such negotiations also sometimes reveal intraparty conflict between the state and federal level. When federal policymakers know that an idea has little support not only in the political opposition, but also within their own party regionally, they can try to adapt legislative proposals with an eye to long-term implementation. For the Energiewende, this ability to implement gradual change and prevent legislative rollbacks has been crucial.

In contrast, such US political experts as Francis Fukuyama have spoken of a "vetocracy" in America, whereas vetoes (and the mere threat of the "nuclear option") have led to a legislative stalemate. Given the composition of the Senate, less populated rural states—where fossil fuel interests are often based—enjoy disproportionate power. In filibusters (another word that does not exist in German), a single senator can delay or even prevent a vote on a bill unless 60 out of 100 senators oppose. The incentive to compromise is small; the incentive for individuals to grandstand, large. There is also less party discipline than in Germany, as shown by the numerous politicians across the country who refuse to associate themselves with the incumbent president from their party during election campaigns. For US politicians, the financial support of donors, often lobbyists and corporations, is more important than the support of their party.

German parties are tightly run organizations that finance election campaigns, nominate candidates, exact membership fees from their members, and subject members in parliament to strict caucus rules. Individual politicians play a minor role in this system. The parties receive government funds and are subject to modest disclosure requirements. In most parties, platforms are a matter of intense negotiation and discussion before any election. Out of the tradition of coalition governments, they become the foundation of the coalition treaty. It is thus not hard for voters to predict what they will roughly get from a particular coalition. The strong role of parties over individual politicians means that politicians usually stick with their party's position. For lobbyists, it simply does not pay to target and "buy" individual parliamentarians if they will vote along party lines anyway.

In this comparison of the US and German political systems in particular, we run the risk of making Germany look like a model democracy. Of course, the German government is open for industry lobbyism; just think of how it rejected stricter EU pollution limits for cars on behalf of German car companies. Naturally, the Germans themselves have plenty of bones to pick about their politicians and their political system, and political scientists also talk about the "crisis of democracy" in Germany. They do not, however, generally speak of a German political gridlock, nor do the Germans generally speak of their political system as dysfunctional.

Furthermore, nothing about the German character guarantees democracy any more than the American character predisposes the US political system to whatever condition it is currently in. Rather, German democracy continually has to be defended. After several decades of success in marginalizing neo-Nazi groups, for instance, German democracy now faces a threat from a new political party called the AfD. In the midst of the current upsurge in migration and in the wake of the recent financial crisis, this party reached 13 percent of the polls in January 2016. That month, the head of the party suggested that refugees entering Germany from war-torn regions should be shot at "as a last resort." Once again, German society will have to decide whether such opinions are anti-democratic—and then stand up and fight to keep such voices marginalized. Nothing about the German political system rules out the success of the AfD without a struggle. But the German political system does provide a platform where the majority can decide whether a minority's views should be taken into account or fended off as uncivil and anti-democratic.

This chapter hopefully reveals several starting points for political scientists to pursue further—and lay readers to consider. In Germany, decisions are based on negotiations and compromise, not strict majority rule; there is no "tyranny of the majority" in Germany. The focus is not on exerting political power alone, but on formulating policies that can garner so much support across party lines that legislation survives elections with relatively minor tweaks. The German political system is designed to produce consensus, but it would all come to naught if politicians, the media, and the public did not demand respectful discourse. Finally, wherever Germany is located on the spectrum between technocracy and democracy, it seems to be moving away from the former. Scholars in the post-democracy field should find the Energiewende to be a case study worth investigating.

11

Utilities Bet on Gas and Coal and Renewables Boom (2005–2011)

After Gerhard Schröder's "suboptimal" alpha-male moment on television, the Social Democrats followed standard procedure and sat down with the Christian Union to negotiate a governing coalition—with Angela Merkel as Chancellor. By the time the next regular parliamentary elections are held in 2017, she will have been in office for 12 years. If reelected, she could remain in office longer than any other German Chancellor before her (the current record holder is Helmut Kohl at 16 years).

By all appearances demure in the beginning, Merkel has evolved into the country's most popular politician. In 2015, she still enjoyed staying out of daily political bickering, so much so that a new verb—*merkeln*—has been coined to mean "not saying anything when people expect you to say something." But when she speaks, it is sometimes what the Germans call a *Machtwort* (literally, a word of power). Merkel generally lets the alpha males around her fight things out in public for weeks, waiting to see who keeps the upper hand—and then quietly hands down the coalition's direction like a decree. The buck stops with Chancellor Merkel.

Her position on refugees is the best recent example. In 2015, she put all her political clout behind a simple statement on open borders for refugees: "Wir schaffen das" (we can handle this). Her stance has brought her

© The Editor(s) (if applicable) and The Author(s) 2016
C. Morris, A. Jungjohann, *Energy Democracy*,
DOI 10.1007/978-3-319-31891-2_11

accolades internationally, including being named Person of the Year by Time Magazine in 2015. But her unwavering kindness—"I'm not going to take part in a competition to see who is the meanest to refugees," she stated in October 2015—has also cost her support within *Realpolitik* proponents in her own party and among nationalistic Germans—and may still prevent her from becoming the longest serving chancellor in history.

In her first chancellorship from 2005 to 2009, in contrast, she remained rather demure. In those early years, the alpha males still thought Merkel would do what they wanted.

In the fight over nuclear power, however, there simply was no consensus within the new governing coalition in 2005. Because the Social Democrats were signatories to the nuclear phaseout, they made it clear during the coalition negotiations that it would not be reneged.

Still, the former Environmental Minister quickly began sharpening her profile on climate issues. The German press even called her the Climate Chancellor after her efforts to get climate action on the agenda at the G8 meeting in 2007.[1] But the moniker did not stick; as Chancellor, she found it difficult to push the climate agenda consistently within her own country.[2] And as the Volkswagen scandal of 2015 shows, her ministers had always worked to protect German industry at the EU level, even at the expense of climate and the environment. Some international observers may take Chancellor Merkel to be a climate leader, but those familiar with the ongoing in Brussels do not. And the grassroots Energiewende movement in Germany is also repeatedly at odds with her policies.

Her stance on nuclear power starting in 2005 is a case in point. Merkel and her party not only remained committed to keeping existing nuclear plants in operation longer, but also contained members calling for new nuclear plants. Minister-President of Bavaria Stoiber called for a public debate on the issue "without ideology." The Minister-President of Hessen, Roland Koch, wanted the discussion to include building new

---

[1] Thalman, Ellen. "The making of "Climate Chancellor" Angela Merkel." Clean Energy Wire. 26 November 2015. Accessed February 7, 2016. https://www.cleanenergywire.org/factsheets/making-climate-chancellor-angela-merkel.

[2] Glazek, Christopher. "The World From Berlin: 'Climate Chancellor' No More." Der Spiegel Online. 9 December 2008. Accessed February 7, 2016. http://www.spiegel.de/international/the-world-from-berlin-climate-chancellor-no-more-a-595362.html.

nuclear plants, as did two other Minister-Presidents: Christian Wulff of Lower Saxony (who would later briefly become German President) and Günther Oettinger of Baden-Württemberg (who would later become the EU's Energy Commissioner).

Behind the scenes, numerous politicians and the nuclear industry itself therefore worked to ensure that no nuclear plants would need to be shut down before the next scheduled elections in 2009. They hoped that the Christian Union would then be able to form a center-right coalition with the libertarian FDP. And then, the phaseout could be rolled back. They banked on Merkel's support.

Delaying Nuclear Plant Shutdowns

The FDP and the Christian Union had gone into the 2005 elections promising to extend the service lives of nuclear power plants. When this coalition did not come about, share prices of German nuclear power operators temporarily fell. A financial analyst explained that the market had already priced in extended service lives for the plants. Now, that decision had been postponed until the next elections in 2009.[3]

The phaseout specified that each plant could produce a certain number of kilowatt-hours within a 32-year lifespan. The number of kilowatt-hours was therefore more important than the number of years. While the coalition agreement between the Christian Union and the SPD stated that the fundamentals of the phaseout would remain unchanged, the agreement left open the possibility that younger nuclear plants might be able to transfer some of their kilowatt-hours to some of the older plants. The new coalition now stated that this option would be considered, but the government had the final say. By 2009, four nuclear plants would otherwise have needed to be shut down: Biblis A and B, Neckarwestheim 1, and Brunsbüttel—assuming they continued to run properly. But even if they did, the firms could always try to find something that needed fixing so the plants could stay off-line.

---

[3] Leuschner, Udo. "Energiechronik: Wahlausgang drückt Aktienkurse von KKW-Betreibern." September 2005. Accessed February 7, 2016. http://www.energie-chronik.de/050903.htm.

The idea of transferring allotted kilowatt-hours from a newer plant to an older one was nothing new. In 2002, the then Chancellor Schröder had personally—and secretly—promised EnBW's CEO that he would allow 5.5 billion kilowatt-hours to be transferred from Phillipsburg 1 to Obrigheim.[4] The deal caused a public scandal at the time, partly because of speculation about what the government may have gotten in return. In the end, the transfer of kilowatt-hours was not even fully utilized; Obrigheim could have remained open until 2004, but it was closed in 2003 for economic reasons. The deal therefore mainly set a precedent.

In 2005, Sigmar Gabriel became Environmental Minister and was in charge of these nuclear decisions. At the time, he was nicknamed as "Siggi Pop"; in an effort to bring him to the national stage after he lost a state election in 2003, the Social Democrats created an office for Pop Culture, with Siggi at the helm. But the Social Democrats knew what they were doing; never a lightweight, Gabriel has become a formidable politician.

In 2006, RWE applied for a transfer of kilowatt-hours from the newer Biblis B to the older Biblis A. The full story is even more complicated: the firm was technically asking for 30 billion kilowatt-hours, 21 billion of which would have come from the Mülheim-Kärlich plant—which ran from 1987 to 1990.[5] The Nuclear Phaseout Act of 2002 explicitly ruled out the transfer of this allotment to Biblis A. Would the government make an exception under the new coalition agreement?

In September 2006, Gabriel rejected the application to transfer kilowatt-hour allotments from Biblis B to A. RWE responded with a second petition to transfer all the 30 billion kilowatt-hours from the Emsland reactor. In April 2008, Gabriel also rejected this proposal, specifically because Biblis A did not have as many safety reserves as Emsland.

While the applications were being reviewed, RWE conveniently found thousands of improperly installed dowels in both Biblis A and B. The two plants were therefore ramped down in October 2006, so that the dowels could be properly installed. The process was completed at the newer Biblis B in November 2007 (13 months later), leaving this reactor with

---

[4] Leuschner, Udo. "Energiechronik: Kernkraftwerk Obrigheim darf bis Ende 2004 betrieben werden." October 2002. Accessed February 7, 2016. http://www.energie-chronik.de/021002.htm.
[5] Leuschner, Udo. "Energiechronik: RWE beantragt Verlängerung der Laufzeit für Biblis A." September 2006. Accessed February 7, 2016. http://www.energie-chronik.de/060901.htm.

enough remaining kilowatt-hours to stay in operation until 2010—and hence, long enough for the 2009 elections.

But time was running out for Biblis A, where work was finished in February 2008 (17 months later). Even in previous years, RWE had kept Biblis A off-line for extended periods: 151 days in 2000 to 2001, 110 days in 2002, 107 days in 2003, and 94 days in 2005.[6] In February 2008, when Biblis A finally went back into operation with its new dowels, RWE announced five months of further revisions for 2009—ending in September, the month before the next scheduled parliamentary elections. With those final months of downtime, Biblis A would go into the next elections with enough power production left in its allotment to stay online.

EnBW pursued a slightly different tactic at Neckarwestheim 1. The company petitioned the government to have kilowatt-hours transferred from the newer Neckarwestheim 2 reactor within the same complex. In June 2008, Gabriel rejected this proposal as well—once again because the younger reactor had better safety reserves. EnBW responded by running the plant at lower capacity. By the time Gabriel reached his decision, it was already clear that the older reactor—which would otherwise have been shut down for good in mid-2009—would be able to keep running possibly into 2010.[7]

Finally, there was Vattenfall's Brunsbüttel reactor. The firm—a wholly owned subsidiary of the Swedish state—wanted to transfer kilowatt-hours from its Krümmel plant. This time, it was hard to tell which reactor performed worse. From June 2007, when Krümmel had to shut down immediately after a transformer station caught fire, the reactor repeatedly made headlines for minor incidents up until July 2009, when there was another emergency shutdown due to a transformer failure.

Brunsbüttel also went off-line in June 2007 after failing to ramp up properly after an emergency shutdown to accommodate for a grid failure. While making repairs, technicians also discovered that the wrong dowels had been used, requiring longer downtime. This reactor never went back

---

[6] "Dübel-Panne in Biblis könnte für RWE zum Segen werden." FAZ. 10 July 2008. Accessed February 7, 2016. http://www.faz.net/aktuell/rhein-main/wirtschaft/nach-erfolgter-nachruestung-duebel-panne-in-biblis-koennte-fuer-rwe-zum-segen-werden-1665699.html.

[7] Pross, Steffen. "GKN I bleibt länger am Netz." Ludwigsburger Kreiszeitung. 28 February 2008. Accessed February 7, 2016. http://www.lkz.de/home_artikel,-GKN-I-bleibt-laenger-am-Netz-_arid,7699.html.

online again, though it had kilowatt-hours left over from all the down-time. Gabriel's rejection of allotment transfers from Krümmel—once again based on safety concerns—in May 2009 thus remained moot.

The strategy to delay the shutdown of reactors was not without risks for the nuclear power firms. The public repeatedly heard of failed nuclear plants going off-line for repairs and maintenance, suggesting that the reactors were unreliable. Even worse for the firms, power reliability remained exception-ally high at the time, indicating that the failing reactors were not needed at all to provide enough electricity. Reactor operators were succeeding in their desperate attempt to keep their plants running, but the strategy only further convinced the public that nuclear plants could and should be closed.

Nuclear plant operators thus prevented the shutdown of additional nuclear plants before the 2009 election. During these years, the German debate took place within a largely pro-nuclear EU—but also in the midst of nuclear failures in Europe. In August 2006, an incident at the Forsmark 1 reactor in Sweden was initially reported as minor—until a Vattenfall employee told the press that a nuclear meltdown had only just barely been avoided. Though Vattenfall itself tried to discredit the employee, the head of Sweden's nuclear power authority spoke of "the worst event in the history of nuclear power" in the country. As with Brunsbüttel and Krümmel, Forsmark 1 had failed to react properly to an external disturbance on the grid.[8]

Meanwhile, France not only remained committed to its EPR reactor design, but also aimed to export nuclear technology, including to Libya. In July 2007, French President Nicholas Sarkozy signed an agreement with Moammar Gadhafi for the construction of a nuclear plant outside Tripoli. Hoping for similar deals elsewhere, the French founded an export agency for nuclear technology in May 2008: *France Nucléaire International.* The list of interested buyers was long: Algeria, Morocco, Egypt, the United Arab Emirates, the UK, and Italy. Russia also planned a new nuclear plant in the exclave of Kaliningrad, partly to sell electricity to the EU.[9]

At the time, the Germans must have felt surrounded by foreign gov-ernments in support of nuclear. Even Belgium, which had resolved

---

[8] Leuschner, Udo. "Energiechronik: Störfall im schwedischen Kernkraftwerk Forsmark 1 wurde zunächst unterschätzt." August 2006. Accessed February 7, 2016. http://www.energie-chronik. de/060807.htm.

[9] Leuschner, Udo. "Energiechronik: Italien will wieder Kernkraftwerke errichten." May 2008. Accessed February 7, 2016. http://www.energie-chronik.de/080508.htm.

to phase out nuclear by 2025 in 2003, saw a new governing coalition elected in August 2007 agree to do exactly what nuclear proponents in Germany were calling for. That month, the new Belgian coalition of Christian Democrats and Libertarians—the same pro-nuclear parties as in Germany—said nuclear plant commissions would be extended.[10]

But as so often in the history of nuclear power, little came of all these plans. Though the deal between Sarkozy and Gadhafi made news in 2007,[11] the idea passed out of public favor when a civil war broke out in 2011. After Gadhafi's death that fall, Sarkozy denied having further pursued the agreement.[12] What's more, France did not export any nuclear reactors to the countries listed above. Indeed, few of those countries went forward with nuclear plans at all, and none of them with French technology. As of 2015, Russia is building a new reactor in Kaliningrad, and one is also going up in the UAE—but the latter is being built by Koreans, not the French. None of the plants had been finished as of mid-2015. The fate of the Hinkley plant in the UK, which would come from French engineering firm Areva, also remained up in the air as of 2015.

Finally, though the event is forgotten today, an earthquake damaged the world's largest nuclear plant in July 2007, when a quake of magnitude 6.8 struck the seven reactors at the Kashiwazaki facility in Japan. It was not the next Chernobyl, however. On the other hand, had the right lessons been drawn, there might not have been a second Chernobyl.

The Rollout of Emissions Trading

In January 2005, the pilot phase of the EU's Emissions Trading Scheme (EU-ETS) also began. Like the nuclear phaseout, it captured a lot of utility attention. Judged in terms of its outcome, emissions trading has been

---

[10] Leuschner, Udo. "Energiechronik: Belgische Parteien wollen Laufzeiten für KKW verlängern." August 2007. Accessed February 7, 2016. http://www.energie-chronik.de/070812.htm.

[11] "Sarkozy Meets Gadhafi: France to Build Nuclear Reactor in Libya." Der Spiegel Online. 26 July 2007. Accessed February 7, 2016. http://www.spiegel.de/international/world/sarkozy-meets-gadhafi-france-to-build-nuclear-reactor-in-libya-a-496711.html.

[12] "French Sarkozy denies hawking nuclear reactor to Gaddafi." Reuters. 17 April 2012. Accessed February 7, 2016. http://www.reuters.com/article/us-france-election-nuclear-idUSBRE83G0FF20120417.

a failure. In combination with other legislation, it perversely led to the construction of a slew of coal plants throughout Europe.

Yet, if properly designed, it could help clean up conventional industry—which is exactly why countless lobbyists continue to make sure that the EU-ETS remains toothless. Every time a higher carbon price and a lower ceiling for emissions is proposed, industry representatives say such an outcome would simply drive them out of the EU.[13] But there was hope in 2005 that the platform could be made to work.

The concept of emissions trading—or cap and trade as it is called in the US—is especially attractive to neoliberal leaders in Brussels, for whom the price-fixing in feed-in tariffs seems anti-market. Essentially, the emissions platform sets a limit on emissions and divvies out allowances to large emitters. Initially, large power providers were forced to participate, as were large industrial firms; later, the aviation sector was added, and the scope of industry firms covered was expanded. Heat and transportation remain a matter for national governments to decide; the EU-ETS does not cover emissions in those sectors. Overall, the EU-ETS covers an estimated 45 percent of greenhouse gas emissions within the EU.

If an emitter approaches the limit of its allocated allowances, it has two basic options: purchase additional allowances from another emitter who has some left over; or invest in technologies that will reduce emissions. The decision will be based on whatever option is the cheapest, and hence the entire platform was expected to lead to lower emissions at the lowest possible price—moreover, one determined by the market. Allowances are limited and tradable—hence the popular name "cap and trade" (Box 11.1).

What exactly went wrong on the EU-ETS? The pilot phase ran from 2005 to 2008. The main purpose of Phase 1 was not primarily to reduce emissions considerably, but rather to put the price of carbon "on the board room agenda," as the International Emissions Trading Association

---

[13] See Dierig, Carsten. "Stahl vor dem großen Exodus." Die Welt Online. 29 July 2015. Accessed February 7, 2016. http://www.welt.de/print/die_welt/wirtschaft/article144559601/Stahl-vor-dem-grossen-Exodus.html.

> **Box 11.1 Emissions trading and renewables**
>
> Proponents of emissions trading have sometimes charged that the platform makes additional policy support for renewable energy superfluous—that the money devoted to renewables is wasted as long as the emissions trading platform exists. However, emissions trading could not have ramped up a mix of renewable energy, where each technology type required a different level of policy support. In supporting only the cheapest solutions, emissions trading would give the biggest push to the technology that is already the most competitive and therefore needs the least support (generally, onshore wind). In return, no renewable energy policy can ensure a transition from older to newer, more efficient power plants and from high-carbon (such as coal) to lower-carbon fuels (such as gas). Support for renewable energy is clearly needed alongside emissions trading—an effective trading platform, that is.

puts it.[14] Basically, the companies forced to participate first had to learn to play the game. From now on, carbon emissions would need to be monitored, reported, and traded in verifiable fashion, something none of them had done before. In trying to convince firms that "being included in the pilot ETS is <u>not</u> a disadvantage," UK-based NGO Sandbag argued that companies would "acquire new knowledge and skill" and start talking more in-house about improving efficiency.[15]

So far so good—except that far too many allowances were handed out in Phase 1, as everyone now agrees. The bureaucrats who believed markets set prices better than governmental officials can were not very good at setting market volumes, either. The result was a carbon price so low that it had little impact.

At the beginning of trading, a ton of carbon dioxide (CO_2) emitted cost around 20 euros. The price briefly rose to around 33 euros, but it fell to below 25 euros for most of the trading phase—and finished at a

---

[14] IETA. "The EU Emission." Now offline. Accessed August, 2015. http://www.ieta.org/index.php%3Foption%3Dcom_content%26view%3Darticle%26catid%3D54:3-minute-briefing%26id%3D324:the-eu-emissions-trading-system.

[15] Elsworth, Rob. "The EU Emissions Trading Scheme (ETS). A critical perspective." Sandbag. Wuhan Workshop. July 2013, July 2013. Accessed February 7, 2016. http://ccap.org/assets/Rob-Elsworth_Sandbag.pdf.

miserable 14 euros. A much higher carbon price would be needed for the platform to be effective: around 30 euros for a transition from hard coal to gas in the power sector, for instance, and at least 50 euros for a switch from European lignite to gas. And the worst was yet to come.

In 2008, the second phase began with the aim of a stronger harmonization of allocation rules among different member states. The option of investing worldwide in the least expensive emissions-reduction projects was taken up. Now, firms could offset emissions at home with far cheaper, but equally effective credits from investments in, say, Africa and Asia. Since the planet does not care where carbon is emitted, it makes sense to spend money cleaning up relatively inefficient infrastructure in the developing world rather than using the same amount to bring already relatively efficient OECD infrastructure just a small step forward. Again, what sounded good on paper proved difficult in practice. Carbon offsets came under fierce criticism both for their ineffectiveness and for the underlying philosophy. It was hard to say which foreign project would not have happened anyway, so there was concern that the European trading scheme simply provided windfall profits to projects already underway. In the end, it was the unclarity of such wheeling and dealing that led critics of emissions trading, such as Pope Francis, to depict it as, in the worst case, a "new form of speculation" and a "ploy" that does not "allow for the radical change which present circumstances require."[16]

The price of carbon continued to drop. Although the subprime lending market in the US had already begun to collapse in 2007, few anticipated the nearly global recession it would eventually lead to starting at the end of 2008. As European economies shrank in 2009, so did carbon emissions, leaving the trading platform once again flooded with allocations no one needed. As of March 2011, the end of the period covered in this chapter, the price of CO_2 hovered at an ineffectively low level of around 16 euros per ton. Perversely, the first two phases of EU-ETS flooded utilities with liquidity, leading to the massive construction of new fossil-fired power plants, including coal plants, throughout Europe.

---

[16] Vatican. "Encyclical Letter Laudato Si' of the Holy Father Francis on Care for Our Common Home." 24 May 2015. Accessed February 7, 2016. http://w2.vatican.va/content/francesco/en/encyclicals/documents/papa-francesco_20150524_enciclica-laudato-si.html.

The Boom in Coal Plant Construction

The ETS provided some of the funding, but not the focus on new coal. Partly, the interest in constructing new coal plants stemmed from the Industrial Emissions Directive (IED), which took effect at the beginning of 2011 but was discussed starting at the end of 2007.[17] By 2016, it would force a number of old polluting coal plants to close. Utilities were convinced that this capacity would need to be replaced, and they aimed to do so not only with natural gas turbines (whose emissions were two thirds lower), but also with new coal plants that complied with the IED. It did not occur to them that wind, solar, and biomass might grow so quickly that the decommissioned coal power capacity would not need to be replaced. Likewise, European utilities did not anticipate the drop in power demand resulting from the financial crisis. Furthermore, the decade had begun with a relatively low level of investments as utilities took a wait-and-see approach during the beginning of liberalization. Europe's power plant fleet was aging. The general feeling was that the time had come for new plants.

Utilities could have invested in renewables, but they did not believe these sources worked. By 2009, Germany's four biggest utilities only made up 1 percent of investments in renewables. "How can fluctuating wind and solar ever replace reliable coal power?", practically every conventional energy expert and consultant asked rhetorically.

For example, German energy agency Dena produced its first Grid Study in 2005. It failed to see photovoltaics coming altogether, subsuming it under "Other" in forecasts. To be fair, even expert proponents of renewables—such as a joint study by the DLR, Heidelberg's IFEU and the Wuppertal Institute in 2004—also failed to see solar coming by the end of the decade, including it only under the small category of "Other."[18]

The DLR, IFEU, and Wuppertal at least called for new capacity construction to focus on *flexible* plants to backup wind power. The Grid

---

[17] European Commission. "The Industrial Emissions Directive: Summary of Directive 2010/75/EU on industrial emissions (integrated pollution prevention and control)." Accessed February 7, 2016. http://ec.europa.eu/environment/industry/stationary/ied/legislation.htm.

[18] Jungjohann, Arne, and Craig Morris. "The German Coal Conundrum." Accessed February 6, 2016. https://us.boell.org/sites/default/files/german-coal-conundrum.pdf.

Study called for new capacity to be built in order to replace upcoming decommissions of less flexible baseload—both nuclear and fossil. This poor advice would be repeated throughout the 2000s in studies from business consultants, whose word utility companies were likely to heed. For instance, three conventional consulting firms warned in 2008 that Germany would become "a net power importer" even if the nuclear phaseout was postponed—while in reality German net power exports reached a record level in 2015. Likewise, in 2007 the *Leitstudie* (the official governmental roadmap for the Energiewende) expected 38.6 GW of new capacity, 15.6 GW of which was to be hard coal and lignite. To build that amount, some 20 large coal plants would need to be constructed.

When trying to understand such recommendations, it helps to remember governmental targets for renewable electricity at the time. The 2004 EEG aimed to reach 20 percent green electricity by 2020. Around that time, the last nuclear plants were also expected to be shut down—and they covered closer to 30 percent of power supply. You didn't have to be a math whiz to calculate in 2004 that an additional 10 percent of German electricity would need to come from fossil plants. But with each passing year, you did have to be hardheaded not to accept that the equation was changing as renewables clearly shot past all targets; in 2010, the target for 2020 was set at 35 percent renewable power.

Naturally, some proponents of renewables forecast much stronger growth for both wind and solar—and therefore focused on the need for conventional plants to react flexibly. For instance, German renewable energy organization BEE and Eurosolar came closer to predicting the growth of solar than business consultants did, and Greenpeace's *Energy [R]evolution* studies also proved more accurate in forecasting global markets than, say, the International Energy Agency (IEA) and other such organizations did. But conventional energy sector experts were still likely during those years to dismiss such optimistic forecasts as the wishful thinking of lobbyists from the other camp.

Convinced that the best minds advocated new gas and coal construction, Europe's biggest utilities set forth to do just that—and windfall profits from the free allocation of emission allowances ironically gave them the cash to do so. The power sector had the nerve to pass on the value of these free allowances to consumers. When consumer advocates and politicians asked why the theoretical price of free allowances was being passed on to consumers, power firms spoke of opportunity costs; the companies simply priced in the value for these emission permits.

The amounts involved were staggering. In the first phase of emissions trading alone, German utility RWE—the coal power giant—posted some 5 billion euros in windfall profits. The second phase only added to that amount, so that the WWF estimated windfall profits from emissions trading for Germany's five biggest utilities at a minimum of 31 billion euros from 2005 to 2012.[19] For years, this criticism mainly came from environmentalists, but it was undeniably true. In fact, in January 2016 the CEO of Eon publicly confirmed the analysis[20]:

> We as corporations—all of us without exception—received a lot of carbon certificates at the beginning of the EU's emissions trading for free. The exchange price was included in power prices and made us money. There was nothing wrong about that, but it did not reflect any good performance on our part either.

Because the phenomenon was Europe-wide, the effects were seen in numerous countries. Europe's eight biggest utilities invested just under 20 billion euros annually in new facilities from 2003 to 2005. But as the windfall profits from ETS began rolling in, investments skyrocketed to around 50 billion euros annually from 2008 to 2012. In Germany, these windfall profits were one reason why the country's four big utilities tripled their profits from 2003 to 2007; likewise, the stock value of these four companies quadrupled.[21]

During these years, a slew of coal plant projects began. By the beginning of 2009, nine out of ten utility projects in the pipeline were fossil-

---

[19] Leuschner, Udo. "Energiechronik: "Stromkonzerne kassieren 20 Milliarden Euro Windfall-Profits"." March 2006. Accessed February 7, 2016. http://www.energie-chronik.de/060303.htm.

[20] Teyssen, Johannes. ""Wir erleben eine Achterbahn der Gefühle"." Die Zeit Online. 7 January 2016. Accessed February 7, 2016. http://www.zeit.de/wirtschaft/unternehmen/2016-01/johannes-teyssen-energie-e-on-energiewende/seite-4.

[21] Leuschner, Udo. "Energiechronik: Energiekonzerne konnten ihre Gewinne in fünf Jahren verdrei-fachen." December 2008. Accessed February 7, 2016. http://www.energie-chronik.de/081202.htm.

fueled.[22] Utilities were not the only ones convinced that such projects were necessary. In 2009, Environmental Minister Sigmar Gabriel stated during a visit to the grounds of a new coal plant project near Mainz that "we will need 8 to 12 new coal plants if we want to phase out nuclear." A year earlier, Chancellor Merkel herself visited Hamm, where construction was beginning on two new coal plants. After placing the symbolic cornerstone, she explained to the cameras that Germany, as an industrial country, would need "new, more powerful power plants." She also accused those who opposed new coal plants of putting the German economy, the German labor market, and the future of the country "at serious risk."[23] She called the investments "future-proof."[24] The project had a price tag of 2 billion euros; RWE could pay for it with a fraction of the windfall profits it made from emissions trading up to 2008.

Renewable energy advocates railed against these projects, but to little avail. Hermann Scheer, coauthor of the 2000 Renewable Energy Act, wrote in 2005 that "no new fossil plants will be needed to replace old ones." By the end of the decade, he was putting a finer point on the argument, specifically stating that there would soon be "no need for baseload."[25]

The most flexible backup generation was clearly what was needed, as the government's Advisory Council on the Environment (SRU) stated in 2009, arguing that "baseload power plants will be of limited use" when renewables make up a large share of power supply. That same year, the head of the SRU explained separately that flexible backup "practically is impossible with nuclear and coal baseload plants." Then, renewable energy professor Volker Quaschning produced a chart in 2010 that went viral.[26] It extrapolated current renewable energy production into the

[22] Leuschner, Udo. "Energiechronik: Neue Kraftwerke werden zu 94 Prozent fossil befeuert." January 2009. Accessed February 7, 2016. http://www.energie-chronik.de/090110.htm.

[23] Magoley, Nina. "Merkel fordert weitere Kohlekraftwerke." WDR. 29 August 2008. Accessed February 7, 2016. http://www1.wdr.de/themen/archiv/sp_kohle_abschied/grundstein_hamm100.html.

[24] Morris, Craig. "New German coal plant worth one euro." Renewables International. The Magazine. 29 July 2015. Accessed February 7, 2016. http://www.renewablesinternational.net/new-german-coal-plant-worth-one-euro/150/537/89142/.

[25] Jungjohann and Morris, *German coal conundrum*.

[26] Quaschning, Volker. "Grundastkraftwerke: Brücke oder Krücke für das regenerative Zeitalter?" In. Sonne Wind & Wärme. 05/2010, p. 10-15. Accessed February 7, 2016. http://www.volker-quaschning.de/artikel/grundlast/index.php.

future based on projections from renewable energy organization BEE (estimates that conventional utilities doubted). Known as the "dental chart," it showed fluctuating wind and solar cutting like long teeth into baseload power so much that demand for these plants disappeared completely for hours at a time, only to roar back up by tens of gigawatts a few hours later. Everyone knew that baseload power plants could not switch off for a couple of hours and then ramp up again so quickly. The scariest thing was the year of the projection: 2020. Ten years away.

Energiewende critics like to point to the aforementioned statements made by Gabriel and Merkel, who argued that new coal plants would be needed to replace nuclear, to demonstrate that the government intended all along to switch from nuclear to coal. Rarely are the simultaneous voices of renewable energy proponents quoted in the same context. In reality, there was a debate in Germany about whether coal would be needed to replace nuclear in the years from 2005 to 2011. Indeed, as far back as 1991—when the nuclear phaseout was still just an unlikely proposal— market analysts at Prognos argued that lignite in particular would be needed if nuclear were ever phased out. Some politicians—unfortunately, the leading ones—agreed for decades.

Protests Grow Against Coal

The experts were wrong—drastically so, in fact. The Hamm coal plant where Merkel expressed her support for new coal power was cancelled in December 2015 after facing setbacks due to construction flaws overlooked by French engineering firm Alstom. By that summer, there was so much surplus capacity on the German power market that low electricity exchange prices would have made the nearly finished coal plant unprofitable anyway. The 23 municipal utilities that had purchased stakes in the project told RWE, the main investor, that they wanted out. RWE offered them one euro each—for the investment Merkel had called "future-proof."[27]

---

[27] Morris, *New German coal plant worth one euro.*

Likewise, the plant near Mainz where Gabriel expressed his support for new coal plants was never constructed. No power shortage resulted. German anti-coal activists prevented the construction of numerous coal plants throughout the country during these years. Protests against coal power went back decades in the country at least to the 1960s, but they largely focused on coal mines. At the beginning of the 1960s, coal power was still closely associated with economic progress—so much that Social Democrat Willi Brandt was laughed at when he called for "blue skies over the Ruhr Area" in the 1961 elections.[28] Pollution had become obvious; people in the area could no longer hang their clothes outside to dry—they got dirty before they were dry. But with no clean alternative in sight, what were the options?

The protest therefore focused on the expansion of lignite mines. When a coal field was to be expanded, destroying a number of neighboring villages in the process, locals would protest, but they were often on their own. Other neighboring communities not yet affected were happy to take up their relocated neighbors and see the economy grow further. The case of Erkelenz is illustrative. Located on the edge of giant coalfields between Cologne and the Dutch border, this town supported the expansion of these coal mines from the 1950s into the 1970s—until 1980, when a top politician mentioned in passing (and in public) that "by 2013, Erkelenz will have disappeared from the map." At that point, the town of some 45,000 took up arms against lignite mining.[29]

Gradually, the focus turned toward legal challenges, but the option also divided anti-coal activists into two camps by the late 1990s: those who still fundamentally opposed coal, and those who wanted to spend their limited resources on getting the most out of legal challenges—even if that meant a compromise in the middle.[30]

By 2007, church groups had come together to form the Climate Alliance, specifically to protest against a coal plant proposed in Neurath. It was a battle the campaigners would lose. In 2012, the then Environmental

---

[28] See https://www.wikiwand.com/de/Blauer_Himmel_%C3%BCber_dem_Ruhrgebiet.

[29] Jansen, Dirk, and Dorothea Schubert. *Zukunft statt Braunkohle: 30 Jahre Widerstand gegen den Braunkohlentagebau Garzweiler II*. Düsseldorf: BUND, Landesverband NRW, 2014.

[30] Jansen and Schubert, *Zukunft statt Braunkohle*, p. 124.

Minister Peter Altmaier would personally cut the ribbon for the opening of the world's largest lignite power plant in Neurath.[31] But overall, the German anti-coal movement was highly successful in preventing coal plant projects after 2005. For instance, of the 23 major projects initially announced between 2005 and 2011, a whopping 15 were blocked entirely. Court challenges against Datteln 4 began in September 2009, when the German chapter of Friends of the Earth (BUND) tipped off a local farmer who opposed the project but did not know that the firm behind it (Eon) had not stuck to the exact planning specified in the permit. By then, German courts supported citizen protests against coal plants so much that this challenge had held up the plant until the writing of this chapter in early 2016.

For at least the past decade, no coal plant has been built in Germany without significant protests. What's more, the coal firms themselves probably wish more of these projects from 2005 to 2011 had been blocked. For instance, the new Moorburg hard coal plant went into operation in 2015. Originally planned in 2006, when energy markets looked fundamentally different, the project went through an odyssey against fierce opposition from citizens of nearby Hamburg but with crucial support from top politicians. Vattenfall, the utility behind the coal plant, eventually tried to step away from the project but had already invested too much money.[32] In preparation for the grand opening at the beginning of 2015, Vattenfall's CEO called the project "a fiasco."[33]

The reason was that Germany's top energy sector thinkers had failed to see the boom in renewables happening right under their noses.

---

[31] Wetzel, Daniel. "Ein Kohle-Koloss soll die Energiewende sichern." Die Welt Online. 15 August 2012. Accessed February 7, 2016. http://www.welt.de/wirtschaft/article108636663/Ein-Kohle-Koloss-soll-die-Energiewende-sichern.html.

[32] Morris, Craig. "Germany opens giant new coal plant." Renewables International. The Magazine. 27 February 2015. Accessed February 7, 2016. http://www.renewablesinternational.net/germany-opens-giant-new-coal-plant/150/537/85919/.

[33] "Ny mångmiljardsmäll för Vattenfall." SvD Näringsliv. 16 January 2015. Accessed February 7, 2016. http://www.svd.se/ny-mangmiljardsmall-for-vattenfall_4259675.

Renewable Energy Becomes Unstoppable

The growth of onshore wind continued more or less unabated from 2005 to 2011 at a level just below 2 gigawatts onshore. In 2006, the symbolic threshold of 20 gigawatts of installed capacity was crossed. That year, the average turbine newly installed had a capacity of 1.85 megawatts—11 times bigger than in 1991—and some 50 percent larger than the average from 2001.[34]

As wind power grew, it also needed to help ensure grid stability. Switching off gigawatts of wind power at once when the grid became unstable was likely to worsen whatever problem was occurring. During the major blackout of 2006, Europe got a taste of what not accounting for wind power would mean. The final report by European grid experts on the blackout found that the German grid operator Eon had not taken into account wind power in particular when an otherwise controllable grid malfunction occurred.[35] The result was the largest blackout in European memory, stretching from Poland to Greece and even Morocco. So at the beginning of 2009, new wind turbines were required to have special equipment added for "ancillary services": maintaining voltage and frequency levels. That year, wind turbines took a major step forward in becoming equal partners with conventional power plants toward grid stabilization.

However, new turbines were not to be installed in large numbers onshore. Rather, the new coalition announced in November 2005 that the focus was to be on "repowering existing wind farms and expanding offshore wind farms."[36] Off the German coast, some 30 wind farms were planned as of 2007, with a cumulative capacity exceeding 40 gigawatts—

---

[34] Leuschner, Udo. "Energiechronik: Windkraft wächst im Inland wieder langsamer - weltweit hält der Boom jedoch an." January 2008. Accessed February 7, 2016. http://www.energie-chronik. de/080112.htm.

[35] Union for the Co-ordination of transmission of electricity. "Final Report. System Disturbance on 4 November 2006." Accessed February 7, 2016. https://www.entsoe.eu/fileadmin/user_upload/_ library/publications/ce/otherreports/Final-Report-20070130.pdf, p. 5.

[36] Leuschner, Udo. "Energiechronik: Das Energie-Kapitel und weitere die Energiewirtschaft berührende Themen im Koalitionsvertrag zwischen CDU, CSU und SPD vom 11. November 2005." November 2005. Accessed February 7, 2016. http://www.energie-chronik.de/051102d.htm.

approximately twice as much as was installed on shore at the time.[37] These projects required giant transformer stations that engineering firms were accustomed to providing to large utilities. Finally, the conventional players had their own renewable playground: wind over water. The success of citizen-owned onshore wind farms was not viewed as a model for the future, but as something that needed reigning in.

Biomass projects, also largely community-driven, began to boom in these years as well. From 2004 to 2011, power from biomass quadrupled to 5 percent of demand. Numerous communities also began recovering waste heat from generators powered with bioenergy to tap a source of renewable heat. The main concern was increasingly on competition between food crops and energy crops (such as corn for biogas), but for the years under review in this chapter, bioenergy growth was sustained. This focus on local biomass means that roughly half of the feedstock used for energy production is recovered from waste. Though there has been concern in North America about Europe's focus on the biomass imports, Germany is a net exporter of wood pellets, mostly made from waste wood from forestry. Germany hardly imports any wood pellets from the US,[38] nor does the country use fresh timber for co-firing in coal plants.[39]

While the community projects moved forward, the announcements about gigantic offshore wind projects produced more hot air than energy. In 2007, the largest renewable energy developers, such as Prokon, joined forces with France's Areva, for instance. The two never built any offshore turbines together. At the end of 2008, RWE announced plans to build a single offshore wind farm in the North Sea with a capacity of 1000 megawatts.[40] Called Nordsee 1, it was to be completed in 2015, consisting

---

[37] Leuschner, Udo. "Energiechronik: E.ON läßt Offshore-Windpark für 300 Millionen Euro ans Netz anschließen." September 2007. Accessed February 7, 2016. http://www.energie-chronik. de/070910.htm.

[38] Morris, Craig. "Germany is not burning US forests." Energy Transition. The German Energiewende. 5 March 2015. Accessed February 7, 2016. http://energytransition.de/2015/03/ germany-is-not-burning-us-forests/.

[39] Morris, Craig. "0.07 percent of German electricity comes from fresh timber." Energy Transition. The German Energiewende. 14 October 2015. Accessed February 7, 2016. http://energytransition. de/2015/10/0-07-percent-of-german-electricity-comes-from-fresh-timber-2/.

[40] Leuschner, Udo. "Energiechronik: RWE plant Nordsee-Windpark mit tausend Megawatt." December 2008. Accessed February 7, 2016. http://www.energie-chronik.de/081216.htm.

of some 150 to 180 turbines with a rated capacity of 5 to 6 megawatts each. By 2015, RWE had finally begun construction, but the project had already shrunk to 54 turbines with roughly the same rated capacity. The problem was that these projects proved to be far more expensive than originally hoped. At the end of 2014, Germany just barely had a single gigawatt of wind capacity in operation offshore—far short of the 40 gigawatts planned seven years earlier. During all those years, Germany had to make do with the moderate, but steady growth of boring old onshore wind farms, which somehow were not sexy enough to be celebrated as the revolution they certainly were.

The focus on offshore wind is the only plausible explanation for the lack of further growth on shore. After the record year of 2002 with 3.2 gigawatts, onshore installations continue to come in at or below 2 gigawatts. There were no major obstacles to onshore projects—no specific laws slowing them down, just a lot of wasted effort on offshore. In fact, to investigate the reasons for stagnation in onshore capacity, we spoke with Peter Ahmels for this book. Ahmels was the head of German wind power association BWE during those years. In 2015, he said he could not remember any other reason why the onshore market slipped from 3 to 2 gigawatts annually in the decade after 2002 (Box 11.2).

In contrast, the growth of photovoltaics was nothing short of breathtaking from 2005 to 2011. The market was still quite small, but in 2005 and 2006 cloudy Germany alone generated more than 40 percent of global solar power production. Still, we are only talking about five terawatt-hours of solar power worldwide for 2006. In 2014, Germany alone generated

---

Box 11.2 Misleading focus on the Big

As the example of offshore wind in Germany shows, the problem with the "focus on the big" is that it slows down the energy transition by draining momentum from countless small projects.

> **Box 11.2 (continued)**
>
> Industry experts are used to thinking in terms of large projects. In 2009, when distributed photovoltaics began to boom in Germany, the gigantic Desertec project made it on the German nightly news. At a press conference, the Dii industry consortium announced a mere feasibility study; Europe might get inexpensive green electricity from the Sahara. Before the consortium disbanded in 2014, Germany had already reached its 2050 targets for photovoltaics at home. Desertec may still happen; it would be good for North Africans. But in those years, the unrealistic project merely cast doubt on the need for photovoltaics within Germany. Experts, such as the head of RWE's renewable energy division, thought photovoltaics should be installed in deserts, not cloudy Germany; famously, the CEO of RWE also stated in 2012 that photovoltaics in Germany makes about as much sense as growing pineapples in Alaska. In 2010, the Big Four utilities only accounted for 0.2 percent of photovoltaics arrays in Germany. Fortunately, German citizens disagreed and continued to build solar at home.
>
> The Pew Charitable Trust's report entitled "Who's winning the clean energy race?" also used to overlook small renewables by disregarding projects smaller than 1000 kilowatts (a homeowner array has around 3 or 4 kilowatts). For 2009, Pew estimated that Germany had spent 4.3 billion US dollars in clean tech, but the figure for solar alone was closer to 14 billion.
>
> Those who doubt that small projects can add up to a fast transition should consider German experience. From 2010 to 2015, the share of renewable electricity increased by three percentage points annually on average, largely with community projects. In 2014, the government implemented the first limit on growth: no more than 45 percent renewable power by 2025. To stay within that limit, annual growth will have to be cut by a roughly third to just over one percentage point from 2016-2025. Community renewable projects moved too quickly for the German government.

seven times as much, while globally 185.9 terawatt-hours was produced—37 times more.[41]

But in 2007, a new kid showed up on the solar block: Spain. The Spaniards installed 544 megawatts of photovoltaics that year. The Germans remained King of the Hill with 1,270 megawatts, but the Spanish were not finished. The next year, they would install more than 2,500 megawatts—and

---

[41] Chabot, Bernard. "Analysis of the Global Electricity Production up to 2014 With a Focus on the Contribution From renewables and on CO2 Emissions." Accessed February 7, 2016. http://cf01.erneuerbareenergien.schluetersche.de/files/smfiledata/4/8/7/3/4/7/119belec2014.pdf.

then disappear almost entirely. The first photovoltaics market had boomed and gone bust.

What happened? In the English-speaking world, commentators have often assumed that the financial crisis, which roughly coincided with Spain's solar bust, must have been one reason. In reality, it played a marginal role. The main problem was that solar is a different beast. Utilities and policymakers did not understand what they were dealing with.

In 2007, a solar budget went into force in Spain, with a target of 371 megawatts of photovoltaics by 2010. Once 85 percent of the target had been installed, the feed-in tariffs on offer were to be continued only for another 12 months while the government reviewed the policy for a possible extension.[42] The budget was intended to prevent uncontrolled growth. It actually produced that uncontrolled growth unintentionally.

You see, 12 months is a short time frame for utilities. When implementing that grace period, policymakers did not sense that a year was long enough to start a run on the market. The purpose was simply to allow projects underway to be completed. As we see from the history of coal plants above, it takes a good six years to complete construction of them. Nuclear plants generally take longer. A gas turbine takes closer to two years. Ditto for wind farms, which could go up in a year, but only if all the permits are immediately available and in order. But for solar, 12 months is generous.

Spain was the first country to make this realization, and it learned its lesson the hard way. When project developers realized time was running out, the Spanish market soaked up global production for exactly 12 months. After that, the financial crisis probably stopped Spanish policymakers from extending policy support, but the crisis did not cause the initial bust of 2008—the speed with which solar arrays can be built did.

The German market also grew by more than 50 percent in 2008 to reach nearly 2 gigawatts—not quite enough to stop the Spaniards from being world champions once.

The German solar market looked on with dismay at the Spanish boom and bust, of course. They worried that German policymakers might

---

[42] del Río, Pablo, and Pere Mir-Artigues. "A Cautionary Tale: Spain's Solar photovoltaics Investment Bubble." Accessed February 7, 2016. http://www.iisd.org/gsi/sites/default/files/rens_ct_spain.pdf.

also get cold feet, especially in light of the cost impact. *Photon*, the monthly photovoltaics magazine founded in the 1990s, then began calculating what solar power actually cost. Around 2005 to 2007, there was a severe shortage of solar silicon, which had originally been made largely from waste products in the semiconductor sector. But semiconductors are made by the square millimeter; solar panels, by the square meter. As photovoltaics grew, demand for solar silicon quickly outstripped the supply of waste semiconductor silicon. Solar firms that did not make silicon in-house already were left forced to purchase silicon at high prices on the spot market, so they began scrambling to set up silicon production facilities.

In late 2005, Merkel's coalition announced plans to review feed-in tariffs in 2007. German solar firms began explaining that feed-in tariffs for photovoltaics were not too high because skyrocketing silicon prices were cutting into their margins. That year, *Photon* took the bold step— after all, the firm lived partly off of advertising—of calculating that silicon made up less than 10 percent of the price of an array and began openly calling for lower feed-in tariffs for solar. For instance, in April 2007 the magazine calculated that solar power from new systems in 2006 cost as little as 24 cents in Germany, and that level was expected to drop to 15 cents by 2010.[43] The article caused outrage in many circles; after all, as much as 52 cents was paid for a kilowatt-hour in the feed-in tariffs from 2006, and the maximum was still nearly 35 cents by 2010.

The German photovoltaics community was then split into two camps: those who spoke mainly for consumers and citizen investors; and those who represented installers and manufacturers. The former were mainly concerned about excessive profits leading politicians to overturn support for solar entirely. The latter focused on the profitability of German photovoltaics manufacturers and protecting domestic jobs.

Photon lost advertisers, some of whom migrated to a new competing magazine which addressed the disgruntled: *Photovoltaik*. (The present author Craig Morris cofounded *photovoltaics Magazine*, the English version of *Photovoltaik*, in 2008.) A subsequent analysis from 2014 showed how

---

[43] Podewils, Christoph. "Billiger als Braunkohle." *Photon*, 2007, 40–44; April 2007. Accessed February 7, 2016. http://www.photon.de/presse/mitteilungen/hintergrund_truecost.pdf.

right Photon was (see Chap. 9).[44] German Solar Industry Association BSW openly doubted *Photon*'s calculation at the time. ("We were just repeating what our members told us," a member of the BSW stated off the record in 2015.) Thus, the solar sector argued for several years starting in 2007 that feed-in tariffs should not be reduced for photovoltaics. By the end of 2011, solar power made up half the cost of feed-in tariffs—but only a sixth of renewable power production.

The year 2007 marked the pinnacle of German industry leadership in the photovoltaics sector, but China also appeared on the global stage as a photovoltaics manufacturer in these years. Historically, Japan's Sharp had been by far the biggest producer of solar cells, but Germany's Q-Cells briefly took the pole position in 2007.[45] That year, only four Chinese firms were in the top 10. By 2011, eight of the top 10 were from China, and Q-Cells no longer made the list. Likewise, Q-Cells competed with Sharp for the lead in the list of photovoltaics panel producers in 2007 and 2008, with Germany's SolarWorld also making the list.[46] Both had disappeared from the top 10 list for panels by 2010, which no longer contained any German manufacturers.[47]

The German solar sector argued during these years that German feed-in tariffs should not be lowered, lest even more sales go to the cheaper Chinese suppliers, but the argument never made any sense. Assume that a Chinese panel costs 200 euros; a German one, 220 euros (all other

[44] Morris, Craig. "Germany overpays solar relative to wind." Renewables International. The Magazine. 29 July 2014. Accessed February 7, 2016. http://www.renewablesinternational.net/germany-overpays-solar-relative-to-wind/150/537/80598/.

[45] Doe, Paula. "Explosive Growth Reshuffles Top 10 Solar Ranking." Renewable Energy World. 12 September 2008. Accessed February 8, 2016. http://www.renewableenergyworld.com/articles/2008/09/explosive-growth-reshuffles-top-10-solar-ranking-53559.html.
Note that other sources have the firm in second place for 2007 but first place for 2008, the difference being production versus shipment: Mints, Paula. "Top Ten photovoltaics Manufacturers from 2000 to Present: A Pictorial Retrospective." Renewable Energy World. 21 January 2014. Accessed February 8, 2016. http://www.renewableenergyworld.com/articles/2014/01/top-ten-pv-manufacturers-from-2000-to-present-a-pictorial-retrospective.html.

[46] "Top 10 producers of photovoltaics modules by installed capacity in 2007 and 2008 in megawatts." Accessed February 8, 2016. http://www.statista.com/statistics/264509/largest-producer-of-pv-modules-by-capacity-in-2007/.

[47] Mehta, Shyam. "photovoltaics News Annual Data Collection Results: 2010 Cell, Module Production Explodes Past 20 GW." Greentechmedia. 9 May 2011. Accessed February 8, 2016. http://www.greentechmedia.com/articles/read/pv-news-annual-data-collection-results-cell-and-module-production-explode-p.

aspects being equal). Now reduce feed-in tariffs. Guess what? The Chinese panel is still cheaper. Feed-in tariffs are prices for power production, not equipment. And German feed-in tariffs never made a distinction between domestic and foreign products.

In contrast to solar industry organization BSW, the older DGS, which represents solar pioneers and the public more than it represents manufacturers, opposed any policy protectionism, arguing in 2009 that "German manufacturers have to compete with quality, longer warranties, and better service."[48] German consumer advocate Holger Krawinkel, a seasoned supporter of the Energiewende from the 1980s, joined forces with *Photon* at the time to prevent requirements for "domestic content" from being included in German feed-in tariffs, as had happened in other countries already.[49]

On the other hand, around 2007 to 2008 six German firms were among the top 10 suppliers of photovoltaics manufacturing equipment, and German engineering firms are still leaders in this market segment today.[50] In other words, the Chinese are making inexpensive solar panels on production lines largely from Germany.

Amendments to the Renewable Energy Act (EEG) of 2009

In 2008, the review of prices led to an agreement that took effect in 2009. The law now had 66 sections instead of 22 from 2004—and 12 from 2000. Specified in the 2004 EEG, the goal of 12.5 percent renewable energy by 2010 had been surpassed in 2007. Likewise, the old goal of 20 percent renewable electricity by 2020 would clearly be reached sooner (it was surpassed in 2011), so a new goal was set at 35 percent of power supply by 2020 (as of 2015, it is just below 35 percent). Here, we again see how the Germans did not let renewable energy growth collapse when

---

[48] Morris, Craig. "German Solar Association opposes protectionism." Accessed February 8, 2016. http://notesfromotherside.blogspot.de/2009/10/german-solar-association-opposes.html.

[49] Such as in Ontario's Green Energy Act of 2008.

[50] "Top 10 suppliers of photovoltaics Manufacturing Equipment." Cleaninvest. Accessed February 8, 2016. https://cleaninvest.files.wordpress.com/2009/08/vlsi-pv-mfg-equipment-top-10-suppliers.gif?w=1000.

targets were met; instead, they raised their targets years in advance so that markets could continue growing.

A debate between the German Economics Ministry and the Environmental Ministry leading up to these amendments illustrates how the German policy landscape operated in these years with split mandates for energy across different ministries.[51] In April 2007, the Economics Ministry produced proposals for changes to the EEG based on a report contracted from independent market analysts; it met with fierce criticism from the Environmental Ministry—which produced its own annual review of renewable energy that month.[52] The arguments were aired in public and in the media and widely commented on in the press, which explained the discrepancies to laypeople.

One of the proposals in the Economics Ministry's study was later taken up: inclusion of "avoided costs" in determining the renewable energy surcharge.[53] Remember, this surcharge is the flip side of the feed-in tariff coin. Grid operators pay FITs to producers of renewable energy. The grid firms then sell the electricity on the power exchange. Today, the difference between FITs paid out and revenue taken in on the exchange is then spread across ratepayers as the surcharge. At the time, however, the surcharge did not take account of revenue from the power exchange. The Economics Ministry wanted to limit the cost impact, so it proposed that money income from the sale of green electricity be subtracted from the surcharge.

The goal was to reduce the surcharge based on the assumption that prices on the exchange would increase. For instance, the 2007 *Leitstudie* (the master plan for the Energiewende) expected carbon trading to help keep the cost of wholesale electricity high (around 7 cents), so that the added expense of renewable power would be negligible (less than 1 cent).[54] Furthermore, the cost of fossil energy was expected to continue rising; the

---

[51] After the elections of 2013, the Economics Ministry officially took charge of all energy matters, putting an end to the sharing of competence—at least in the power sector—with the Environmental Ministry which had been responsible for renewable energy.

[52] Leuschner, Udo. "Energiechronik: Streit zwischen Glos und Gabriel um Gutachten zu erneuerbare Energien." April 2007. Accessed February 8, 2016. http://www.energie-chronik.de/070411.htm.

[53] Leuschner, *Energiechronik: Streit zwischen Glos und Gabriel.*

[54] See Nitsch, Joachim. "Leitstudie 2007. Ausbaustrategie Erneuerbare Energien.: Aktualisierung und Neubewertung bis zu den Jahren 2020 und 2030 mit Ausblick bis 2050." Accessed February 8, 2016. https://www.clearingstelle-eeg.de/files/Leitstudie_2007_Langfassung_0.pdf, p. 64.

experts did not see the shale gas boom in the US coming. As the US power market shifted from coal to gas, the US flooded Europe—and Germany—with cheap hard coal. As a result, wholesale prices begin dropping in 2011, thereby inflating the renewable energy surcharge once the Economics Ministry's proposal from 2007 took effect in 2010. And when the "Compensation Mechanism" (as the new surcharge item was called) was finally presented in May 2009, another rationale was given in addition to the hope that it would shrink the surcharge as wholesale prices rose: the new calculation was considered less complex for the utilities involved.[55]

For solar, the amendments in the 2009 EEG included a minor innovation that would later take center stage as a bone of contention. The change was designed in preparation for "grid parity"—the point at which solar power from household rooftops would cost the same or less than retail electricity. The German policy was called *Eigenverbrauch*, meaning literally "own consumption." It differs from net-metering, which is common in the United States, in a crucial respect. In the US, the meter simply runs backward if you produce more electricity than you consume, and the difference is tallied at the end of whatever billing period the law defines (such as monthly or annually). In contrast, *Eigenverbrauch* means that the electricity has to be consumed behind the meter—either immediately or from a battery storage system without ever touching the grid.

Grid parity was reached in Germany in 2012. Though largely held to be the point at which solar power would continue to grow without any further governmental support, at the point of grid parity the German photovoltaics market shrank by nearly 75 percent in 2013 and 2014. In the US, where grid parity is increasingly also being reached, utilities are imposing—or at least proposing—special charges for households with solar roofs. Some of Germany's top executives in the solar sector failed to see this pushback coming when heralding grid parity as the tipping point to "unsubsidized" solar growth.[56] By 2014, the German government had

---

[55] Leuschner, Udo. "Energiechronik: Neue Windenergieanlagen müssen Systemdienstleistungen erbringen können." May 2009. Accessed February 8, 2016. http://www.energie-chronik. de/090507.htm.

[56] Morris, Craig. "FITs needed after grid parity." Renewables International. The Magazine. 17 February 2011. Accessed February 8, 2016. http://www.renewablesinternational.net/ fits-needed-after-grid-parity/150/511/30197/.

also implemented a special charge on arrays that took advantage of the incentives for *Eigenverbrauch*. Though the government had specifically incentivized the direct consumption of solar power, German Energy Minister Sigmar Gabriel stated that the storage of solar power was "absolute madness" in 2014. His statement was widely circulated among critics of the Energiewende as an admission of German policy failure without any indication of the underlying irony: Gabriel was disparaging a policy he himself had implemented as Environmental Minister in 2009.[57]

When the amendments for 2009 were finally adopted, the rates for solar were reduced considerably, though not as much as originally proposed. The Christian Union was already divided over the issue; top politicians in Berlin wanted photovoltaics feed-in tariffs reduced faster, but local party members saw how engaged rural communities in particular had become in renewables. The libertarian FDP was the only party that opposed the 2009 EEG altogether, with party whip Gudrun Kopp calling it "political madness" that could be done away with now that the emissions trading platform had been established.[58] Perhaps the most important message from the amendment process that year was that a Christian Union coalition with the FDP spelled trouble for renewables. And that is exactly the coalition Germany got in the 2009 fall elections.

2009 Elections and the Nuclear Phaseout

The grand coalition between the Christian Union and the Social Democrats had not been good for the SPD as a junior partner. The party fell to its lowest share of the vote since World War II. The party began losing more votes to other left-of-center newcomers, and the public attributed the successes under Merkel's coalition more to the CDU. The FDP also managed to convince the public that it would be a more effective "corrective" to the CDU than the SPD had been. Voter turnout was also

[57] Morris, Craig. "Germany's incentivized solar failure." Renewables International. The Magazine. 12 May 2014. Accessed February 8, 2016. http://www.renewablesinternational.net/germanys-incentivized-solar-failure/150/537/78765/.

[58] Leuschner, Udo. "Energiechronik: Bundestag verabschiedete neues EEG und drei weitere Gesetze des Klimapakets." June 2008. Accessed February 8, 2016. http://www.energie-chronik.de/080601.htm.

the lowest in history at 71 percent. But the libertarian FDP had managed to get almost 15 percent of the vote, the highest level ever.

The financial crisis of 2008 overshadowed the elections. Once again, the energy transition was a minor issue, as talk focused on tweaking taxation to prevent the recession from becoming a prolonged depression. And the main issue with an energy policy was not renewables, but the nuclear phaseout.

The new coalition of the Christian Union and the FDP under the continued leadership of Chancellor Merkel came into office with a pledge to extend the commissions of German nuclear plants, which were now called a "bridge technology" that would give renewables more time to grow—a bold claim after half a decade of renewables blasting past targets.

Just days before the election, a 108-page strategy paper was leaked to Greenpeace. Drawn up by a PR agency, it showed how German utility Eon could manipulate public opinion in favor of nuclear power during the election campaign.[59] It contained lists of specific journalists and their political orientation, and also included names of leading energy experts who frequently appeared in the media supporting "renewables in combination with nuclear power." German economist Claudia Kemfert of the DIW, who became a leading critic of nuclear after Fukushima, was still included in that group of prominent energy experts.[60] Overall, the PR agency recommended that Eon not support nuclear in public too much, lest the effort backfire. Rather, more discreet communication was recommended, with a focus on independent data that "suggests neutrality." Germany's broad landscape of independent analysts was to be relied on: "Studies conducted by renowned institutes are expensive, but they pay for themselves because the institute's credibility cannot fundamentally be doubted." Also, bloggers were recommended as a new channel. When Greenpeace published the leaked strategy paper, Eon claimed it had not contracted the study at all, explaining that the agency merely submitted it in the hope of getting future work. The

---

[59] PRGS. "Kommunikationskonzept Kernenergie: Strategie, Argumente und Maßnahmen." Accessed February 8, 2016. https://www.energieblogger.net/public/via_GreenPeace_Kommunikationskonzept_Kernernergie.pdf.

[60] Leuschner, Udo. "Energiechronik: PR-Agentur erstellte Strategiepapier für die Atom-Lobby." September 2009. Accessed February 8, 2016. http://www.energie-chronik.de/090907.htm.

cover read "for Eon's nuclear power division," and the recommendations directly addressed Eon repeatedly.

It wasn't the only leak about nuclear strategy behind closed doors. For instance, EnBW produced a 26-page document about "shadow planning" to keep the Neckarwestheim plant running until after the elections. This paper was also leaked to Greenpeace just weeks after the elections. It detailed how power production was reduced to 4.7 terawatt-hours in 2007 at the reactor—and 3.8 terawatt-hours in 2008. From 1990 to 2006, in contrast, the plant generated between 5.4 and 6.3 terawatt-hours annually.[61] The paper expressed the hope for a better political constellation that would roll back the nuclear phaseout after the elections.

By September 2010, the new coalition had made a deal to extend the lifetimes of Germany's nuclear plants. Plants that had gone into operation before 1980 were given another eight years; those after 1980, another 14. The last nuclear plant in Germany would now not go off-line until 2036— instead of around 2022 under the Red–Green phaseout law. The political opposition, of course, vociferously opposed the extension of nuclear plant commissions. The term "phaseout of the phaseout" (*Ausstieg aus dem Ausstieg*) was coined, which clearly overstates the case; no new nuclear reactors were to be built. The phaseout had merely been postponed, not reversed.

The deal gave even more windfall profits to the nuclear plant owners, who were still awash in liquidity from emissions trading. But the arrangement did not fare well with the public. The government therefore imposed a tax on nuclear fuel to recover some 30 billion euros, but Merkel's new coalition had not even asked the firms to meet her halfway; the total windfall profits from the extension were estimated at 100 billion euros.[62] On 13 December 2010, the arrangement became law after being signed by German President Christian Wulff, whose signature was required after a review of constitutionality. Nonetheless, German states governed by the SPD immediately announced they would challenge the agreement in

[61] Messina, Sarah. "EnBW: Laufzeitverlängerung selbstgemacht." Klimaretter. 13 October 2009. Accessed February 8, 2016. http://www.klimaretter.info/energie/hintergrund/4062-enbw-laufzeitverlaengerung-selbstgemacht.

[62] "Atomkraft, ja bitte." Süddeutsche Zeitung. 28 September 2010. Accessed February 8, 2016. http://www.sueddeutsche.de/politik/energiepolitik-kabinett-beschliesst-laengere-laufzeiten-fuer-atomkraftwerke-1.1005340.

Constitutional Court. In particular, approval from the *Bundesrat* normally would have been required because each German state is liable for its own nuclear waste. But in a clever move the Nuclear Power Act was amended at the same time so that now only the federal government was liable. Merkel's new coalition had lost its majority in the *Bundesrat* after state elections in May 2010. Though the chamber did not have a right of veto, it could have asked parliamentary questions for clarification, thereby slowing down the process considerably. The tax on nuclear fuel rods was also destined for the federal budget, not state budgets, so Chancellor Merkel promised to revisit the general sharing of tax revenue with the states if the *Bundesrat* (which represents states' rights) did not slow down the adoption of the nuclear deal.[63]

Municipal utilities and others without nuclear facilities also opposed the deal not only because it transferred profits to their competitors, but also because the German power market already had too many power plants. Most of all, the deal energized the anti-nuclear community and mobilized large demonstrations. In April 2010 (during the negotiations), some 120,000 citizens had formed a 120-kilometer chain of people holding hands between the Brunsbüttel and Krümmel reactors. Another 10,000 surrounded Biblis, with 5000 also demonstrating at Ahaus, an interim repository site.[64] The popularity of the Greens reached a national record level at 21 percent in November 2010 in the first polls taken after the new nuclear deal had been announced.[65] Likewise, polls found that a majority of Germans opposed the extension of nuclear plant commissions.[66] Nonetheless, Merkel herself and the CDU remained popular, largely because of her handling of the economic crisis, which Germany

---

[63] Leuschner, Udo. "Energiechronik: Bundesrat läßt Atom-Gesetzgebung ohne Einspruch passieren." November 2010. Accessed February 8, 2016. http://www.energie-chronik.de/101105.htm.

[64] Leuschner, Udo. "Energiechronik: Massenkundgebungen gegen Revision des Atomausstiegs." April 2010. Accessed February 8, 2016. http://www.energie-chronik.de/100409.htm.

[65] "Politbarometer November 2010.: CDU/CSU erholt sich weiter - die SPD verliert Mehrheit will Steuermehreinnahmen für Schuldenabbau nutzen." 12 November 2010. Accessed February 8, 2016. http://www.forschungsgruppe.de/Umfragen/Politbarometer/Archiv/Politbarometer_2010/November/.

[66] Brost, Marco. "Schon wieder Ärger mit dem Volk: Die Mehrheit der Deutschen ist gegen die Atomkraftpläne von Union und FDP." Die Zeit Online. 22 July 2010. Accessed February 8, 2016. http://www.zeit.de/2010/30/Atomausstieg.

was weathering relatively well—and that issue overshadowed nuclear negotiations.

The way the deal was announced embarrassed the government. Only the indiscretion of RWE's executive Martin Schmitz to a question posed by Greenpeace's Tobias Münchmeyer at a conference revealed the finalization of the deal, which had not yet been properly announced. Under the assumption that the details were still being negotiated, Münchmeyer asked Schmitz how the firms could be prevented from passing on the nuclear fuel tax to power consumers. Schmitz answered that those details had been signed the previous night. Perhaps tired from an all-night negotiations session, Schmitz indiscreetly revealed that the deal had been signed at 5:23 AM on September 6—and he added, "we even got Environmental Undersecretary Jürgen Becker out of bed for the signing."[67] The impression was that secret negotiations were being held behind closed doors between top industry representatives and middle-level governmental officials—and that top policymakers had only been called in to sign what utilities wanted.

Corporate watch groups like LobbyControl pointed out that the "compromise" reached was based on a study produced by the EWI, an energy research institute largely funded by RWE and Eon.[68] The EWI had also previously called for the EEG to be done away with completely, arguing that the existence of the emissions trading platform made it redundant—even as the carbon price remained ineffectively low.[69] After several days of public pressure, the government finally published the signed agreement on September 9, but it was too late—already, everyone was calling it the "secret agreement."[70]

---

[67] Offline as the FTD closed in 2012: http://www.ftd.de/unternehmen/industrie/:energiekompromiss-der-geheimvertrag-mit-den-konzernen/50166716.html

[68] "Geheimabkommen zwischen Regierung und Atomlobby – LobbyControl fordert Offenlegung." LobbyControl. 8 September 2010. Accessed February 8, 2016. https://www.lobbycontrol.de/2010/09/geheimabkommen-zwischen-regierung-und-atomlobby-lobbycontrol-fordert-offenlegung/.

[69] Leuschner, Udo. "Energiechronik: Stromkonzerne wollen EEG durch europaweite "Harmonisierung" ersetzen." April 2010. Accessed February 8, 2016. http://www.energie-chronik.de/100408.htm.

[70] Leuschner, Udo. "Energiechronik: Deutsche Kernkraftwerke dürfen bis zu 14 Jahre länger laufen." September 2010. Accessed February 8, 2016. http://www.energie-chronik.de/100901.htm.

The EWI had also been involved in the Monitoring Report of 2010. In 2008, this report had been drawn up for the first time to inform policy-makers during preparations for the 2009 amendments to the EEG. That first report found that Germany would "probably have sufficient genera-tion capacity up to 2020"—with only a few nuclear power plants remain-ing online by then under the original nuclear phaseout plan from 2002. With renewable electricity having grown faster than expected, there was no reason for the second Monitoring Report to come to a different con-clusion only two years later. Indeed, the major change in the power sector was lower power demand in the wake of the financial crisis that started in September 2008. But an official study drawn up for the Economics Ministry could hardly find that nuclear plant commissions need not be extended now that the official line was different.

As a result, there was some maneuvering. First, publication of the study was postponed. The law required a new study by 31 July 2010, and when none appeared Greenpeace sued to have the report released.[71] The Economics Ministry then promised to produce the study in the fall; it was finally published in January 2011, months after the agreement for nuclear plant commissions had been reached.[72] To save face for the government, EWI took the blame for the delay, with the Economics Ministry explaining that EWI had been contracted to produce too much material by the deadline—a depiction that the Institute itself duly called "accurate."[73]

Across a flimsy 25 pages (roughly one page for each week of delay), the Monitoring Report on Electrical Supply Security included two sce-narios up to 2020: one (called "conservative") with the nuclear phaseout of 2002; the other (called "probable"), with the proposed reactor permit extensions.[74] "Unless the service lives of nuclear plants are extended," the

---

[71] Leuschner, Udo. "Energiechronik: Verzögert die Regierung absichtlich den "Monitoringbericht 2010"?" August 2010. Accessed February 8, 2016. http://www.energie-chronik.de/100808.htm.

[72] "Monitoringbericht 2010." Bundesnetzagentur. 30 November 2010. Accessed February 8, 2016. http://www.bundesnetzagentur.de/SharedDocs/Pressemitteilungen/DE/2010/101130Monitoring bericht2010.html.

[73] Leuschner, *Energiechronik: Verzögert die Regierung absichtlich den "Monitoringbericht 2010"?*

[74] Bundesministerium für Wirtschaft und Technologie. "Monitoring-Bericht zur Versorgungssicherheit im Bereich der leitungsgebundenen Versorgung mit Elektrizität." January 2011. Accessed February 8, 2016. http://www.bmwi.de/BMWi/Redaktion/PDF/M-O/monitor-

experts wrote, "Germany will need to import relevant amounts of electricity (around six percent) because neighboring countries (France and Eastern European countries) will be able to offer carbon-free electricity (mainly nuclear power) less expensively." The key to understanding this passage is "carbon-free." The experts were still assuming that the emissions trading platform would make coal power expensive. The actual outcome was the opposite of what the experts predicted: when Germany switched off 40 percent of its nuclear power capacity just three months later, it should have immediately become a net power importer, with high power prices on the exchange. In reality, as we will see below, its net power exports reached a record level, and its wholesale prices were lower than in neighboring countries.[75] Furthermore, France remained a net importer of German electricity. The report was clearly not worth the wait; it could hardly have been more off the mark.

In fact, current events already belied the study's conclusions at the time. In December 2010, the month before the Industry Ministry's second Monitoring Report warned about German reliance on French electricity, France nearly had a blackout during a cold spell, when power demand outstripped supply. France was home to roughly half Europe's electric heating systems at the time. Built initially to offset oil with nuclear power in the heat sector, the combination proved troublesome. French nuclear plants—like all nuclear plants everywhere—like to run at a constant level near the maximum. They therefore cannot ramp up further to cover spikes in power demand caused by the use of electric heaters. Indeed, the authors of the second Monitoring Report only needed to look at the capacity utilization levels of nuclear reactors in Germany's neighboring countries to realize that German reliance on carbon-free nuclear power from next door was a physical impossibility all along. These neighboring reactors were all running as close to maximum

ingbericht-bmwi-versorgungssicherheit-bereich-leitungsgebundene-versorgung-elekrizitaet,proper ty=pdf,bereich=bmwi2012,sprache=de,rwb=true.pdf.
[75] Morris, Craig. "The German switch from nuclear to renewables – myths and facts." Renewables International. The Magazine. 9 March 2012. Accessed February 8, 2016. http://www.renewablesinternational.net/the-german-switch-from-nuclear-to-renewables-myths-and-facts/150/537/33308/.

as possible already. They would not be able to ramp up further for exports to Germany during times of peak demand.

To make matters worse, when French power demand peaked in December 2010, seven of the country's 58 nuclear plants were off-line for planned and unplanned maintenance.[76] Power demand surpassed 96 gigawatts, roughly 50 percent above the country's total nuclear generation capacity even when all the reactors are running. As a result, grid operator RTE asked the public to reduce its power consumption "especially between 5 and 8 PM." Power plant operator EDF assured the public that "19 degrees Celsius [66 degrees Fahrenheit] suffices as an indoor temperature." The situation of power shortages was new for France, having begun in 2009. That October, France imported more electricity than it exported on a monthly basis for the first time in 27 years. Nuclear power had made France a power exporter for years, but demand had continued to rise while the country was unable to add new reactors. In October and again in December 2009, shortfalls in power supply brought prices on the French electricity exchange up to the maximum tradable level (an artificial cap) of three euros per kilowatt-hour (usually, kilowatt-hour prices are counted in cents). France imported nearly 7.2 gigawatts of electricity during those hours, close to the maximum line capacity of 9 gigawatts.[77] In 2009, German power exports covered roughly 2.5 percent of French power demand net. In 2010, French reliance on German electricity was only half as high—but Germany was the only country that France had a major negative power trading balance with.[78] This trend has continued until 2016.

In other respects, the Economics Ministry's second Monitoring Report also revealed a political signature. It called for an end to priority grid access for renewable electricity, arguing that the grid frequency would otherwise be endangered. It claimed that new power lines were needed because "stability limits" were already being reached in some parts of the

---

[76] Leuschner, Udo. "Energiechronik: Alle Jahre wieder: Frankreich leidet unter Stromknappheit." December 2010. Accessed February 8, 2016. http://www.energie-chronik.de/101207.htm.

[77] Leuschner, Udo. "Energiechronik: Stromknappheit in Frankreich ließ Börsenpreis explodieren." December 2009. Accessed February 8, 2016. http://www.energie-chronik.de/091202.htm.

[78] Réseau de transport d'électricité. "The French Electricity Report 2010." Accessed February 8, 2016. http://www.rte-france.com/uploads/media/pdf_zip/publications-annuelles/rte-be10-fr-02.pdf.

grid.[79] It called on renewable power generators to provide "ancillary services," meaning they would not only generate electricity, but also help stabilize grid frequency. Yet, wind turbines had been able to do so for years already. German manufacturer Enercon proved back in 2004 that its turbines could "ride through a fault," to use utility jargon.[80] The wind sector itself would later (selflessly) call for the bonus paid to the wind turbines for ancillary services to be done away with altogether because the bonus provided no incentive once the service became mandatory: "Wind turbine owners should cover the additional technical costs themselves, which would save [consumers] some 30 million euros per year," wind energy organization BWE argued in 2013.[81] The report was calling for something the wind sector itself was more than willing and able to provide. Likewise, at the beginning of 2011 the solar sector produced a proposal so that solar could help stabilize the grid.[82] In 2012, the industry came up with equipment that could do so around the clock, even in the dead of night. The solar sector then called on the government to require large arrays to provide this service.[83]

Confusingly, other studies are also called Monitoring Report. The one produced for the German Network Agency (the grid regulator) at the end of November 2010 (while everyone was waiting for the other one) is also illustrative. Focusing on trends in the power and gas markets, this report at least had a thickness Germans expect from experts: 305 pages. (In comparison, in 2015 critics called a 62-page proposal for new cogeneration rules "flimsy.") In order to justify the need for nuclear

---

[79] Leuschner, Udo. "Energiechronik: "Stabilitätsgrenzen des elektrischen Systems bereits punktuell erreicht"." January 2011. Accessed February 8, 2016. http://www.energie-chronik.de/110114.htm.

[80] Morris, Craig. "Intelligent innovations in wind power." Renewables International. The Magazine. 12 March 2014. Accessed February 8, 2016. http://www.renewablesinternational.net/intelligent-innovations-in-wind-power/150/435/77417/.

[81] Morris, Craig. "BWE's proposals for new energy policy." Renewables International. The Magazine. 16 April 2013. Accessed February 8, 2016. http://www.renewablesinternational.net/bwes-proposals-for-new-energy-policy/150/505/61969/.

[82] Morris, Craig. "Switching off gigawatts of solar inverters." Renewables International. The Magazine. 15 March 2011. Accessed February 8, 2016. http://www.renewablesinternational.net/switching-off-gigawatts-of-solar-inverters/150/407/30446/.

[83] Schwarzburger, Heiko, and Craig Morris. "photovoltaics to stabilize grid around the clock?" Renewables International. The Magazine. 16 February 2012. Accessed February 8, 2016. http://www.renewablesinternational.net/pv-to-stabilize-grid-around-the-clock/150/537/33107/.

plant extensions, the Network Agency's new Monitoring Report argued along similar lines as the Economics Ministry's Monitoring Report would later. "Demand for electricity has to be covered at all times," it stated. In 2009, power demand had peaked at 73.0 gigawatts, with 92.8 gigawatts of "hourly secure net dispatchable capacity" available at the time. The difference was 19.2 gigawatts—and Germany had 20.5 gigawatts of nuclear. Only an additional 4.7 GW of non-nuclear dispatchable capacity was in the pipeline net by 2020 (after the closure of numerous existing plants). It was thus not hard to read into the document the insinuation that a nuclear phaseout would endanger supply security, thereby leading to a reliance on power imports. But the document itself took a more diplomatic stance, stating instead that "no detailed conclusions can be drawn for the next few years in light of the discussion about extensions to nuclear power plant commissions during the drafting of this report."[84]

In addition to questions about corporate influence from EWI, nuclear critics raised the usual safety issues in arguing against extended nuclear plant commissions. In particular, Germany still had no final repository for nuclear waste, and the nuclear reactors were not required to beef up safety measures in return for longer plant commissions. But increasingly, a new concern was being voiced—inflexible nuclear is incompatible with fluctuating wind and solar. As Gabriel put it, as long as nuclear plants continue to run "you can't even support renewables because there won't be any space for the electricity on the grid."[85]

The reason for this concern was the unprecedented level of solar power production. In 2009, solar made up just over 1 one percent of total power supply, but it was already peaking at a tenth of demand at noontime on sunny days. To reach a mere 5 percent of total supply, solar would soon be peaking at half of demand—a level already reached on exceptional days in 2013. At 10 percent of total supply, solar will peak at

---

[84] "Monitoringbericht 2010." Bundesnetzagentur. 30 November 2010. Accessed February 8, 2016. http://www.bundesnetzagentur.de/SharedDocs/Pressemitteilungen/DE/2010/101130Monitoring bericht2010.html, p. 30.

[85] "Atomkraft, ja bitte." Süddeutsche Zeitung. 28 September 2010. Accessed February 8, 2016. http://www.sueddeutsche.de/politik/energiepolitik-kabinett-beschliesst-laengere-laufzeiten-fuer-atomkraftwerke-1.1005340.

close to 100 percent of demand in Germany—a level not yet reached as of 2016, but foreseeable on the horizon.[86]

The German nuclear sector responded to this new discussion by stressing the flexibility of nuclear reactors. For instance, in the spring of 2010 RWE claimed that German nuclear plants only ran full-blast as much as possible for economic reasons; they could, the firm argued, adjust power output "to a larger extent and faster than almost all other power plant types."[87] The German Physics Society also chimed in with a 123-page study claiming that nuclear plants can change their power output by 20 percent in only two minutes "and, within 10 minutes, from full load to half load and vice versa"[88]—but physicists are not power plant engineers.[89] After the major power outage of 2003 in the US, it took some nuclear power plants up to 12 days to go back into operation.[90] The data from Germany also clearly reveal that nuclear plants remain the least responsive to power demand and price signals of all generator types on the German market.[91]

In other words, Minister Gabriel had put the focus on a conflict that is still not properly understood in countries with lower levels of fluctuating solar and wind, not to mention less publicly available data than Germany has. Large shares of wind and solar power will require flexible backup generation capacity. Baseload has to go. And nuclear is the most inflex-

---

[86] These levels do not apply for all countries. In sunny parts of the United States, for instance, solar power production is more even across the year, meaning the peaks are smaller relative to the total annual contribution. More importantly, solar can make up a far greater share of supply in countries with greater power demand in the summer than in the winter—the opposite of the German situation.

[87] Leuschner, Udo. "Energiechronik: Atomlobby entdeckt die "Flexibilität von Kernkraftwerken"." March 2010. Accessed February 8, 2016. http://www.energie-chronik.de/100312.htm.

[88] Deutsche Physikalische Gesellschaft e.V. "Electricity: The key to a sustainable and climate-compatible energy system." Accessed February 8, 2016. http://www.dpg-physik.de/veroeffentlichung/broschueren/studien/energy_2011.pdf.

[89] Leuschner, Udo. "Energiechronik: Vorteile der Kraft-Wärme-Kopplung werden überschätzt." June 2010. Accessed February 8, 2016. http://www.energie-chronik.de/100612.htm.

[90] Lovins, Amory B., Imran Sheikh, and Alex Markevich. "Forget Nuclear." Rocky Mountain Institute. 2008. Accessed February 8, 2016. http://www.rmi.org/Knowledge-Center/Library/E08-04_ForgetNuclear.

[91] Moris, Craig. "A bad bank for German hard coal?." Energy Transition. The German Energiewende. 28 February 2014. Accessed February 8, 2016. http://energytransition.de/2014/02/bad-bank-for-german-coal/.

ible type of baseload. So you have to decide whether you want nuclear or wind and solar.[92]

Wheeling and Dealing in the Power Sector

In addition to extended nuclear plant commissions, the business situation for big energy firms seemed bright in general in 2010. Low carbon prices on the emission trading platform allowed them to continue old coal power plants without constraints, and strong lobbying efforts paid off to prevent a substantial reform of the overall trading system.

Meanwhile, attention turned to the German electricity spot market. Increasingly, there were charges of manipulation—artificial power shortages leading to price spikes. Germany and Austria had formed a single trading zone (called Phelix), and along with Switzerland and France the Epex market had been created in Paris in late 2009. But no one had bothered to create a proper regulator; French authorities were still only responsible for trading within the country, while German and Austrian authorities could not reach across the border into France.[93] State prosecutors would therefore not have been able to look into charges of price manipulation. In January 2011, the German Cartel Office, which had investigated the matter, announced that it had not been able to find any evidence of Germany's four biggest utilities withholding generation capacity in order to drive up prices.[94] Nonetheless, the European Commission had previously found enough suspicious activity at Eon forced the firm to sell 5 gigawatts of generation capacity at the end of 2008. Suspiciously, Eon had broken the seal on rooms containing documents the Commission wanted to investigate further. The seal was to prevent the company from destroying evidence. When

---

[92] Leuschner, Udo. "Energiechronik: Atomlobby entdeckt die "Flexibilität von Kernkraftwerken"." March 2010. Accessed February 8, 2016. http://www.energie-chronik.de/100312.htm.

[93] Leuschner, Udo. "Energiechronik: Strombörse betreibt Spotmarkt seit einem Jahr ohne behördliche Aufsicht." November 2010. Accessed February 8, 2016. http://www.energie-chronik.de/101108.htm.

[94] Leuschner, Udo. "Energiechronik: Kartellamt kann Konzernen keine Strompreismanipulation nachweisen." January 2011. Accessed February 8, 2016. http://www.energie-chronik.de/110106.htm.

it was broken, Eon unconvincingly claimed that humidity and residue from cleansers on the door had caused the breach.[95]

Overall, the biggest three German utilities had unusually high profits during these years. Collectively, RWE, Eon, and EnBW posted between 12 and 14 billion euros in profits annually between 2003 and 2006, but that level rose to nearly 25 billion in 2009.[96] A study conducted for the Green Party in 2010 found that Eon was making more profit from derivatives than from power sales, suggesting that the firm was speculating more than hedging against risks.[97,98]

Despite the tremendous growth in profits, the firms continued to raise retail rates—Eon by as much as 6.5 percent in 2010.[99] In their defense, the firms stated that their costs were increasing, partly because of renewable electricity. But another study produced for the Greens found that RWE had raised its rates that year even though the company had *lower* procurement costs—both in terms of wholesale power prices and fuel costs.[100]

In 2010, "energy poverty" began drawing more interest. It should be noted that the term is not used per se in German; in contrast to the UK, for instance, German policymakers have never bothered to define the level at which *Energiearmut* (a word not used in German policy) occurs.[101] Instead, Germans see energy poverty as a subset of poverty, and

---

[95] European Commission. "Antitrust: Commission imposes € 38 million fine on E.ON for breach of a seal during an inspection." 30 January 2008. Accessed February 8, 2016. http://europa.eu/rapid/press-release_IP-08-108_en.htm?locale=en.

[96] Leuschner, Udo. "Energiechronik: E.ON, RWE und EnBW erzielten in sieben Jahren über 100 Milliarden Euro an Gewinn." October 2010. Accessed February 8, 2016. http://www.energie-chronik.de/101003.htm.

[97] In April 2008, the Swedish government had taken over all the shares at Vattenfall, the other Big Four utility in Germany, so it no longer had to report its financial figures as a publicly traded firm.

[98] Leprich, Uwe, and Andy Junker. "Stromwatch 3: Energiekonzerne in Deutschland. Kurzstudie im Auftrag der Bundestagsfraktion Bündnis 90/Die Grünen." Accessed February 8, 2016. http://www.energieverbraucher.de/files_db/1287647417_3999__12.pdf.

[99] "E.ON erhöht Strompreise im Mai 2010 um bis zu 6,5 Prozent." Check24. 15 February 2010. Accessed February 8, 2016. http://blog.check24.de/strom/strompreise-erhoehung-e-on-mai-2010-461/.

[100] Harms, Gunnar. "Gerechtfertigte Strompreiserhöhung? Kurzgutachten im Auftrag der Fraktionsgeschäftsführung der Bundestagsfraktion Bündnis 90/Die Grünen. Berlin, July 2010." Accessed February 8, 2016. http://www.baerbel-hoehn.de/fileadmin/media/MdB/baerbelhoehn_de/www_baerbelhoehn_de/studie_strompreiseharms.pdf.

[101] Morris, Craig. "Energy poverty still hard to compare." Renewables International. The Magazine. 25 February 2014. Accessed February 8, 2016. http://www.renewablesinternational.net/

they focus on reducing general poverty. Up to the economic crisis of 2008, most of the concerns revolved around the cost of oil and gas. Germany therefore covers the cost of heat, but not electricity, for welfare recipients.[102] One statistic put the number of German households that had their power cut off because they couldn't pay their bills at 600,000 in February 2012, but the number was extrapolated from Germany's most populous state of North Rhine-Westphalia.[103]

In 2010, Eurostat began producing annual comparisons of retail power prices in the EU, which drew more attention to Germany's relatively high rates—second only to Denmark's. The impression was that renewable electricity was making German power expensive. Historically, little had changed, however; Germany had been in the top four in terms of retail rates going back to the 1990s (along with the Netherlands, Italy, and Denmark), primarily because of high taxes and levies.[104]

Few Changes for Renewables

Along with the new nuclear legislation produced in the fall of 2010, a year after the elections, the new coalition also published an Energy Concept.[105] The previous coalition had produced a draft of this concept at the Meseburg Summit in 2007. As such, it is a good example of the strong continuity of policymaking across consecutive government coalitions. With the 2010 Energy Concept, the government set out Germany's energy policy goals until 2050 and specifically laid down measures for the development of renewable energy sources, power grids,

---

energy-poverty-still-hard-to-compare/150/537/77095/.

[102] Leuschner, Udo. "Energiechronik: Energiepreise werden immer mehr zum Politikum." July 2008. Accessed February 8, 2016. http://www.energie-chronik.de/080709.htm.

[103] Leuschner, Udo. "Energiechronik: Jährlich wird 600.000 Kunden der Strom abgestellt." February 2012. Accessed February 8, 2016. http://www.energie-chronik.de/120211.htm.

[104] Leuschner, Udo. "Energiechronik: Haushalts-Strompreise in den europäischen Ländern in Euro/kWh, einschließlich Steuern (Jahresverbrauch von 3.500 kWh)." August 2006. Accessed February 8, 2016. http://www.energie-chronik.de/060812d2.htm.

[105] Leuschner, Udo. "Energiechronik: Energiekonzept für eine umweltschonende, zuverlässige und bezahlbare Energieversorgung." September 2010. Accessed February 8, 2016. http://www.energie-chronik.de/100902d1.htm.

transportation, and energy efficiency. The concept addressed the sorely overlooked heat sector, with incentives for the weatherization of buildings and renewable heat. Requirements for energy renovations met with considerable real-world obstacles. The proposal to require building retrofits and the renewal of heating systems after a certain number of years, for instance, was widely perceived as governmental expropriation of private property—building owners were to be forced to spend money on insulation, better windows, and more efficient heating systems, which would only have paid for themselves over decades. These requirements were therefore watered down. Nonetheless, even the weaker requirements met with fierce criticism, with the *Frankfurter Allgemeine* claiming that making "the German economy an ecological example to follow" would "cost more than German reunification."[106]

In the wind sector, the focus for new incentives remained on offshore projects. Now, the government aimed to accelerate the construction of new power lines, including a PR campaign for high-voltage power lines. Germany's Energy Agency, Dena, published its second Grid Study at the beginning of 2010. Drawn up largely by the EWI (the same institute that had produced the nuclear compromise for longer reactor commissions), the study called for up to 3,600 kilometers of new transmission power lines in the base scenario. Another scenario had only 1,700 kilometers of new high-voltage power lines, but in turn 5,700 kilometers of existing power lines would have needed to be upgraded. The total—up to 7,400 kilometers—was substantial in light of the size of Germany's ultra-high-voltage grid: 18,000 kilometers. In retrospect, this estimate is clearly exaggerated; little has been built, and the plans have been trimmed down considerably.

The underlying assumptions for the growth of renewables were also widely off the mark. At the end of 2014, 38.1 gigawatts of onshore wind power capacity had already been built, but the Grid Study II expected only 34.1 gigawatts by 2015. Furthermore, a mere 37 GW was to be built by 2020—a rate of less than 0.6 gigawatts per year for the onshore sector, where citizens were heavily involved in community projects. The experts

---

[106] Leuschner, Udo. "Energiechronik: Bundesregierung will auch Altbauten "klimaneutral sanieren"." September 2010. Accessed February 8, 2016. http://www.energie-chronik.de/100902.htm.

were now telling them that their projects would be cut by two thirds in the second half of the decade.

In return, the expectations for offshore wind—where big energy firms do business—remained exaggerated. By 2015, seven gigawatts of wind turbines were to be in the water; in reality, the number will be closer to 2.7 gigawatts. Likewise, the assumption for 2020 was 14 gigawatts, a figure that had fallen to 6.5 gigawatts by 2015.

Likewise, the assumptions for photovoltaics were strongly underestimated. A mere 13 gigawatts was to be built by 2015, rising to only 17.9 gigawatts by 2020. Yet, 10.6 gigawatts had already been built by the end of 2009, when roughly 3.8 gigawatts was added. And the market had exploded further in 2010, reaching around 7.5 gigawatts—a level that would be sustained in 2011 and 2012 as well. At the time the Grid Study II was published, Germany had already built the amount of photovoltaics expected for 2015, and by the end of 2010 the level expected for 2020 would also be surpassed.

This different outcome greatly reduced the need for power lines, which were not being built anyway. As of 2012, only a few hundred kilometers of new transmission grid lines had been completed, but the reliability of German power supply remained among the highest in the world. The onshore wind farms and photovoltaics arrays that continued to be built exceeded the expectations of experts, but they were distributed—meaning built close to sites of consumption. Much of this electricity was therefore consumed in the vicinity without ever touching transmission power lines.

At the time, however, no one could be sure that the solar boom would continue. With the recent bust in Spain in mind, the German sector feared the worst. There were calls for drastic cuts to feed-in tariffs for photovoltaics not only among opponents of the Energiewende. Consumer advocate Holger Krawinkel called for a 30 percent reduction; solar monthly *Photon*, 20 percent.[107] But most proponents of solar opposed such great reductions. When the government finally announced a 15 percent cut in January 2010, Eurosolar, the organization founded by Hermann Scheer

---

[107] Leuschner, Udo. "Energiechronik: Solarstrom-Förderung soll um 15 Prozent gekürzt werden." January 2010. Accessed February 8, 2016. http://www.energie-chronik.de/100105.htm.

(co-author of the 2000 EEG), accused the government of "endangering jobs within the German solar sector."[108]

The proposal of 15 percent lower feed-in tariffs faced opposition in the *Bundesrat*. Christian Union politicians at the state level saw too many local benefits from photovoltaics, particularly in the economically struggling east. Although the *Bundesrat* could only have slowed down the legal changes, not veto them, the German political penchant for consensus prevailed. Governments of the *Länder* negotiated a compromise. For systems installed on buildings, feed-in tariffs were reduced by 16 percent, while those on brownfields dropped by only 11 percent. In addition, the annual automatic reduction in feed-in tariffs for systems newly connected to the grid was to drop by 9 percent annually, not 5 percent as in previous years. The price of solar arrays was simply falling faster than anyone had expected.

Furthermore, a growth corridor was defined for the first time for photovoltaics. The target range was now 2.5 to 3.5 gigawatts. If more was installed, future feed-in tariffs would drop even further. Likewise, if less was installed, feed-in tariffs would be reduced less. It seemed like a sensible mechanism, though *Photon* expressed concern about a pork cycle: the market might overheat, causing feed-in tariffs to drop so low so quickly that the market would stall, thereby keeping feed-in tariffs stable and producing another boom followed by another bust, and so on. These concerns turned out to be unfounded. In 2011 and 2012, the market exceeded the middle of the corridor—3.0 gigawatts—by 150 percent. Investors could see that rates would be lower in the future, so everyone scrambled to build quickly. The result was a sustained run on the market.

The Renewable Energy Surcharge Starts to Skyrocket

By the end of 2009, the new way of calculating the renewable energy surcharge for green electricity had backfired. In the new arrangement (the Balancing Mechanism), grid operators first sold green electricity on the power exchange, and the difference between feed-in tariffs and the

---

[108] Leuschner, *Energiechronik: Solarstrom-Förderung soll um 15 Prozent gekürzt werden.*

revenue from power sales was then passed on as the surcharge. Effective 1 January 2010, all renewable electricity fell under this new financing scheme, but wind power started to be sold in this way in May 2009. The effects were dramatic.

On 4 October, a surge in wind power production brought the price of electricity on the spot market down to −5 cents, meaning that power generators were paying buyers a tidy sum to take electricity off their hands. The main reason was a combination of fluctuating wind power generation and the inflexibility of German baseload power plants. They preferred to pay the market rather than ramp down their plants. Had Germany's 17 nuclear plants still in operation at the time been as flexible as the German Physics Society claimed, these negative prices would either not have occurred at all or at least not have been so low. The event clearly showed the incompatibility of inflexible baseload with wind power (and, by extension, solar).

In addition, the negative prices showed that the new financing scheme would artificially increase, not reduce, the renewable energy surcharge. In 2000, it started off at a mere 0.19 cents per kilowatt-hour. By 2009, it had risen to 1.33 cents in line with the increase in payments for feed-in tariffs. But starting that year, it exploded thanks to the Balancing Mechanism, rising roughly fivefold to 6.24 cents in 2014—although payments for feed-in tariffs had only doubled (from 10.45 billion euros in 2009 to 21.26 billion euros in 2014).[109]

If you, dear reader, are having trouble following these explanations, imagine the plight of the German public. Amidst reports of a booming photovoltaics market and excessively high solar feed-in tariffs, it was difficult to explain that a policy mechanism—not photovoltaics—was the real reason why the renewable energy surcharge was skyrocketing. It is also counterintuitive that lower prices on the spot market for electricity should raise a surcharge on power rates. But that is exactly how the Balancing Mechanism is designed.

---

[109] Mayer, Johannes N., and Bruno Burger. "Kurzstudie zur historischen Entwicklung der EEG-Umlage. Fraunhofer. Freiburg, 21 May 2014." Accessed February 8, 2016. https://www.ise.fraunhofer.de/de/downloads/pdf-files/data-nivc-/kurzstudie-zur-historischen-entwicklung-der-eeg-umlage.pdf.

Top politicians who opposed renewables pounced on the rising surcharge to argue that poor households could not afford the Energiewende. Reverting to the old way of tallying the surcharge would have fixed the problem immediately, and such changes were within their mandate—but they preferred instead to cherish the moment. The FDP's Christian Lindner reminded everyone of the then Environmental Minister Jürgen Trittin of the Greens comparison from 2004.[110] Trittin had stated that renewable electricity only cost the average household around one euro per month, "not more than a scoop of ice cream." After the financing scheme for the renewable energy surcharge led to much higher monthly expenses for households, Trittin was roundly—and unfairly—ridiculed; after all, he was not in office when the Balancing Mechanism was introduced. Critics of the Energiewende, including politicians responsible for the skyrocketing surcharge, began asking Trittin rhetorically how big that scoop of icecream was going to get.

Numerous pro-Energiewende experts did their best to shed light on the real problem. During the election campaign of 2013, the renewable energy surcharge became synonymous with the cost of the energy transition. Hermann Falk, head of renewable energy organization BEE, clearly stated, "The renewables surcharge is not a price tag for the Energiewende." Felix Matthes, the head energy researcher at Öko-Institut, also fought to make the distinction clear to the public, but as an academic, he struggled to find a wording that the public at large would understand: "The renewables surcharge is not a useful indicator for assessments or steering. It is a technical parameter for the implementation of the Renewable Energy Act, which has a lot of components."[111]

The explanations not only went over most people's heads, but they also came too late. When the calculation method was changed in 2009, no one from the political opposition in parliament pointed out the inevitability of what is now called the "cannibalization effect"—the more solar

---

[110] "Jürgen Trittins Eiskugel für 355 Euro." Liberale. 17 October 2013. Accessed February 8, 2016. http://www.liberale.de/content/juergen-trittins-eiskugel-fuer-355-euro.

[111] "The right price tag for the Energiewende: Interview with Felix Matthes and Hermann Falk." Renewables International. The Magazine. 19 November 2013. Accessed February 8, 2016. http://www.renewablesinternational.net/the-right-price-tag-for-the-energiewende/150/522/74670/.

and wind power are produced, the lower spot market prices will be.[112] This effect had become common knowledge by 2015, but Germany's Network Agency and the Environmental Ministry had nothing to say at the time. On the contrary, when spot market prices once again dipped into the negative in November and December 2009, the Network Agency defended negative prices as "necessary."[113]

One of the few voices who correctly predicted the future at the time were economist Lorenz Jarass and retired Undersecretary Wilfried Voigt, who published a paper explaining that high levels of wind power were incompatible with "the large number of new coal plants and extended service lives of nuclear plants, both of which need to run more than 6000 hours a year for economic and technical reasons" (a year has 8760 hours). The result, they explained, would be "surplus capacity that cannot be sold." They put their finger right on the problem: "When there is a lot of wind, the spot market price will be very low; when there is little wind, it will be high…For that reason alone, selling green electricity on the power exchange is problematic."

Written just before the extremely negative spot market power prices of October, the timing of the publication was perfect; it was newly in circulation when the negative prices drew everyone's attention. Yet, its conclusions were not heeded. The fundamental conflict remains: solar and wind power production responds to the weather, not to spot market prices.

Thanks to growing grassroots support, solar advocates fended off the pushback against photovoltaics from politicians at the national level. In March 2011, it seemed that solar would be able to continue to grow unabated for the time being. So would bioenergy. The German ability to find a consensus had produced a state of relative tranquility. The major groups in the energy transition all had a piece of the pie by the spring of 2011.

---

[112] van Renssen, Sonja, and Hughes Belin. "Dimitri Pescia, Agora Energiewende: "No more baseload in 2030, no case for new nuclear in Europe"" EnergyPost. 23 June 2015. Accessed February 8, 2016. http://www.energypost.eu/interview-dimitri-pescia-agora-energiewende-baseload-germany-2030/.

[113] Leuschner, Udo. "Energiechronik: Bundesnetzagentur will finanzielle Risiken des EEG-Ausgleichs begrenzen und verteidigt Negativpreise für Windstrom." February 2010. Accessed February 8, 2016. http://www.energie-chronik.de/100201.htm.

Except perhaps for opponents of nuclear power. "We felt that decades of successful protest had come to naught" when reactor service lives were extended, remembers Greenpeace's Tobias Münchmeyer.[114] On the other hand, the extensions gave Germany greater credibility in the eyes of several governments in neighboring countries. In March 2011, Italian Prime Minister Silvio Berlusconi still planned to build four new nuclear reactors in his country. For his part, French Prime Minister Nicolas Sarkozy had made it clear in 2009 how silly he found the German nuclear phaseout. He reminded the Germans that the wind generally blew from France to them: "And then the problem is the same whether the reactors are here or there."[115] With their extension of nuclear power in 2010, the Germans had at least rehabilitated themselves in the eyes of the somewhat tactless French leader.

And let's not forget the larger context. The financial crisis overshadowed all these aspects of the Energiewende for the general German public. But by the beginning of 2011, it was already clear that Germany was weathering the crisis relatively well. So on 10 March 2011, the Germans went to bed optimistic that their jobs were safe and that everything could continue more or less as expected. The Germans had what they wanted most: security.

[114] Personal communication with Interview Tobias Münchmeyer.

[115] Leuschner, Udo. "Energiechronik: Personalia." December 2009. Accessed February 8, 2016. http://www.energie-chronik.de/091217.htm.

12

From Meitner to Merkel: A History of German Nuclear Power

On 1 January 1939, Lise Meitner was hiking through the snow with her nephew Otto Frisch in a forest near Stockholm. Meitner was in Sweden involuntarily, for she was a Jew. While the younger Frisch had fled Germany in 1933, his aunt Meitner had only just barely managed to get out of Nazi Germany in time—with only 90 minutes of preparation, a single handbag, and 13 deutsche marks in her pocket.

Like so many Jews in the German-speaking world, Meitner failed to see the need to escape Nazi Germany in time. When Adolf Hitler became Chancellor on 30 January 1933, she wrote to her colleague Otto Hahn, a famous chemist, "Hindenburg [the old Chancellor] spoke a few short sentences and then introduced Hitler, who spoke very powerfully, tactfully, and personally. I hope things continue to develop this way."[1]

While Meitner may not have been an astute judge of the historical events unfolding before her, she was a genius in physics. That was why Otto Hahn had written her the letter she was brooding over during her

---

[1] Quoted in Sime, Ruth L. *Lise Meitner: A life in physics.* A centennial book 13. Berkeley: Univ. of California Press, 1996, p. 136.

© The Editor(s) (if applicable) and The Author(s) 2016
C. Morris, A. Jungjohann, *Energy Democracy*,
DOI 10.1007/978-3-319-31891-2_12

hike through the snow with her nephew. For his part, Hahn was a genius in chemistry. The two had worked together for several years in Berlin, especially since Meitner read about Enrico Fermi's work on "transuranium"—elements larger than uranium, the heaviest natural element on Earth. In the mid-1930s, Fermi bombarded uranium with neutrons, creating new elements in the process.

But Fermi was at a loss to explain exactly what was happening, and this inability enticed Meitner. As a physicist, she had her own ideas about Fermi's results. Eventually, she convinced Hahn to help her experiment. He had the skills in chemistry both to bombard a nucleus and isolate the products created in the process without any impurities—under the quite primitive lab conditions of the day.

After she went into exile, Hahn and Meitner continued writing letters, and in November 1938 their correspondence took on historic import. The experimenting in Berlin had continued, and Hahn asked Meitner if she understood his findings. Fermi logically suspected he had created elements with atomic numbers 93 and 94, even heavier than uranium. After all, uranium was being "pumped up" in the experiments, so the element must get bigger, right?

He was wrong. But Hahn had chemical skills enough to know what he was looking at when he bombarded atomic number 92: barium (atomic number 56) and krypton (atomic number 36). Still, the outcome made no sense to Hahn. So he wrote Meitner in exile.

Meitner had not been investigating the idea of a nucleus "bursting." But as she discussed the matter with her nephew, who was also a physicist, on their snowy hike, she realized that Einstein's $E = mc^2$, the formula for the theory of relativity, might be the explanation.[2] Armed with just a pencil, a pocket notepad, and her physicist nephew as a sounding board next to her on a log in the snowy landscape, Meitner and her nephew talked about how quite a lot of energy would be required to break apart an atom, but some mass would also be lost when ura-

---

[2] Einstein had won the Nobel Prize himself in 1922, but not for the formula everyone associates with him today; rather, it was for a previous paper on the "photoelectric effect"—essentially, what we now know as photovoltaics.

nium is broken up into barium and krypton. Maybe a uranium atom bombarded with a neutron breaks in two, like a drop of water that has gotten too big?

On a scrap of paper, she began calculating. Roughly 200 million electron volts would be needed to break apart the atom. Then, she calculated—using $E=mc^2$—how much energy would have been released from the amount of mass lost in this particular process: around 200 million electron volts. And sitting on a log in the snow with a pencil and a scratchpad in Scandinavian exile, Meitner made a breakthrough discovery of a lifetime—the one every scientist dreams of. She explained that a uranium nucleus doesn't just go up a few notches on the periodic table when bombarded with neutrons. It falls apart.

Hahn and his colleague Fritz Strassmann immediately published a paper reporting their findings from the experiment but left the explanation for their exiled friends Meitner and Frisch to publish in February. In that paper, Meitner chose to speak of "nuclear fission" (otherwise, we might talk about "nuclear bursting" today).

Meitner did not win the Nobel Prize for her explanation of Hahn's experiment, an oversight many feel is an injustice. She also would have been a good candidate for the Nobel Peace Prize. During the war, when prominent physicists from Fermi to Oppenheimer went to work on the Manhattan Project, Meitner wanted no part of the research on nuclear bombs, though she was invited to join. Hahn also refused. Later, she also refused to work on nuclear power, whereas Hahn promoted "atoms for peace." Perhaps Meitner felt a bit like Mary Shelley's scientist Victor Frankenstein, who regretted unleashing the beast he created.

The Germans first split the atom. Their skepticism of nuclear weapons and energy begins at the moment of the discovery of nuclear fission. It starts at the top level of those who understand the technology best and continues until today. In 2011, Christian Democrat Chancellor Angela Merkel put an end to her party's support for nuclear with the country's second nuclear phaseout. Merkel holds a PhD in quantum chemistry.

The Initial Years of Atomic Hype

After World War II, "atomic" had a bad name. Everyone associated it with the terrible bombs that destroyed the Japanese cities of Hiroshima and Nagasaki. In the mid-1950s, President Eisenhower announced a plan: nuclear reactors would also be built to produce cheap electricity. His "Atoms for Peace" speech to the UN General Assembly led to the 1954 Atomic Energy Act, which incentivized utilities and foreign countries to start investing in nuclear power.

In the US, private firms were to develop the technologies, which would then be sold to foreign countries, especially Europe. Millions of dollars in foreign aid was made available to countries like France and Germany for the purchase of nuclear technology made in the US, and the US shared its expertise with experts from these countries so that skills could be built up locally.

European governments willingly accepted the offer from America. In 1957, Euratom was created alongside the European Economic Community, the predecessor of the EU. Its purpose was to create a market for nuclear power and distribute the technology to member states; thus, nuclear has been at the heart of the EU since the outset. Oddly, all EU member states are still Euratom members, includes those without nuclear power or with nuclear countries phaseout plans (Austria, Belgium, Denmark, Germany, and Italy, for instance).

But the link between nuclear power and nuclear weapons remains until this day, as the conflict in Iran demonstrates; the Iranians claimed they were merely building a nuclear power plant, while the West accused the country of ulterior motives. As US author Mark Hertsgaard put it in his seminal history of nuclear power from 1983 entitled *Nuclear, Inc.*, "There is no firm line dividing the military and non-military uses of atomic energy."[3]

The German equivalent of Hertsgaard's book is Joachim Radkau and Lothar Hahn's *Rise and Fall of the German Atomic Industry* (*Aufstieg und Fall der deutschen Atomwirtschaft*, not translated into English). This his-

---

[3] Hertsgaard, Mark. *Nuclear inc: The men and money behind nuclear energy*. 1. American ed. New York: Pantheon Books, 1983, p. 8.

tory of German nuclear power does not mince words: "nuclear physicists needed to believe in the blessings of peaceful atoms to protect their elite status, lest the world see them as henchmen of death."

Nuclear was hyped up at the time. In 1954, Lewis Strauss, head of the US Atomic Energy Commission (AEC), famously stated that atomic power would soon be "too cheap to meter." Other overblown claims are less known today. People were told they would one day drive nuclear-powered cars. We never got those, but we did get a handful of cargo/passenger ships. The one made in Germany was even named after Otto Hahn. Built in the 1960s, it was a technical success but a commercial failure. The ship was not allowed to pass through the Suez or Panama Canal. Eventually, the nuclear equipment was removed and replaced with a conventional diesel engine. Nuclear has made more progress in military applications, such as submarines, Russian icebreakers, and France's aircraft carrier.

In 1956, in the run-up to the Euroatom agreement, an expert from the German Health Office stated, "Every major city is expected to try to get its own nuclear reactor." These "municipal nuclear plants" would have been quite small at only around 50 megawatts, compared to 1,400 megawatts today. Such small reactors would have been everywhere. Who would mind, with a nuclear car in every driveway?

Many of the people behind the nuclear hype were not physicists, but politicians, journalists, and even philosophers. In the 1960s, when the first major wave of nuclear builds was about to get started, the Social Democrats in Germany had argued that the amount of energy in uranium and thorium is "endless by human standards." One Social Democrat wanted to have "atomic boxes" of just a few cubic feet; with just a cable sticking out, a single unit was expected to serve a few thousand households.[4] The great energy density of uranium and thorium initially led people to expect nuclear to be used in small applications everywhere. The gigantic nuclear plants we have today—the largest power plants built— were not the obvious outcome in the mid-1950s.

---

[4] Radkau, Joachim, and Lothar Hahn. *Aufstieg und Fall der deutschen Atomwirtschaft*. München: oekom Ges. für ökologische Kommunikation, 2013, l. 973 ff.

In the late 1950s, German philosopher Ernst Bloch wrote one of the most influential books in his field, *The Principle of Hope*. In it, he speaks of nuclear energy "turning deserts into cropland, ice into spring; a few hundred pounds of uranium and thorium would suffice to make the Sahara and the Gobi deserts disappear and would turn Siberia and North America, Greenland and the Antarctic into the Riviera." (Sea level might have also risen several meters, which was not mentioned.) Worldwide, such ideas were taken seriously; Israeli President Ben Gurion was interested in nuclear research in order to irrigate the desert with desalinated ocean water.

But one German politician was not interested in the hype: Chancellor Konrad Adenauer. Adenauer wanted to use Euratom so that Germany could make its own nuclear weapons as quickly as possible.[5] The German Armed Forces had only been created anew in 1955, a year after Germany signed an agreement promising not to develop any weapons of mass destruction. The discussion about equipping the Armed Forces with nuclear weapons met with criticism among leading German nuclear researchers, including Otto Hahn and Warner Heisenberg. They were two of 18 leading scientists who signed the *Göttinger Manifesto* in 1957, which spoke out against nuclear weapons, but encouraged the use of nuclear as an energy source: "At the same time, it is extremely important for us that the peaceful use of atomic energy be promoted to the utmost, and we aim to continue to participate in this task." Nuclear weapons would eventually be stationed in Germany, but by US forces under US military control.

Nuclear power seemed to many progressive thinkers like the obvious path to take—the human mind prevailing over the scarcity of natural resources. Prudent utilities were not so quickly convinced, however. In both the US and Germany, power providers were sitting on enormous coal resources; they still are today. Their experts understood fossil fuels well, but they had no experience with nuclear energy. Indeed, practically no one did initially outside the military.

There was also a conflict of interest: what impact would new nuclear power plants have on existing coal plants? How would the value of coal

---

[5] Radkau and Hahn, l. 138.

reserves be affected? Cheap coal would be "alarming news for the US coal industry [and] ominous for the railroads that depend heavily on coal hauling fares," Forbes wrote in the early 1960s.[6] Not surprisingly, RWE—the German utility with the most coal assets—was the least interested in nuclear initially.

While people tend to think of big utilities today as pro-nuclear, utilities actually first had to be convinced to build nuclear plants in the 1950s. Utilities demanded two things before they got started: money and liability protection. First, the construction of the first reactors was to be funded with tax money because the projects constituted research; and second, liability for the risks of an accident needed to be limited.

In the US, the second demand led to the Price-Anderson Nuclear Industries Indemnity Act of 1957. It limited the power company's liability to 60 million dollars per reactor at the time, with the state covering an additional 500 million—a paltry sum given the 7 billion dollars in damage that a 500 megawatt nuclear reactor was estimated to be capable of in the Brookhaven Report published that year.[7] But the exact figures never mattered anyway; liability protection simply needed to be provided at some level so that the issue could be set aside and construction could begin.

International estimates of damage on which liability was based varied widely. In their Atomic Act of 1960, the Germans did not bother to produce a real estimate of potential damage either. They simply took the figure of 500 million from the Americans and switched the currency from US dollars to deutsche marks.[8] One expert described the discussion about whether corporate liability should be set at 10, 25, 50, or 100 million deutsche marks as "a crapshoot."[9]

For the liability of "third parties" (as the nuclear plant owners were misleadingly called), Europe came together in 1960 to agree on the Paris Convention and later on the Vienna Convention. Most European countries have signed and ratified one or the other, if not both. The

---

[6] Hertsgaard, p. 144.
[7] Wikipedia. "WASH-740." Accessed February 4, 2016. https://en.wikipedia.org/wiki/WASH-740.
[8] Radkau and Hahn, l. 4468.
[9] Radkau and Hahn, l. 4453.

Conventions make operators of nuclear plants liable for any damage and require them to be insured. At the same time, the Conventions also limit liability, currently to no more than US$15 million.

In the United States, each owner of a nuclear power plant pays a premium annually for each reactor to cover damage up to $375 million. If an accident causes greater damage, a second combined insurance of all operators comes into play. For a major accident, each licensee would contribute up to $112 million to pay for the damages. With 100 reactors currently licensed to operate, this combined insurance amounts to $12 billion overall. Neither utilities nor insurers would have to pay for damages exceeding this sum. Taxpayers would—or the people affected would be left to their own devices.

In Germany, the operator's insurance covers up to €2.5 billion in damages. If the insurance is insufficient, the operator is liable with its own corporate equity, which might be billions. If the damage exceeds both the insurance coverage and the company's equity, the extra damage coverage is passed on to society.

While these numbers might sound quite high, the damage caused in Fukushima dwarfs them. The Fukushima bills are not all in yet, but by May 2014 Tepco—the owner of the Daiichi reactors—had paid $38 billion alone in compensation to people and businesses in the affected area by the accident. Expecting ever-increasing compensation claims and clean-up costs, the government has raised the upper limit of its financial assistance to Tepco to $86 billion, estimating it would take 40 years to clean up the Fukushima site. Total cost estimates range from around $200 billion to about $500 billion—many times the entire equity value of any nuclear power corporation.[10] The estimate is in line with a French governmental agency, which put the cost of a Fukushima-scale accident in France at up to 430 billion euros, equivalent to around 20 percent of French GDP and "more than 10 years' economic growth."[11] The nuclear

[10] World Nuclear Association. "Fukushima Accident." Accessed February 4, 2016. http://www.world-nuclear.org/info/safety-and-security/safety-of-plants/fukushima-accident/.

[11] Morris, Craig. "French governmental agency puts price tag on nuclear disaster." Renewables International. The Magazine. 8 February 2013. Accessed February 4, 2016. http://www.renewablesinternational.net/french-governmental-agency-puts-price-tag-on-nuclear-disaster/150/537/60424/.

sector is only insured for a fraction of that and can only cover a small additional amount from its own equity. Everything else is passed on to society. The insurance sector itself would probably require so much money to cover the full cost as to make nuclear power prohibitively expensive. After the accident in Fukushima, a German insurer was asked how much it would require in provisions to cover the full liability of nuclear power, and it put the figure at 6 trillion euros. The term "maximum credible accident" (MAC) came into common use in the 1960s to ensure a higher level of safety. The term now used is "design basis accident," which more clearly expresses the intent: to account for and therefore rule out accidents stemming from a nuclear plant's specific design.

But as Radkau and Hahn argue, the MAC was primarily a bureaucratic fiction that allowed authorities to sign permits for reactors.[12] The definition was neither based on an analysis of actual accidents, nor further developed as experience was gathered with additional incidents. Here again, actual safety was not the point. The point was to remove obstacles so that construction could begin soon.

The Nuclear Race Begins

When firms first started building reactors, they faced a wide range of design options. Although some of them were tried and found wanting, all those still on the drawing board were already there in the 1950s.

One of them, surprisingly, was nuclear fusion. As opposed to fission, in which energy is released when a nucleus breaks up into smaller nuclei, nuclear fusion provides energy when two nuclei fuse to form a larger one. Our Sun is essentially a fusion reactor in which hydrogen (atomic number 1) nuclei fuse to form helium (atomic number 2). In that respect, solar panels (and, indirectly, wind turbines) are receivers of nuclear fusion energy.

At present, a prototype fusion reactor is being constructed in southern France. It is hoped that the facility, which is called International Thermonuclear Experimental Reactor (ITER), will provide proof of the

---

[12] Radkau and Hahn, l. 4650.

concept in the 2020s. If so, a second unit will have to be built to actually provide power as a fully operational fusion plant, and that may take until midcentury—if everything goes well.

In 1954, the German journal *Außenpolitik* (Foreign Policy) claimed that nuclear fusion would be "practically ready to use in two years."[13] In 1955, German nuclear minister Franz-Josef Strauss declared at the Atomic Conference in Geneva that fusion "might be" the conference's only real sensation.[14] He was referring to Indian nuclear physicist Homi Bhabha, president of the conference, who had stated that the world could expect nuclear fusion in around 20 years, putting us in 1975. In 1960, Germany founded the Institute of Plasma Physics (IPP) in Garching, which focused on nuclear fusion under the direction of Werner Heisenberg—who, ironically, stated in 1956 that "nuclear synthesis [meaning 'fusion'] is a long way off."[15] But if the government is handing out money, why say no?

GE stepped away from fusion research in 1969, and around that time the head of the IPP called fusion "adventuresome...100 million degrees is a fantastical temperature."[16] And so, 1975 came and went without nuclear fusion, which still remains adventuresome. The headline of a recent article in the New Yorker on the prototype plant being built in France put it well: "A star in a bottle."[17]

The great thing about fusion is that the plants would not produce any nuclear waste (though the facility itself becomes radioactive). Another technology option known as the "breeder reactor" also promised a solution to the waste problem; these plants would simply reuse the waste created as additional fuel. The term "breeder" refers to the production of further fuel in the process of generating electricity.

German nuclear researcher Wolf Häfele, who headed the nuclear research center at the town of Jülich, was fond of claiming that the "fast breeder" would "generate more new nuclear fuel than it consumes." The

---

[13] Radkau and Hahn, l. 1089.

[14] Radkau and Hahn, l. 864.

[15] Radkau and Hahn, l. 875.

[16] Radkau and Hahn, l. 892.

[17] Khatchadourian, Raffi. "A Star in a Bottle." The New Yorker. March 3, 2014 Issue. Accessed February 4, 2016. http://www.newyorker.com/magazine/2014/03/03/a-star-in-a-bottle.

idea was to produce a closed fuel cycle, with these plants essentially recycling their own waste. The technology continues to be worked on, but the idea of a closed fuel cycle has been abandoned. But in the 1960s, the Germans and the French competed fiercely over breeder technology. In December 1966, the Germans raced to complete a breeder plant in Karlsruhe before the French finished theirs in Cadarache (now the site of ITER).[18] Two months earlier, a breeder reactor near Detroit named after Enrico Fermi partially melted down, an event that might seem to warrant greater caution. The nuclear sector apparently took the meltdown in Detroit as the starting gun in a sprint. The French beat the Germans by a few hours in this "race," but few cared, and the entire event is largely forgotten today.

Still, the idea of waste-less or at least low-waste nuclear power is enticing. The Germans went on to complete a larger fast breeder in Kalkar in 1985. The price had skyrocketed from 500 million deutsche marks at the end of the 1960s to 7 billion marks when it was completed. With fluid sodium flowing through the cooling circuit, the plant was ready to go, but the state government of North Rhine-Westphalia refused to issue the operating permit. The federal government could have intervened, but the power provider's heart wasn't behind the project any longer. Power demand simply had not risen as expected, so the power sector was sitting on overcapacity—even though far fewer nuclear plants were built than expected. The plant was gutted and turned into an amusement park; you can now ride a carousel within the cooling tower.

The US stance on fast breeders may have also played a role in ending German interest. In 1977, President Carter put an end to the technology. Essentially, breeder reactors produce a lot of plutonium, which can be fed back into nuclear reactors for further use as a fuel. The problem is that plutonium is used to make nuclear bombs as well, so Carter decided that the technology posed too great a risk to national security.

---

[18] Radkau and Hahn, l. 2059.

A Range of Nuclear Design Options

At this point, you might think we are running out of nuclear options. In fact, we've hardly gotten started. Comparisons of different nuclear technologies do not make for light reading, either. Critics of nuclear have often complained that the technology's complexity makes it hard for the public to understand, and hence to assess. Some nuclear proponents have tried to turn that argument around. In 1980, Swiss nuclear researcher Walter Seifritz wrote that the knowledge required for solar technology is "simple," whereas nuclear requires "an active mind" to deal with the "much more complex intellectual challenges" than solar poses. With that encouragement in mind, we delve further into the niceties of various nuclear technologies.

In 1959, German researchers produced a small high-temperature test reactor in Jülich, leading to headlines in the German press reading, "Germany has won the first nuclear battle!"[19] Two similar reactors (also called "pebble bed reactors") were built, but this technology has remained marginal.

In addition to uranium, thorium can also be used as a fuel. Such plants have also been built, and a few are still in operation. Germany was initially a leader in this field. Two of its high-temperature reactors ran on thorium. A small test reactor led to a larger facility, which went into operation near the town of Hamm in 1985—after 15 years of construction.

Often, proponents of nuclear argue that unfounded or at least exaggerated concerns among environmentalists prolonged construction and caused the cost increases of nuclear. The reactor in Hamm is a good example of how things actually went. In the week before construction was to begin, one of the firms in the consortium, engineering giant Krupp, abandoned the technology altogether.

Power firms weren't interested in the technology either. They only offered to cover 50 million deutsche marks of the 690 million the plant was expected to cost initially, with the rest having to be covered by research funding; after all, it was the government that wanted the technology tested. With the plant, engineers entered new territory—so

---

[19] Radkau and Hahn, l. 2029.

much so that the project was called an experiment, not a demonstration. Disinterest from the power sector in this technology continued after it was completed; no such reactors were ever ordered.

The experience in Hamm was typical. Power plant builder Siemens stated that delays at another reactor were "doubtlessly not the fault of industry, which had serious concerns from the beginning about building such a complicated facility without clear technical instructions."[20] Environmentalists didn't cause these delays either. Rather, each nuclear plant was unique—a different technology, a different size. There was no standardization.

The Germans really wanted to build the heavy-water reactor Werner Heisenberg originally worked on. What sets this reactor apart is the fuel: natural, unenriched uranium. Most other reactor types require enriched uranium, which meant a dependence on supplies from the United States. One of the goals of nuclear power for European countries (especially France) was energy independence, so these countries should have rejected a plant design that would have made them dependent on the US.

Yet, the Germans only built two heavy-water reactors. The first was a relatively small "multipurpose research reactor" in Karlsruhe, which made history mainly for leaks—and the cover-up of such incidents, as we will see below. The other was a larger, but still relatively small unit in Bavaria, which ran for around 200 hours across its service life of 18 months in 1973–74. The plant worked so poorly that the operator shut it down without even asking the government for permission. Indeed, the reactor never actually passed the buyer's inspection.

The Nuclear Reactor Design Most Similar to Coal Plants Wins

Given all the problems described above, it is perhaps no surprise that the nuclear power plant design that became the most popular—or at least the most familiar to power plant builders—was the one that most resembled a coal plant: the light-water reactor (LWR).

---

[20] Radkau and Hahn, l. 2955.

Though this technology had a few strikes against it (not intrinsically safe, produces nuclear waste), it also had a couple of things going for it. First, it had the stamp of approval from the US government and nuclear power sector. It was also being built successfully in numerous places and was supported by large industrial firms. Investors increasingly liked it.

Whereas other technologies would see a small number of plants tested in various countries, with few follow-up projects, one LWR led to another, and they were getting bigger all the time. The high-temperature reactor in Hamm completed in 1985 had a capacity of 300 megawatts. In contrast, Germany had several light-water nuclear reactors with a capacity of more than 1,200 megawatts in the mid-1970s. LWRs made other technologies look insignificant before they had even been properly tested.

The growth in power plant size also had a political downside. Municipal utilities in Germany originally wanted their own small nuclear plants. Large utilities reacted by ordering ever larger units—and convincing politicians that bigger was better. The intent was partly to drive out competition; only the big boys could build nuclear. The focus would eventually become the Achilles's heel of the German nuclear sector; instead of numerous municipals with close contact to citizens, it created a small group of large corporations that could easily be demonized and corralled. On the other hand, small reactors would have meant nuclear everywhere—and in close proximity to populated areas.

LWRs prevailed in the end not because they were found to be the best option in open, objective discussions among experts. Rather, there was little discussion. "The technology won based on relatively low upfront costs and the influence and prestige of the US atomic industry and Atomic Commission," argue Radkau and Hahn.[21] The two analysts seem to regret the lack of discussion the most: "It was a kind of reasoning that did without public debate."[22]

Indeed, although phrases like "psychosis" and "henchmen of death" make Radkau and Hahn sound quite anti-nuclear themselves, they

---

[21] Radkau and Hahn, l. 5492.
[22] Radkau and Hahn, l. 1661.

repeatedly reveal a fascination for various nuclear technologies. At times, they seem to regret that more experimentation was not possible:[23]

> The transition to commercial production came so quickly that it put a definitive end to the experimental phase. In fact, the results of the various experiments were hardly even analyzed, making the development of nuclear power unique in its hectic pace, its high risks, its enormous cost, and the pressures brought about by the sluggishness of bureaucracy and large industry. Because promising concepts were repeatedly thrown out for a lack of time, there is reason to believe that a longer phase of experimentation would have produced other solutions for nuclear problems.

The two nuclear experts also implicitly defend the heavy-water concept:[24]

> The fate of heavy-water technology in Germany is thus less the result of basic drawbacks of the technology relative to the light water concept and more related to flaws in the decision-making process for nuclear policy, especially the short timeframe and inability to plan in one common direction.

They also seem to lament the demise of fast breeders:[25]

> Experience in the history of technology would suggest that breeders should have been developed in increments and connected to proven technologies to the extent possible.

Like a number of nuclear proponents today, they are not particularly fond of the LWRs that currently dominate the global nuclear fleet:[26]

> There was a wide range of theoretically promising alternatives to light water reactors, but almost none of them were properly investigated.

---

[23] Radkau and Hahn, l. 5517.
[24] Radkau and Hahn, l. 3463.
[25] Radkau and Hahn, l. 3494.
[26] Radkau and Hahn, l. 7669.

Indeed, they admit in their book that they once had a certain weakness for German nuclear expert Rudolph Schulten and his pebble bed reactor.[27] Schulten partly wanted high-temperature pebble bed reactors as a source of process heat and space heat. He realized early on that nuclear would be relegated to a minor role in energy supply if it remained stuck in the power sector. His concern was prescient; nuclear has never covered as large a share of global energy supply (including heat and transport) as biomass alone. But the world did not get Schulten's vision of nuclear plants integrated in power and heat supply. It got the simplest nuclear technology that was scalable the fastest—because that's what utilities wanted.

Although Radkau and Hahn criticize the overstatements made about nuclear in the 1950s, they also write, "Anyone who experienced the hype of the nuclear age in their youth (and shared some of the excitement) will want to be careful about making emotional prophecies about the world's energy future."[28] These two Germans educated themselves and followed the topic for decades. In the end, they reached a sobering conclusion.

The fundamental problem, they argue, lies in the nature of nuclear technology itself. The history of technological experimentation is one of trial and error. We made better coal plants, developed diesel engines, and invented airplanes by testing something—and closely studying why it failed. Explosions and accidents were commonplace. Progress in engineering consists in incremental learning from mistakes due to technology improvements. Nuclear risks are a true handicap when it comes to trial and error. We cannot afford to make many mistakes, and certainly not many big ones, with nuclear. Radkau and Hahn underwent a learning process—and concluded that it is hard for nuclear power to undergo such a learning process itself.

Surging Costs—Wall Street, Not the Greens, Killed Nuclear

Around 1970, the age of experimentation with different nuclear technologies slowly came to an end, and the die was ultimately cast for LWRs, which exist in a two basic designs: pressurized water reactors and boiling

---

[27] Radkau and Hahn, l. 7574.
[28] Radkau and Hahn, l. 7503.

water reactors. Decisions were also made not to have complex designs—no waste heat, no constant feed line for fuel, no connection to gas turbines, and so on. In fact, utilities wanted to use components from coal plants as much as possible. Simplicity was to facilitate construction and keep costs down, but it also came at the expense of safety.

As Radkau and Hahn put it, "The rule of thumb was not to add up multiple risks, but to use trusted components from coal plants. Though the security philosophy was rarely stated outright, it is demonstrated by actual events…The strategy reveals the helplessness of engineers in light of the confusing array of potential dangers." As early as 1974, a report in a German nuclear journal found that the use of "conventional" equipment in nuclear plants to reduce risks "led to the opposite result." [29]

President Nixon had a goal of 1000 nuclear plants by 2000 in the United States. For a few years at the beginning of the 1970s, the US seemed poised to come close to that goal. From 1970 to 1974, US utilities ordered 140 reactors, 28 per year. Orders went into overdrive in 1973 due to the oil crisis. In the US alone, utilities ordered a record number of nuclear plants that year: 41. The number is especially large when we consider that the US only had only around 100 commercial reactors in 2014.

Not all those ordered in 1973 were built, however. Indeed, a whopping 138 nuclear plants ordered had actually been canceled in the US alone up to 2002. While most people will probably tell you that Chernobyl was a major setback, if not a turning point, for nuclear power, in reality, nuclear was already dead when Chernobyl happened. Indeed, it started dying before the accident at Three Mile Island in 1979. In 1983, energy expert Daniel Yergin wrote, "There have been no new orders for domestic plants since 1977, and, according to the government, at least 100 existing orders have been canceled in the last 10 years." [30]

Nuclear proponents like to claim that it was environmentalists demanding excessive and retroactive changes to nuclear plants to increase safety levels excessively. But a study published by the RAND Corporation in 1977 came to a much different conclusion. It confirmed that,

---

[29] Radkau and Hahn, l. 1420.

[30] Yergin, Daniel. "Troubles of the Atomic Brotherhood." The New York Times. 31 July 1983. Accessed February 4, 2016. http://www.nytimes.com/1983/07/31/books/troubles-of-the-atomic-brotherhood.html?pagewanted=all.

What has often been characterized as a 'ponderous' licensing procedure continues to be widely credited with retarding the growth of what would otherwise have been a 'healthy' nuclear industry. Yet, more than three fourths of the 111 commercial-scale power reactors operating somewhere in the world in early 1976 were of light water design. Few were of other than American design; most were types conceived, developed, and successfully demonstrated by the US government and private American firms between 1950 and 1976.[31]

The RAND study finds that the US Atomic Energy Commission in particular did not grant engineers enough time to develop plant designs. In other words, the costly ex post facto requirements that nuclear plants under development increasingly had to fulfill resulted not from excessive protesting on the behalf of uninformed citizens overly concerned about radioactivity. Rather, experts themselves had not taken the time to come up with "sensible, objective regulations," thereby necessitating "lengthy and costly reviews of each individual license application," as nuclear historian Mark Hertsgaard put it in his history of nuclear from the early 1980s.[32]

While the RAND study puts the blame quite firmly on the AEC, Hertsgaard believes the four reactor vendors competing for buyers share some of the responsibility.[33] Each time a plant of a particular size was built, a competitor would offer the market one slightly larger. Even after utilities had settled on the LWR design, the constantly growing plant size led to a lack of standardization that plagued the industry well into the 1970s. Today, the largest nuclear reactors being built come in at around 1.6 gigawatts, but in 1970 reactors with a capacity of 2.0 gigawatts were envisioned. But no further growth in reactor size was possible; the sector simply wasn't receiving new orders.

It thus wasn't the Greens who killed nuclear; the Greens were not even founded in Germany until the 1980s. And it wasn't Chernobyl. It was Wall Street, and the murder happened in the 1970s.

---

[31] Perry, Robert, A. J. Alexander, W. Allen, P. deLeon, A. Gandara, W. E. Mooz, S. Siegel, and K. A. Solomon. *Development and Commercialization of the Light Water Reactor, 1946-1976.* Santa Monica, CA: Rand, 1977; Prepared under a grant from the National Science Foundation. Accessed February 4, 2016. www.rand.org/content/dam/rand/pubs/reports/2007/R2180.pdf.

[32] Hertsgaard, p. 62.

[33] Hertsgaard, p. 61.

Arguably, the murder was more a slow death of natural causes. In the second half of the 1970s, US utilities ordered no more than a dozen new nuclear plants. In the 1950s and into the 1960s, there was no lack of governmental support. In addition to limiting utility liability to a small fraction of the potential damage, the US government had offered "cost-plus contracts," which very much resemble the feed-in tariffs later offered in Germany for renewables in the form of cost-covering compensation. The Germans were simply offering renewables the policy support nuclear originally had in the US. Back in the 1950s, the US government pledged "to absorb all costs and pay the company [building a nuclear plant] an additional fixed fee" to ensure a profit—even as the entrepreneur ran no risk.

The same support was offered for nuclear in Germany. In 1960, RWE told the German government in no uncertain terms that it was not going to be forced to adopt nuclear as a part of any "national task." When the reactor at Gundremmingen was built, the government promised the firm that the plant would cost it a maximum of 100 million marks no matter what. The government ended up having to pay more than three times that much as costs spiraled out of control. Government officials began to suspect that the deal might not be startup funding, but a precedent. They were right—the reactors at both Kalkar and Hamm were 90 percent financed with taxpayer money. [34] Long into the 1970s, German utilities would have preferred not to order any nuclear plants at all; without subsidies, they wouldn't have.

The RAND study explains that until 1963, no reactor was sold under free-market terms in the US. By the mid-1960s utilities were supposed to purchase nuclear plants from engineering firms under more market-like conditions. It wasn't working. Strauss's "too cheap to meter" proved too good to be true. "Those and some associated assumptions proved to be hollow," the analysts at RAND concluded, adding that bid prices considerably understated manufacturing costs. By the early 1970s, the cost estimates from the 1960s for nuclear plants had already doubled. By the end of the 1970s, they had tripled. As long as different technologies were tested, no standardization was possible. But even when LWRs had pushed back competing options, ever increasing plant size meant there were no economies of scale.

---

[34] Radkau and Hahn, l. 2325.

A Long Process of Eroding Public Faith in the Nuclear Complex Begins

The Germans were willing to build nuclear plants very close to populated areas. During the Cold War, West Berlin was an island within communist East Germany. Against this backdrop, the idea to build a nuclear plant on Pfaueninsel, an island within the city limits of Berlin but on the outskirts of town, was born. One main selling point was the potential to use waste heat from nuclear reactors as process heating for industry and space heating in homes. American officials reacted nervously to the proposal and pleaded with West Germany to ask East German officials whether they minded having a nuclear reactor so close to their border (entire Berlin was close to the border, but Pfaueninsel was within shouting distance).[35]

The plant was never built, but West Germany went on to build numerous reactors very close to densely populated areas, and repeatedly the Americans were more concerned about the risks to the public than German officials were. One German expert tried to explain to the Americans that, by their standards, entire Germany is densely populated, so US standards for minimum distances simply could not be applied. The leading journal of Germany's nuclear industry wrote in 1969 that it "is especially interesting that West Germany is the country with the most projects" close to cities.[36]

The first commercial nuclear reactor built in Germany (with a capacity of only 15 megawatts) went into operation in 1960 some 20 kilometers from Frankfurt near Kahl. Although it had the greatest "population factor" of any German reactor, the location was chosen because a nearby heat plant's cooling system was available to be co-used. There were no protests. Indeed, there was little concern about radioactivity at the time. The alarm system consisted partly of a herd of sheep grazing on the grounds. Each year, one was slaughtered and tested for contamination.

In 1965, construction began on a second nuclear reactor nearby, this time with a slightly different design (using superheated steam). It went into operation in 1969, but only briefly; in 1971, after a year and a half, it was shut down again because the fuel rod elements were improperly

---

[35] Radkau and Hahn, l. 5003.
[36] Radkau and Hahn, l. 5034.

designed. By that time, the first reactor from 1960 had already had two major incidents. In 1966, the fuel rod elements overheated and started melting; and in 1968, the entire facility lost power for two minutes. Radioactivity escaped from the reactor numerous times, and the dismantling of the facility starting in 1988 had to be done with remote-controlled vehicles because the structure itself was so contaminated.

Officials knew that the Kahl reactor had been improperly built. Construction began even without a permit, which was provided after the fact; engineering auditors spoke of a "fait accompli." The rush to be successful with nuclear power was simply more important than paperwork, and Kahl would not be the last such example. When construction began on Block A at Gundremmingen (roughly 15 times bigger at 250 MW) in 1962, the safety report was submitted late and did not even contain an engineering drawing of the reactor. The plant owner, RWE, was not in such a hurry itself, but the government insisted that the plant be finished quickly—and promised to cover 90 percent of any losses incurred. The plant went into operation before it even had a proper permit. Roughly 25 kilometers from the midsize towns of Ulm and Augsburg and around 100 kilometers from Stuttgart and Munich, Block A at Gundremmingen ran for 11 years from 1966 until 1977, when an accident caused irreparable damage. In 1961, construction began on the "multi-purpose research reactor" (MZFR) 12 kilometers from downtown Karlsruhe, where the Germans aimed to demonstrate the benefits of the heavy-water reactor, the German nuclear flagship.

While German officials may have been lax about safety, events at the Karlsruhe research reactor revealed that researchers and firms could not be trusted either. On 2 March 1967, an accident at the MZFR released radiation, but the event was kept secret for five weeks—and even then, it was only made public because of an indiscretion. Evidence suggests that Siemens wanted the information held back, but executives at the research facility were also implicated. One of them explained:[37]

We should try to prevent mishaps in our facilities from becoming a public knowledge unless public safety is affected. Today, it will be a problem with

---

[37] Radkau and Hahn, l. 5388.

a loading machine, tomorrow a leak in a heavy water line, the day after tomorrow some valve will get stuck, and if we keep going we will eventually have to publish information about a door lock getting stuck...If we don't talk about these things publicly ourselves, the media will forget about them eventually.

The researchers in Karlsruhe were merely following the advice of US reactor expert Theos Thompson, who held a lecture in the city in 1964, in which he concluded from US experience that the public should not be told too much. In his view, it would be "psychologically clever" to "prevent a public discussion at an early stage of a project" because "otherwise objections and demands for (unrealistic) guarantees and other complications" would arise. Radkau and Hahn comment on the cover-up in Karlsruhe in their history of German nuclear power thus: "The long process of eroding public faith in the atomic complex had begun."[38]

RWE was the reluctant utility behind the research facility in Karlsruhe. The director was none other than Heinrich Schöller, who had been the CEO of the company from 1945 to 1961—at a time when money was so easy to make that executive board meetings were repeatedly not even held. Awash in revenue from coal power, Schöller rejected nuclear in 1957 because disposing of nuclear waste would make it too expensive "for the time being." His reason for taking the helm at the nuclear research center in the 1960s is therefore illustrative: "If the government wants to do something stupid by building nuclear plants too soon, then we might as well do the stupid things ourselves so we can keep them under control." This line of thinking was common throughout the conventional utility and engineering world, with one US executive of a reactor builder commenting around the same time, "In order to protect our birthright as a company, we had to go" into the nuclear field.[39]

Until the 1970s, safety philosophies mainly served to make it easy for authorities to grant permits, not to increase safety. Indeed, the attempt to streamline processes had made safety features an externality

---

[38] Radkau and Hahn, l. 5346.

[39] Radkau, Joachim. "Wie ausgerechnet RWE sich gegen Kernkraft wehrte." Zeit Online. 12 June 2014. Accessed February 4, 2016. http://www.zeit.de/wirtschaft/2014-06/atomkraft-rwe/komplettansicht.

that added to the cost and conflicted with profitability. Concerns were not taken seriously anyway. Engineering auditors reviewing Block A at Gundremmingen concluded in writing that "if the emergency cooling system fails partially for 10 seconds, the fuel rod containment vessel will explode," causing thousands of deaths. The project continued as though experts had given no such warning.[40]

Public officials were equally irresponsible in their search for a nuclear waste repository. In 1967, the German government purchased a disused salt mine called Asse for the disposal of nuclear waste. Work at the repository ended in 2004, but the site has been in the news repeatedly since 2008, when it was discovered that water was leaking into the facility, and numerous barrels were already clearly damaged. Photos of the site reveal that the barrels were not neatly stacked, but dumped onto each other from above—and thus were probably damaged from the moment they landed in the repository.

In the 1960s, the government purchased the salt mines for this purpose because Wintershall, a large German natural gas provider, was willing to sell at a low price—but wanted the deal to go down immediately. Government experts had little time to properly inspect the facility, but Wintershall assured them that "under normal conditions, there is no risk of water entering"—meaning that nothing will go wrong unless something goes wrong. And, the firm added, if anything did go wrong, it would not happen "in the immediate future." Governmental officials may have interpreted that to mean "not during my career."[41]

Experts in the US at the time were equally unconcerned about the seriousness of waste disposal. A "top nuclear executive" quoted in Mark Hertsgaard's history of nuclear (but not mentioned by name) wonders why people worry about what happens hundreds of years later: "To me, it's the craziest thing. Neither they nor their descendants are going to be there at the time when anything could conceivably go wrong. If you do a halfway decent job of disposing of it, it's at least a few hundred years before anything could go wrong, and they won't even be there then." John West, Vice President of Combustion Engineering's nuclear division, is as

---

[40] Radkau and Hahn, l. 4907.
[41] Radkau and Hahn, l. 3925.

quoted in that book, saying there are too many good solutions to choose from (and the language is reminiscent of the early 1980s): "It's kind of like you have a blonde, a brunette, and a redhead, real glamorous gals all lined up for action, and you can't decide which one you'd like to go to bed with. They're all good." ("Rarely do these men speak with reporters at all, much less as frankly as they did with me," Hertsgaard later admitted.)

Of course, some German officials honestly wanted to know how to make nuclear safer. During the era when different nuclear reactor designs were being tested, politicians asked what the outcomes were, but researchers were not about to rat on each other. As one Social Democrat complained, "the safety problem is always the same, and all reactors always have the same level of safety." When a parliamentary commission asked one expert working on a fast breeder what he thought about another design, he said he would not answer the question because "it wouldn't be polite." The politician responded, "That's exactly the problem!"

A Nuclear Convoy—Attempts to Standardize Nuclear Plant Design

At its headquarters in Ludwighafen, BASF considered building a reactor for its chemicals plant (the largest in the world), and the discussion led to talks about "residual risk."[42] Because a populated area was close by, the plant was to have "burst protection." This decision then set a precedent for the plant in Wyhl, when a court ruled that rural people deserved the same protection as those in cities. In the end, BASF never built the 600 MW reactor; instead, RWE built Biblis. And that nuclear plant set the course for what was supposed to become a breakthrough in a "convoy" of nuclear plants based on a similar design.

Though it was a bit further from a conglomeration, Biblis was also the largest nuclear power plant in the world at that point with a capacity of 1.2 gigawatts, twice as large as the reactor proposed by BASF. Traditionally, Germany's largest coal power provider, RWE had been skeptical of nuclear all along. In late 1966, the firm told Germany's research minister

---

[42] Radkau and Hahn, l. 5098.

in writing that "power consumption would have to continue to grow at the current pace or even faster" for there to be any need for nuclear plants in Germany.[43]

The German government made that promise; indeed, the promise was made everywhere nuclear plants were built in the 1960s and 1970s. RWE then chose to build such a gigantic plant to show its competitors who was the boss. The firm was feeling pressure from competitors like Preussenelektra, which had built a 670 MW nuclear plant in Würgassen, and Eon, which built a reactor of nearly the same size in Stade. The former was completed within three years; the latter, in four. The nuclear race was in full swing.

Stade has the distinction of being the first reactor shutdown as a result of the nuclear phaseout of 2002 ("for commercial reasons," as Eon stated at the time). It was one of the few nuclear plants in Germany whose waste heat was recovered for use as process heat, here for a nearby salt production facility. Würgassen didn't even stay open long enough to be phased out officially. The plant was shut down after 19 years of operation when microfissures were discovered.

Located some 20 kilometers from Mannheim and Darmstadt, Block A at Biblis set the stage for what was to come. The plant was just what big utilities wanted. It was gigantic and straightforward, with normal (light) water as the heat medium and with a pressurized reactor (Pressurized Water Reactor [PWR]). This design currently makes up a majority of plants worldwide, including roughly two thirds of those in the US—and most of the ones in France.

Construction began in 1970, and the plant went into operation five years later in 1975. In 1972, construction began on Block B. Slightly larger at 1.3 GW, it contained only slight tweaks from Block A. Block B was also completed in roughly 5 years. It seemed that standardization might be within reach. Two additional blocks, C and D were also planned.

They were never built. Around 100 million marks had already been invested in Block C by the time it was canceled. Nuclear plant costs had not dropped as expected, nor had power consumption increased in line

---

[43] Radkau and Hahn, l. 2450.

with expectations. The 1950s and 1960s were a time of rapid economic growth in Western countries, and they were also a time of "electrification," with an increasing number of household appliances and industrial processes running on electricity. But the 1970s were a time of relative economic stagnation that left US utilities with excess generation capacity at the beginning of the 1980s. By the time Biblis A and B began generating electricity in 1975 and 1977, Germany was clearly building more power plants, including nuclear reactors, than it needed.

The nuclear "convoy" of reactors with similar designs came about, but it was more a lonesome trio than a proud line of tradition. Two different firms had built Stade (Siemens) and Würgassen (AEG), but those two competitors decided competition was not good for business after completing those plants, so they merged to form a new corporation, KWU. The manufacture of nuclear plants was now firmly in the hands of a monopoly in Germany, and the size of the new power plants also restricted new construction to an oligopoly of the biggest utilities. This group set out to dominate the future of nuclear plants. Three were built, each with 1.3 to 1.4 gigawatts of capacity: Isar 2, Emsland, and Neckarwestheim 2. Each was completed within around six years, and all three have been in operation since 1988 to 1989.

Neckarwestheim 2 was the last nuclear plant built in Germany. Construction began in November 1982, three and a half years before Chernobyl. Even at this late date, radioactivity did not seem to be a major concern. Neckarwestheim 2 has the distinction of being quite possibly the only nuclear plant in the world with an official hiking path leading across it. When it was built, it needed access to the Neckar River near Heidelberg for cooling water. The mayor of the village had one condition: the historic hiking path along the riverbank must be respected. Though the path has fallen out of public memory, it is still possible today to walk past a sign reading "do not enter—industrial plant" in the woods. Walk up to the three-meter tall concrete wall topped with razor wire, approach a door, and press a button next to a small red nameplate reading "ring here." When the reactor's security staff answers, simply request passage. An armed security guard with a trained German shepherd will be happy to accompany you to the other side. But you will need to hurry to take the hike. The three convoy plants will be among the last to be closed after some 33 years of service at the end of the nuclear phaseout. Neckarwestheim 2 is scheduled to shut down for good as the last plant in December 2022.[44]

By the time the accident at Chernobyl happened, utilities had already lost interest in nuclear. Chernobyl thus did not spell doom for the German nuclear sector, which was essentially already over at the beginning of the 1980s—not only in Germany, but in the US as well. Indeed, Hertsgaard starts off his history of nuclear from 1983 with two provocative questions: "What went wrong with nuclear power? How did it come to the verge of collapse in this country?"[45]

But one country stands out as a clear outlier in this respect. France actually built most of its nuclear plants in the 1980s. What made that country different?

Nuclear France: Silencing Critics with Brute Force and Pink Slips

In the early 1970s, France arguably had the strongest anti-nuclear movement in Europe. As in the US, the public always associated nuclear power with the country's aims for nuclear weapons. French critics of nuclear considered the technology to be "a large technical system destined to radically transform society, moving it violently towards a technical, centralized, and authoritarian model, with society organized on the basis of consumption and waste," according to sociologist and nuclear historian Sezin Topçu.[46] In her *La France nucléaire*, she traces the country's Messmer Plan of 1974—"all electric, all nuclear"—back to the late 1960s, when the government initially planned to use nuclear as an *oeuvre national* to ensure the country's industrial and military greatness.

In 1971, the government only intended to get 10 percent of the country's total energy supply from nuclear by 1985. Three years later, the new

[44] Petite Planète. *Welcome to the Energiewende*. 2013. Accessed February 4, 2016. http://welcometotheenergiewende.blogspot.de/2013/10/full-movie.html.

[45] Hertsgaard, p. 10.

[46] Topçu, Sezin. *La France nucléaire: L'art de gouverner une technologie contestée*. Paris: Éd. du Seuil, DL 2013, cop. 2013, l. 813.

goal was to get 70 percent of the country's electricity from nuclear by 1985 and eventually 100 percent of all energy, with part of the nuclear transition including electric heating. The plan meant that 80 reactors would need to be built by 1985—and 170 by 2000.[47] But because French power consumption remained stable at around 500 TWh instead of the 1000 TWh envisioned for 2000, the 59 reactors the country now has suffice to cover three quarters of power consumption. France has the greatest share of nuclear power worldwide, but today many think the French "nuclear transition" was more successful than it actually was—only a third of the plants originally envisioned were actually built.

Preparations had started even earlier. In the early 1960s, European countries were forced to relinquish their colonies. In 1961, France signed an agreement with Niger, the Ivory Coast, and Benin (then still called Dahomey) stipulating that:[48]

> the Republic of France shall have the first right of refusal for sales of uranium to meet its domestic demand…In other respects, uranium is to be provided to the French military, while exports to other countries are hereby ruled out. France shall assume control of the use of Nigerian uranium at the moment when it renounces its colonial power in the country.

In the case of Niger, the contract applied until 2004, with Niger forced to sell uranium to the French at prices far below market rates for nearly half a century. In contrast to the US and Germany, France had no major coal reserves. It naturally looked to protect privileged access to natural reserves in former colonies. Without large domestic coal companies, France also lacked a major potential critic of nuclear power from the utility sector.

When the oil crisis struck in 1973, the government presented its Messmer Plan to the world—without thinking to ask the French public for approval first. The government probably suspected an announcement of its nuclear plans before the oil crisis would not have been welcomed. In 1971, Jean-Jacques Rettig—one of the people who initiated major

---

[47] Topçu, l. 429.
[48] Topçu, l. 4542.

protests three years later in Marckolsheim and Wyhl—helped launch the first protest against the reactor in Fessenheim, France, on the German border. In 1972, the French chapter of Friends of the Earth, *Amis de la Terre*, worked with around a dozen other ecological organizations to collect 100,000 signatures calling for a five-year moratorium on new nuclear plants and a definitive end to atomic tests. In November of that year, French protesters did something the Germans would later become famous for—they blocked a shipment of nuclear waste.[49]

Concern was not limited to radioactivity, but also included the impact of nuclear power on democracy. The group *Survivre et vivre* (Survive and live) argued that a discussion focusing on nuclear waste would create "a debate between experts above the heads of the public, thereby excluding those who are truly concerned."[50]

The announcement of the Messmer Plan against the backdrop of the oil crisis did not sway the public, however. The government tried to sell nuclear as a way of becoming independent as well, even though the French national path of graphite-gas nuclear power had been abandoned in 1969, leading to months of strikes. When the government announced its commitment to nuclear power four years later, it was essentially announcing a reliance on licenses from the United States and shipments of uranium from abroad. France never produced significant amounts of uranium within its own borders in Europe and discontinued uranium production altogether around 2000.

While the public continued to debate the technology, the French government rushed into nuclear. By purchasing licenses and enriched uranium from the United States, France could hand over construction to a single firm, Framatom, and get going. The government decreed in March 1973 that construction permits could be granted for nuclear plants even before a "declaration of public utility" has been issued, meaning essentially that the Prime Minister and the *Conseil d'Etat* renounced their right to review projects beforehand. The sector had *carte blanche*.

---

[49] Topçu, l. 849.
[50] Topçu, l. 888.

Environmental impact assessments were not even required until 1978.[51]

As in other countries, safety was viewed as a cost driver, not a necessity. "The goal was not to reduce risks, but rather make them socially and economically acceptable," Topçu writes. She quotes a spokesperson for the then French power monopolist EDF, who argued that "The only reason why nuclear is getting so expensive is because of the growing desire to increase security to an excessive level." [52]

Protests continued at all levels. In 1974, labor union CFGT rejected nuclear power for promoting "technocratic, centralized police state."[53] In April 1975, thousands of people protested in various demonstrations across the country, with the largest one being held in Paris, where 25,000 people came together. A survey held at the time found that 80 percent of locals were against the project proposed in Golfech. In 1977, 90,000 people protested against Superphénix, a fast-breeder reactor that promised a closed nuclear fuel cycle—a breakthrough that should have interested the public. The plant was eventually completed in 1994 and ran for three years before French Prime Minister Lionel Jospin closed it for being "too expensive."

A confidential letter written back in 1977 by the firm building Superphénix reveals the strategy of *faits accomplis* that the sector was adopting: "The best way to counteract the local and national opposition that is developing is to get moving as quickly as possible in order to make the project irreversible." In this respect, the French nuclear sector was quite successful. While it took some 8 to 10 years for a nuclear reactor to be built in the US at the time, the French finished theirs in six on average.[54]

France differs from the US in Germany in another major respect: the way courts handled civil society challenges to nuclear power. In Germany, courts held up some nuclear plant builds for years. In the US, the anti-nuclear movement started off using the courts and

---

[51] Topçu, l. 1303 ff.

[52] Topçu, l. 617.

[53] Topçu, l. 963.

[54] Topçu, l. 1303 ff.

only adopted direct-action tactics after witnessing successes in France and Germany. In France, however, the courts essentially stated they did not understand the material and therefore could not handle cases.[55]

The accident at Three Mile Island marked a setback for nuclear power in most countries, but not in France. French authorities denied that any such accident could happen in the country's nuclear fleet even though French power plants are generally the same type (Pressurized Water Reactors). French experts claimed that the design had been "*totalement francisés*" (completely Frenchified).

Nonetheless, immediately after Three Mile Island a passing visitor might have taken France to be a country destined to abandon nuclear like the United States and Germany did in the 1980s. Nuclear had pushed French power prices through the roof, nearly doubling retail rates between 1975 and 1977. By 1983, a decade of the Messmer Plan accounted for nearly a sixth of the country's total foreign debt.[56] Indeed, Socialist François Mitterrand campaigned on an anti-nuclear platform for the 1981 elections, which he won.[57] One campaign promise was that there would be an 18- to 24-month moratorium on new nuclear plant construction. The position was not even new; in 1978, Mitterrand had helped bring together leading French nuclear critics for a report entitled "*Pour une autre politique nucleaire*" (for a different nuclear policy).

In office, Mitterrand did not slow down nuclear, however; he ensured its success in France during the very decade when the banking sector and eventually Chernobyl would decide the technology's fate in other Western countries. The elections apparently aged Mitterrand quickly, for he announced after entering office that "ecology is an illness that afflicts the young."[58]

Mitterrand got his Socialist Party, the chief remaining an opponent of nuclear in France, to rally around the technology as a national savior. In doing so, he pursued a clever strategy. In the 1970s, the French

---

[55] Topçu, l. 1374.
[56] Topçu, l. 579.
[57] Topçu, l. 124.
[58] Topçu, l. 1677.

anti- nuclear movement covered a wide range of society, but its main proponents were outside of the establishment. The protesters lacked resources, but they benefited from the integrity of their status as impugnable volunteer civil society activists.

Immediately on taking office, Mitterrand began creating new nuclear organizations and hiring leading anti-nuclear protesters to take part in them. Experts critical of nuclear power were broken up and put into different organizations, where they remained a minority opinion. Because their jobs were on the line, they could now also be told what to do. And because the 1980s were a crisis decade for the French economy, researchers in general feared for their jobs.

Crucially, independent research institutes were not created, unlike in Germany. The protests in Wyhl led directly to the creation of the Öko-Institut of Freiburg and Darmstadt and IFEU of Heidelberg. Under Mitterrand, France ensured that its top experts critical of nuclear would not come together in this way. Take two examples: In 1981, the French Agency for Energy Management (AFME; the predecessor of the current Ademe) was founded. By 1986, the year in which the accident occurred at Chernobyl, its budget and staff had been cut by a third. Second, Local Information Commissions (CLIs) were also created in 1981 with the laudable goal of providing locals with the information they desire about nuclear projects that affect them. By 1986, only four of CLIs's 14 chapters originally created were still in business.[59]

Chernobyl Cloud Stops at French Border

The French reaction to Chernobyl was also unique. Readings of elevated radioactivity levels taken across the continent were made public. France was the only western European country that did not publish such data.[60]

Countries throughout Europe warned citizens of the risks. Italy banned the sale of fresh milk for two weeks for children and pregnant women

---

[59] Topçu, l. 1785.
[60] Topçu, l. 2040.

along with the sale of fresh leafy vegetables. Austria temporarily banned the sale of fresh vegetables from Eastern Europe. The German government advised the public to feed powdered milk, not fresh milk, to babies and to wash fresh vegetables thoroughly. It also reassured the public that "no such accident can happen in Germany" because that particular reactor type did not exist in the country.

The fate of two prominent nuclear experts—one German, one French—illustrate the difference between the reactions to Chernobyl in the two countries. Friedrich Zimmermann, Kohl's Interior Secretary, stated in the wake of the accident, "There is only a danger within 30 to 50 kilometers of the reactor. There, the danger is great. We are 2000 kilometers away."[61] Environmental issues fell into Zimmermann's mandate at the time, but Chancellor Kohl did not wish to have a man of such opinions handling nuclear. The public was simultaneously being told not to eat certain foodstuffs from various areas. The government was contradicting itself and losing credibility. When Kohl created the Ministry of the Environment, Nature Conservation, and Nuclear Safety (BMU) roughly a month after the accident at Chernobyl, he took the environmental mandate from Zimmermann and gave it to the new head of the BMU.

In contrast, French medical doctor and radioactivity expert Pierre Pellerin told the French public in the weeks after the accident at Chernobyl that he would be "willing to take a walk a few kilometers from the plant, with my hands in my pockets, without any protection." Pellerin also claimed that "the radioactive cloud [from Chernobyl] did not pass over France." Pellerin remained at the head of an organization called SCPRI from 1959 until 1996—nearly 40 years.[62]

The SCPRI was founded within the French Ministry of Health specifically to handle radiation issues. As the head of this organization, Pellerin was not only the main authority figure in the country, he also was in a position to determine how what information was disseminated—or not. When the accident at Chernobyl happened, initially nothing was known outside the USSR. The Soviets did not report anything. But when

[61] Radkau and Hahn, l. 6698.
[62] Topçu, l. 2040 ff.

radioactivity meters reported unusually high levels in Sweden, European countries began sharing data.

In France, the nuclear cloud that allegedly stopped at the French border also set off alarms at nuclear facilities in Cadarache, Marcoule, and La Hague. Outside of the Soviet Union and the countries of the Warsaw Pact, France was the only European country where this information was kept secret. Indeed, researchers and experts who had this information were explicitly instructed to keep the information from going public. Employees of Pellerin's SCPRI were expected to keep information secret "for their entire professional life." The government forbade other researchers at universities and French scientific research organization CNRS from publishing any measurements taken. Thus, the French press was misled into announcing that France had been "miraculously" protected from the nuclear cloud.

In Germany, as in every other country except France and the Soviet Union, a lively public debate ensued after the accident at Chernobyl. As one French scientist who had been in Munich when the accident occurred put it, "Everyone was talking about it [in Germany]. In France, nothing!"[63] Only in Alsace, the area of France with a German-speaking population, did local officials react—perhaps because they had access to media reports in German—by banning the sale of spinach.

The history of nuclear power in France does not, of course, end with Chernobyl. In the 1990s, we find the emergence of the RSN, and the spelled out acronym says it all: *Regroupement pour la surveillance du nucléaire.* We find the new "re-group" speaking in the 1990s of the concerns from the 1970s—not only about radioactivity, but about technocratic authoritarianism: "first, decisions are made, and then we talk about them—a strange concept of participatory democracy."[64]

France inherited from the Messmer Plan a government that silenced its critics and its own experts initially with the brute force of a police state and later with the more civilized pressure of pink slips. In contrast, Germany inherited from its anti-nuclear movement a healthier democracy in which experts and citizens speak eye to eye, and there is greater transparency. Experts who spoke out in Germany could rest assured that their promi-

---

[63] Topçu, l. 2119.
[64] Topçu, l. 4203.

nence as a whistleblower would get them a good job at one of the numerous independent research institutes in business by the 1980s. Whereas François Mitterrand ended resistance to nuclear in the last major party opposing the technology in France, 30 years later Angela Merkel ended support for nuclear in the last major German party behind it.

To quote Topçu, the French government and nuclear sector used Chernobyl to perfect the "bureaucracy set up to keep critics from participating in the decision-making process" along with "professional tools to control public opinion."[65] The Germans, on the other hand, took Chernobyl as an opportunity to start talking openly about everything. Including a nuclear phaseout.

A Focus on Risks—Nuclear in Post-Chernobyl Germany

Chernobyl finally brought the risks of radioactivity from nuclear plants to the foreground—several years after investors had lost economic interest. It wasn't until the beginning of the 1980s that the public became generally aware of the risks of radioactivity. In the mid-1970s, the German Research Ministry conducted a survey on citizen initiatives against nuclear plants and found that only 123 press reports out of some 20,000 published between 1970 and 1974 expressed concerns about radioactivity.[66]

On the contrary, the Germans seem to be among the last to let go of the notion of radioactivity as having potential health benefits. In 1955, a Social Democrat wrote approvingly of "radium pillows."[67] In the 1960s, Germany investigated uranium mining near the small village of Menzenschwand in the Black Forest. Concern about radioactivity does not occur in documents from that era. Instead, there was a plan to use local radioactivity from the mines to open radioactive spas. The owner of one of Germany's largest publishing houses owned hunting rights in the area and loved the idea, but the amount of radon estimated to be available turned out to be quite low. Locals were divided over the issue, but

---

[65] Topçu, l. 5023.
[66] Radkau and Hahn, l. 5452.
[67] Radkau and Hahn, l. 4396.

not because of concern about radioactivity.[68] Rather, those against the mine and the spa felt that development would ruin their pristine rural environment, much like the villagers around Kaiserstuhl felt about the nuclear plant proposed a decade later. But the radon spa was eventually built and is open today.[69]

Then, US President Ronald Reagan began stationing scores of Pershing II nuclear missiles in Germany in the first half of 1983, which led to the largest protests ever across the country—and finally put nuclear power and nuclear weapons into a single context for the German public. The protests were powerful, because it was not only the peace movement and environmentalists demanding a stop to nuclear; protestors came from all parts of society. By the mid-1980s, the German discussion of nuclear power finally focused on nuclear risks. By that time, the people who would determine the future nuclear phaseout had entered the political scene.

In the State of Hessen, a Green politician named Joschka Fischer had become Environmental Minister. Hessen was the location of the Biblis nuclear power plant. When elevated levels of radiation were measured in 1985, he published them—the exact opposite of what the French government was doing.[70] Then, in December 1987 an accident happened at Biblis Block A. This time, RWE, the plant owner, decided to withhold the information from Fischer.

When the secrecy was revealed 11 months later, however, RWE not only undermined relations remained with Fischer, but also with the new Federal Environmental Minister, a Christian Democrat named Klaus Töpfer. He later revoked one company's permit to transport radioactive waste after it was discovered improper labels and bribes.[71] Töpfer later became head of the United Nations Environmental Program (UNEP).

The emergence of the Green Party gave the protesters a parliamentary voice, and the other parties came under pressure to respond. In 1986,

---

[68] Radkau and Hahn, l. 5725.

[69] http://www.radonrevitalbad.de/.

[70] Radkau and Hahn, l. 5963.

[71] Radkau and Hahn, l. 6603.

the Social Democrats (who were not in the federal governing coalition at that point) called for a nuclear phaseout within 10 years at their annual conference. The Christian Democrats, in contrast, were still largely pro-nuclear, but not without exception. In 1987, Töpfer said, "We have to invent a future without nuclear power."[72]

The German utility sector itself seemed ready for such a future by the end of the 1980s. In 1989, the Germans and the French decided not to further investigate cooperation on the reprocessing of nuclear fuel. Consequently, an end to nuclear power had become a distinct possibility, so plans had to be made. In 1990, German utility VEBA began talking about the future with the newly elected governor of the state of Lower Saxony, a man named Gerhard Schröder. In 1992, after 2 years of the deliberations, VEBA and RWE made a few proposals in writing to Chancellor Helmut Kohl, signaling a principal willingness to start thinking about slowly phasing out nuclear power. In 1993, Kohl launched the "energy consensus talks," in which the discussion focused partly on the future of nuclear power.

Fischer tried to have Biblis shut down starting in 1991. Töpfer's successor as Federal Environmental Minister—Angela Merkel—put an end to the decommissioning process. As a part of the deal, RWE was to increase the safety of Biblis, but the firm did nothing. Other nuclear plant operators, who had upped safety measures, were unhappy about their competitor's inactivity. By the time Merkel's party was voted out of office in 1998, she must have been increasingly unpleased herself. That year, one of her last actions as Environmental Minister was to stop the transport of radioactive waste to a reprocessing plant in Normandy, when 16 of the 68 containers were found to be inadmissibly radioactive.[73]

The new federal coalition that took power in 1998 included the Greens for the first time. Joschka Fischer became Foreign Minister, and hence Vice-Chancellor behind Gerhard Schröder, the former governor of Lower Saxony, a state bordering Hessen to the North. Lower Saxony is also the site of both Asse (the failed nuclear repository) and Gorleben (a site chosen as West Germany's final repository). Over the decades, Gorleben was

---

[72] Radkau and Hahn, l. 7363.
[73] Radkau and Hahn, l. 6760.

the site of repeated protests at least as intense as those in Wyhl—and at least as consensus-building. The site became a rallying point for the environmental movement. Schröder was thus deeply familiar with the anti-nuclear movement from his home state.

Nuclear vs. Democracy

All three of our history books draw similar conclusions about the conflict between democracy and nuclear power in the United States, France, and Germany. Here is Mark Hertsgaard from 1983:[74]

> *Nuclear power requires the type of arrangement between government and private capital known as state capitalism. The state must intervene with money or legal authority, or both, to create and maintain conditions favorable to long-term investment. The industry needs a guarantee of generous and unwavering support for decades to come. It must greatly reduce its vulnerability to the organized public pressure that forces changes in government policy. It must make its deal not with politicians and governments that a capricious citizenry can vote in and out of power, but rather with the permanent institutions of the state itself. The industry must, in short, shield itself from democracy.*

He elsewhere asks whether "a nuclear plant can ever be operated safely in an environment where conflict is the norm, and where the very technology must be designed to sidestep labor-management antagonisms."[75] Before the nuclear phaseout, Germans protested the transport of nuclear waste to France for reprocessing. In doing so, they made these train trips more dangerous. And when Greenpeace activists climb up containment vessels to unfold protest banners, they show the world that nuclear plants are not impenetrable. It's a Catch-22—the risks of nuclear power practically rule out protests against the technology.

Thirty years later, Sezin Topçu's history of nuclear France "pleads for less naivety and a stronger vision of democracy."[76] Topçu is particularly criti-

---

[74] Hertsgaard, Mark. "An indestructible Industry." *Mother Jones*: 27-31, 46; May 1983, p. 46.

[75] Hertsgaard, Nuclear Inc., p. 159.

[76] Topçu, l. 222.

cal of the weight nuclear produces a technocratic, authoritarian government "leading to a police state." As Cécile Maisonneuve, Director of the Center for Energy of the Institut Français des Relations Internationales, puts it, "The French aren't used to discussing energy issues in a public forum."[77]

Also writing in 2013, Radkau and Hahn likewise depict nuclear power as a technology that tends toward secrecy and authoritarian government. The violent protests in Wyhl and Brokdorf show that the technology pitted the public against a police state at one time. But Radkau and Hahn argue—and the present authors agree—that the story of nuclear power in Germany is largely one of the public educating itself: about the actual risks of nuclear power, but also about alternatives to nuclear power.

Radkau and Hahn dismiss out of hand allegations that Germans are overly scared of nuclear: "Jokes about an alleged German angst—standard fare among the critics of the anti-nuclear power movement for decades—are ignorant of history."[78] Rather, they argue that it is hard to compare Germany with other countries because the Germans eventually produced a lot of open data and information on their nuclear sector. "In this respect, the strength of the German anti-nuclear movement demonstrates the benefits of German political culture, especially since the 1970s."[79]

But wasn't the reaction to Fukushima, Chancellor Merkel's nuclear phaseout, an emotional and hasty decision, not based on information and reason? Where the public is concerned, certainly not. Fukushima did not change the German public's opinion on nuclear. It merely brought everyone out for another round of protests after the hatchet had been buried for nearly a decade.

And Angela Merkel? Landlocked Bavaria might not face any risks of tsunamis, but German nuclear plants do indeed lack adequate flood protection and are also built on fault lines. But if we want to know what Merkel thought, we could just quote her directly.

---

[77] Heinrich Böll Foundation. "What we're missing today is a clear and global vision for the future." Interview with Cécile Maisonneuve. 26 April 2013. Accessed February 4, 2016. https://www.boell.de/de/node/277252.

[78] Radkau and Hahn, l. 5931.

[79] Radkau and Hahn, l. 7605.

On 9 June 2011, three months after the accident at Fukushima, Merkel defended her sudden closure of 8 of 17 German nuclear plants and the reinstitution of a nuclear phaseout by 2022:

> You can only accept the residual risk of nuclear if you are convinced that it will not occur as far as it is humanly possible to determine…And that is exactly the point—it's not about whether Germany can ever have such a disastrous earthquake, such a catastrophic tsunami as in Japan. Everyone knows it won't happen exactly the same way. No, after Fukushima we're talking about something else. We're talking about the reliability of risk assessments and the reliability of probability analyses…Fukushima changed my stance on nuclear power.

To understand her stance better, we need look only a few months earlier. When Merkel extended the commissions of nuclear reactors by 8 to 14 years in the fall of 2010, the government issued an official statement entitled, "German nuclear plants are safe." It specifically stated, "An explosion or fire—like the one that happened in Chernobyl—is not possible" in any German nuclear plant.[80] Of course not, she seemed to have realized after Fukushima—if we are prepared, it won't be so bad. So if it's really bad, it has to be a bit different, unexpected.

Unless Angela Merkel publishes her memoirs, we may never know what exactly went on in her mind in the days after Fukushima. But the most likely scenario makes her reaction to Fukushima appear to be quite reasonable.

Immediately after the nuclear accident in Japan, Merkel sat in her office in central Berlin, just a short walk from the building where, more than half a century earlier, Otto Hahn and Lise Meitner had worked to split the first atom. On television, she saw top Japanese nuclear experts and politicians helplessly explaining to the public that they were not exactly sure what was going on. It would have been only human for her to imagine having to make a similar TV appearance, apologizing helplessly to the country she loved.

---

[80] http://www.bundestag.de/presse/hib/2010_08/2010_265/02.html.

We do know that she immediately received a call from Rainer Baake.[81] He had been Undersecretary of the Environment in Hessen during the 1990s, when Fischer (Baake's boss at the time) tried to get Biblis closed—against Merkel's opposition as the then Federal Environmental Minister. He reminded Chancellor Merkel of an almost forgotten review by Germany's Environmental Agency covering the various safety risks at German nuclear plants. The plants the study was most concerned about were the ones Merkel shut down that week.

At that point, Merkel must have remembered the battles she had held with Fischer and Baake. She will also have remembered how RWE failed to make good on its word to her. No doubt, she also knew how such firms had essentially been forced into the nuclear sector by overzealous German politicians, who are no longer among us, sprinting to make Germany a leader in a nuclear race largely forgotten today.

She will have remembered the pressure she had been under, as Chancellor Helmut Kohl's Environmental Minister, to toe the party line. Kohl referred to Merkel at the time as his "little girl" (*mein Mädchen*). She had worked for 15 years to rise up the ladder past the machos in German politics to become Chancellor in 2005. Even when she was reelected in 2009, large utilities, the large pro-nuclear camp within her party, and her new coalition partner pressured her to do what everyone expected when this governing coalition came about—revoke the nuclear phaseout of 2002. In the fall of 2010, she did it.

Half a year later, sitting in the building the Germans so unceremoniously refer to simply as the Chancellor's Office, she may have felt that the pressure was gone. Now, she was in power. And as a woman holding a PhD in quantum chemistry, Angela Merkel will have known about Lise Meitner, one of the first women in physics—someone who had recognized the beast she had unleashed and refused to foster it.

It would only be understandable if, at that point, Angela Merkel had formulated in her mind the same clear, calm, reasonable stance that Lise Meitner had taken a half-century earlier: I'm out.

---

[81] Drieschner, Frank. "Die stille Kraft, seit dreißig Jahren. Rainer Baake." Zeit Online. 27 March 2014. Accessed February 4, 2016. http://www.zeit.de/2014/14/rainer-baake-atomausstieg.

13

Merkel Takes Ownership of the Energiewende (2011–Today)

Overcapacity Instead of Blackouts

In February 2011, just a month before Fukushima, energy sector publisher Montel handed out a special issue at a German trade show. The main story—entitled "German plant closures 'inevitable'"—spoke of the need to shut down between eight and ten gigawatts of generation capacity based on price. The only thing Montel got wrong was the energy source; the article spoke of coal plants and gas turbines.[1]

Only four weeks later, Chancellor Merkel ordered the immediate, but initially temporary closure of eight of the country's 17 nuclear plants with a total capacity of 8.4 gigawatts within a week of the accident in Fukushima. The result, according to Montel, should have been higher wholesale electricity prices and greater profits for power providers—not blackouts.

---

[1] Morris, Craig. "Reality Check: massive overcapacity on German power market." Energy Transition. The German Energiewende. 11 April 2014. Accessed February 8, 2016. http://energytransition.de/2014/04/reality-check-massive-overcapacity-on-german-power-market/.

© The Editor(s) (if applicable) and The Author(s) 2016
C. Morris, A. Jungjohann, *Energy Democracy*,
DOI 10.1007/978-3-319-31891-2_13

Not everyone got the memo. One newspaper article called Merkel's hasty re-implementation of the nuclear phaseout a *Blitzwende*. The journalist revealed that, behind closed doors, power sector experts were not talking about whether a blackout would happen, but when—on a hot day in June, or when power consumption peaks in the winter?[2]

In public, the experts chose their words more carefully to avoid causing a panic. "To my knowledge, Germany has never turned off as much reliably dispatchable electricity so quickly," the head of Germany's Network Agency told the press—adding that the fear of a blackout was "justifiable."[3] But no blackout related to the Energiewende or the *Blitzwende* occurred.

When reviewing the winter of 2011 to 2012, the German Network Agency spoke of a "very tense" situation, which had required considerable actions on the part of transmission grid operators. German power consumers hardly suspected the situation was tense, however. They only experience an extremely high reliability in power supply. From 2009 to 2013, Germans experienced around 15 minutes of grid downtime a year. Then, the number fell even further to 12 minutes in 2014.[4] In comparison, France and the UK tend to have between 50 and 70 minutes of power outages annually; various parts of the US, from 100 to 400 minutes.

Although scarcely reported, the share of nuclear in Germany dropped even further just a few months after Fukushima, when scheduled maintenance led to the shutdown of an additional five nuclear plants, leaving the country with only four online in May 2011.[5] Within two months, Germany had gone from around 21 gigawatts of nuclear to 5.4 gigawatts. The lights stayed on.

---

[2] Iken, Matthias. "Gegen den Strom—Eine Pflichtverteidigung der Atomkraft." Hamburger Abendblatt. 27 May 2011. Accessed February 8, 2016. http://www.abendblatt.de/ratgeber/wissen/article106539997/Gegen-den-Strom-Eine-Pflichtverteidigung-der-Atomkraft.html.

[3] Schultz, Stefan. "Blackout-Debatte: Warum der Stromausfall ausfällt." Der Spiegel Online. 27 May 2011. Accessed February 8, 2016. http://www.spiegel.de/wirtschaft/soziales/blackout-debatte-warum-der-stromausfall-ausfaellt-a-765006.html.

[4] Morris, Craig. "German grid keeps getting more reliable." Renewables International. The Magazine. 21 August 2015. Accessed 08.02.216. http://www.renewablesinternational.net/german-grid-keeps-getting-more-reliable/150/537/89595/.

[5] Morris, Craig. "Can renewables replace nuclear? More nuclear plants off-line in Spain and Germany." Renewables International. The Magazine. 24 May 2011. Accessed February 8, 2016. http://www.renewablesinternational.net/more-nuclear-plants-off-line-in-spain-and-germany/150/537/30998/.

Montel had been right about Germany having more than enough power plants. On the other hand, their prediction about higher wholesale power prices was completely off the mark. From the winter of 2011 to the summer of 2015, German wholesale prices were cut in half[6] as renewable power continued to grow by leaps and bounds, easily making up for the drop in nuclear power production.

The Transdisciplinary Learning Curve

Initially speaking of a "moratorium," Merkel put together an Ethics Commission to advise her on the nuclear phaseout. She did not, however, ask it to review whether the nuclear plants should be closed. Rather, the question she asked the commission read:

> How can I prudently complete the phaseout so that the transition towards the age of renewable energy is practicable and reasonable, and how can I prevent, for instance, risks from the import of nuclear power to Germany, which may be greater than the risks in the production of nuclear power domestically?

The commission consisted of 17 people, many of whom would not have seemed obvious candidates for an assessment of nuclear power. There was a sociology professor, a philosopher, a former education minister, and the current president of the German UNESCO commission. Others seemed less out of place, such as an economist, a former research and technology minister, the current president of the German Academy of Technical Sciences, and other scientists. The ones that raised the most eyebrows outside of Germany were the three representing the Protestant and Catholic churches. The CEO of German chemicals giant BASF was the only businessman present. And there was even an American: political science professor Miranda Schreurs from the Free University of Berlin in her role as a member of the government's Advisory Council of the Environment.

---

[6] Morison, Rachel. "Why do Germany's Electricity Prices Keep Falling?" Bloomberg. 25 August 2015. Accessed February 8, 2016. http://www.bloomberg.com/news/articles/2015-08-25/why-do-germany-s-electricity-prices-keep-falling-.

Against the backdrop of the entire history of the Energiewende, the composition of this Ethics Commission is the culmination of a decades-long learning curve. Germany's energy transition had already clearly required a transdisciplinary approach—not just interdisciplinary across related fields (such as physics, chemistry, and engineering), but across a wider range of studies that truly represent the actual interests of society at large, including philosophy, sociology, theology, and so on. Appropriately, one of the two people heading the Ethics Commission was Klaus Töpfer, the former Environmental Minister and UNEP head, who had just helped found the Institute for Advanced Sustainability Studies (IASS). The IASS defines itself as conducting transdisciplinary research. In terms of political tactics, the Ethics Commission also offered Merkel a way of ensuring her own party's support for her nuclear phaseout.

There is a societal tension between the expert knowledge required for decisions made today in our highly complex civilization and input from the masses affected by these decisions, which they may not completely understand. On the expert end of the spectrum, we find a technocratic approach; on the other end, a transdisciplinary approach. The latter is more democratic; the former, more authoritarian. In putting together such an Ethics Commission, Merkel instinctively acted within the tradition of the Energiewende, which, we will remember, started off as a grassroots movement for more democracy against authoritarian government. Michael Sailer, the spokesperson for Öko-Institut (the research institute produced by the protest movement that started in 1974) stated in 2014 that he and his colleagues were transdisciplinary before the word even existed.[7]

So while international onlookers thought that the German anti-nuclear movement had finally reached its goal with Merkel's *Blitzwende*, in reality something bigger had happened: the German government had become open and answerable to the public. The Ethics Commission was a manifestation of the energy democracy called for way back in 1974 by the conservative rural communities in Kaiserstuhl.

---

[7] Sailer, Michael. "Transdisziplinarität nicht erst seit heute: Das Vorwort von Michael Sailer, Sprecher der Geschäftsführung des Öko-Instituts." Accessed February 8, 2016. http://www.oeko. de/e-paper/gemeinsam-fuer-veraenderung-der-beitrag-der-transdisziplinaeren-nachhaltigkeitsforsc hung/#c3958#c3944.

Remember the children's book *The Little Prince?* A Turkish astronomer presents the discovery of an asteroid, but no one believes him because he is dressed like a Turk. When he comes back in a suit and tie, his European audience listens.

Sailer himself frequently appeared on television during the spring of 2011. He never wore a tie, and it wouldn't have looked right anyway—Sailer has long gray hair, a mustache, and a soul patch. One can easily imagine him as a young hippie in the 1980s. He is also a chemist specializing in reactor safety. A member of Euratom's Scientific & Technical Committee, Sailer was head of the German Reactor Safety Commission when Fukushima happened.

Thanks to the internet, you can go back and compare, say, media coverage of Fukushima in the US with reports on the BBC, and if you speak a foreign language, you could see what was being said in France and Germany (students of media will find a treasure trove in such comparisons). In the US and the UK, only suits and ties were allowed in front of the camera. No one even closely resembling Sailer would have been asked to explain what was happening.

A BBC report entitled "Fukushima—disaster or distraction?" from 18 March 2011 can serve as an example. A "nuclear campaigner" from Greenpeace is quoted in the report, but the journalist foregrounds the brief mention by reminding readers unnecessarily that the organization has "a long history of opposition to nuclear power." In contrast, Richard Wakeford is presented as a "visiting professor in epidemiology at the University of Manchester"—not as someone who spent 30 years working for British Nuclear Fuels Limited in Sellafield,[8] a nuclear fuel reprocessing and decommissioning site with a murky history.[9,10] In other words, Greenpeace is presented as partisan, whereas the man whose career is partly based on downplaying the risks of nuclear power is presented as a neutral expert.

---

[8] "CV of Dr. Richard Wakeford." Accessed February 8, 2016. http://www.icrp.org/cv/%7BCA8DFACB-64B6-4393-B928-FD52B0302BBA%7D/Wakeford_CV.pdf.

[9] Nuclear waste has been dumped into the sea, and contaminated water is also pumped into the ocean near the facility.

[10] See the German documentary Ladwig, Manfred, and Thomas Reutter. *Versenkt und Vergessen. Atommüll vor Europas Küsten. Documentary produced by arte and SWR.* 2013. YouTube. Accessed February 8, 2016. https://www.youtube.com/watch?v=di0hzvK-Rd4.

and the BBC report Phoenix, Éamon. "NI state papers: Files Reveal Secret Dumping of Radioactive Waste." BBC. 28 December 2013. Accessed February 8, 2016. http://www.bbc.com/news/uk-northern-ireland-25470028.

In the spring of 2011, when nuclear proponents outside Germany were accusing the Germans of panicking, Michael Sailer was on television telling them in a calm voice that the radiation released from the destroyed Japanese reactors might not impact the population much if we got lucky and the wind blew it all out to the ocean.[11] When asked whether such an accident could happen in Germany, he reminded everyone of the similar nuclear plant designs and said the outcome would be similar if all safety measures failed at a German plant.

Sailer was never presented as a nuclear critic, but as a top nuclear safety expert. He answered questions "clearly...without making too little or too much out of anything," one media analyst group wrote, adding: "His appearances as an expert are free of ideology, blinders, and arcane technical jargon."[12]

Compare that description with the diatribe hurled against Germany by the pro-nuclear camp abroad. In 2013, for instance, British environmental journalist George Monbiot ridiculed Merkel for protecting Germany against the "imminent risk of tsunamis in Bavaria" (the state is at least 500 km from the nearest saltwater).[13] Such critics do not know that most of the nuclear reactors still in operation in Germany at the end of 2011 lacked basic protection from earthquakes and floods.[14] The cooling tower at the Neckarwestheim reactor has already sunk 40 centimeters into the ground because it was built on porous limestone; in 2002, an 18-meter crater opened up just 5 kilometers away, similar to the craters that form in Yucatán, Mexico.[15] In 2014, a smaller one opened up

---

[11] Tagesschau. *Michael Sailer (Ökoinstitut) über das Ausmaß der Gefahren von Kernkraft in Reaktion auf Fukushima*. ARD. YouTube. Accessed February 8, 2016. https://www.youtube.com/watch?v=tVdrH1EPJR0.

[12] Winterbauer. "Die Erklärer der Japan-Katastrophe." Meedia. 14 March 2011. Accessed February 8, 2016. http://meedia.de/2011/03/14/die-erklarer-der-japan-katastrophe/.

[13] Monbiot, George. "Out of Steam." 4 February 2013. Accessed February 8, 2016. http://www.monbiot.com/2013/02/04/out-of-steam/.

[14] Morris, Craig. "NGO itemizes risks of nuclear in Germany." Renewables International. The Magazine. 7 March 2013. Accessed February 8, 2016. http://www.renewablesinternational.net/ngo-itemizes-risks-of-nuclear-in-germany/150/537/61070/.

[15] Bartsch, Matthias, Andrea Brandt, Michael Fröhlingsdorf, Laura Höflinger, Simone Kaiser, Gunther Latsch, Cordula Meyer, Marcel Rosenbach, Holger Stark, and Gerald Traufetter. "Atomkraft: Landkarte der Schreckens." Der Spiegel 12/2011. Accessed February 8, 2016. http://www.spiegel.de/spiegel/print/d-77531730.html.

around 3 kilometers away.[16] Reactor 1 in Neckarwestheim was shut down in 2011. Reactor 2 will be one of the last three closed at the end of 2022.

Myth-Busting Post-Fukushima

Safely retired, former Chancellor Helmut Kohl called Merkel's decision a mistake, and the odd industry leader or economic advisor tried to explain why the sudden phaseout was a bad idea, but otherwise Merkel was not challenged.

Outside Germany, Merkel's decision was more divisive. Reactions broke down cleanly along people's stance toward nuclear. Germany was either hysterically overreacting or the only country that had the guts to do the right thing. That debate can still be held today, and the answer is still largely a matter of opinion. But not everything discussed at the time remains unanswered. Lots of predictions have already turned out to be false.

Chancellor Merkel's question to the Ethics Commission included the widespread assumption that Germany would simply import more nuclear power from abroad if it shut down its own nuclear plants. Yet, that concern was an obvious impossibility and could have been dismissed immediately. Nuclear plants generally run near full capacity, meaning they cannot be ramped up further. The major exception to this is France, where nuclear reactors do ramp down when demand is low at home— but then, demand is also low in Germany at such times.

A broader look at German power exports and imports reveals what actually happened. Germany exported more electricity in 2011 than it imported. From 2012 to 2015, it posted record levels of net electricity exports each year. By the end of 2013, more new green power was being generated than nuclear power had been reduced in 2011. It took Germans three years to replace eight nuclear plants—nearly half of the

---

[16] Brinkschmidt, Christian. "Erdabsenkung in Kirchheim: Hohlraum mit unabsehbaren Folgen." Bietigheimer Zeitung. 15 January 2014. Accessed February 8, 2016. http://www.swp.de/bietigheim/lokales/landkreis_ludwigsburg/Erdabsenkung-in-Kirchheim-Hohlraum-mit-unabsehbaren-Folgen;art1188795,2399672.

fleet—with biogas, wind, and solar power.[17] In 2015, power exports had already surpassed the record level of the previous year by September, with roughly an eighth of all power generated in Germany produced for neighboring countries.

There was also concern in mid-2011 about Germany switching from nuclear to coal power, and indeed carbon emissions temporarily rose slightly until 2013. But the effect was of limited duration and partially caused by the economy picking up again after the 2009 to 2010 crisis; by 2014, carbon emissions were down again. By 2014, electricity from fossil fuels had reached a 35-year low as renewables cut into power from fossil fuels—despite record high power exports.[18] Carbon emissions then fell further in 2015.[19]

The concern about the impact on the economy was also a hot topic in 2011. Hans-Werner Sinn, a leading German economist, called nuclear power "mankind's only hope" and added that the Energiewende was "committing a crime against future generations."[20] He was especially concerned about the impact on domestic industry. In May 2011, he predicted that "energy-intensive industry will disappear from Germany."[21]

It hasn't happened yet. Since 2011, power prices on the German spot market have known only one direction: down. The price in 2015 was some 50 percent below the level pre-Fukushima—as low as a decade ago, and the downward trend continued on the futures market. By 2015,

---

[17] Morris, Craig. "The wrong lessons at the New York Times." Energy Transition. The German Energiewende. 7 May 2014. Accessed February 8, 2016. http://energytransition.de/2014/05/the-wrong-lessons-at-the-new-york-times/.

[18] Morris, Craig. "Power from fossil fuel drops to 35-year low in Germany." Energy Transition. The German Energiewende. 8 January 2015. Accessed February 8, 2016. http://energytransition.de/2015/01/fossil-fuel-power-at-35-year-low-in-germany/.

[19] Morris, Craig. "CO2: German emissions adjusted." Renewables International. The Magazine. 14 January 2016. Accessed February 8, 2016. http://www.renewablesinternational.net/german-emissions-adjusted/150/537/92698/.

[20] Morris, Craig. "Energiewende a down payment on future affordable energy." Renewables International. The Magazine. 6 February 2014. Accessed February 8, 2016. http://www.renewablesinternational.net/energiewende-a-down-payment-on-future-affordable-energy/150/537/76750/.

[21] Iken, Matthias. "Gegen den Strom—Eine Pflichtverteidigung der Atomkraft." Hamburger Abendblatt. 27 May 2011. Accessed February 8, 2016. http://www.abendblatt.de/ratgeber/wissen/article106539997/Gegen-den-Strom-Eine-Pflichtverteidigung-der-Atomkraft.html.

wholesale prices had reached around 3.2 cents.[22] The biggest industrial power consumers pay rates based on these prices, so the outcome is the opposite of what was feared; these firms pay drastically lower electricity prices than they did before the nuclear phaseout.

But have we not repeatedly read stories about various German firms investing in the US because energy prices are lower there? Indeed, Industry Minister Sigmar Gabriel, a Social Democrat, warned in 2014 that Germany's energy transition is deindustrializing the country. Perhaps we should first mention that Germany at least still has a lot of industry to lose. In 2012, industrial production made up 28.1 percent of the German economy, compared with only 21 percent of the UK's—and 19.1 percent of the US economy.[23]

Still, deindustrialization is nothing to take lightly. After Gabriel conjured up the specter of deindustrialization, the Greens asked him for a list of industrial firms that had left the country because of high energy prices. The official response from the Industry Ministry read, "The government does not have any reliable statistics about the total number of firms that have left." The only example mentioned was Finnish stainless steel conglomerate Outokumpu, which had closed one plant in Germany and planned to close a second, but the firm itself spoke of "global excess capacity" as the reason, not high energy prices in Germany. Indeed, Outokumpu had also sold or closed facilities in Italy, Spain, Turkey, and France. Embarrassingly for the German Industry Ministry, the buyer was Germany's ThyssenKrupp.[24]

At the same time, foreign firms complained about low industry power prices in Germany. The most spectacular case was a Dutch aluminum firm Aldel, which said its German competitors were paying 25 percent less for electricity. It folded at the end of 2013 only to announce a comeback in 2015, when a connection to the German grid (and hence, to German

[22] Morris, Craig. "German renewable energy surcharge to rise?" Renewables International. The Magazine. 17 September 2015. Accessed February 9, 2016. http://www.renewablesinternational. net/german-renewable-energy-surcharge-to-rise/150/537/90334/.

[23] "List of countries by GDP sector composition." Wikipedia. Accessed February 9, 2016. https:// en.wikipedia.org/wiki/List_of_countries_by_GDP_sector_composition.

[24] Morris, Craig. "German industrial firms flee the country – which ones?" Renewables International. The Magazine. 27 March 2014. Accessed February 9, 2016. http://www.renewablesinternational. net/german-industrial-firms-flee-the-country-which-ones/150/537/77853/.

prices) was promised.[25] Also in 2013, a French industry lobby group complained that their German competitors would pay 35 percent less for power in 2014.[26] The announcement came just weeks after a Norwegian aluminum producer invested 130 million euros in a new production line in Germany.[27] Once again, this outcome is the exact opposite of what was predicted. Instead of the nuclear phaseout raising prices for heavy industry, these firms are the biggest winners in the Energiewende.

Yet, there was no scarcity of reports of German firms leaving the country. The Financial Times claimed that high energy prices were driving German carmaker BMW to the US, where the company was investing 100 million dollars to expand a plant that made carbon fiber, a raw material increasingly used to make cars lighter. The article spoke of "a global cost-cutting program" and left readers assuming that money was being pulled from Germany. In fact, the firm announced investments worth nearly 4 *billion* dollars in its German facilities that same year. (So did its German competitor Daimler, incidentally.)[28] The Financial Times also failed to mention that BMW had already installed four giant wind turbines at a production site in Germany to produce its own electricity—and that US workers at the carbon fiber plant make half as much per hour as their colleagues at BMW in Germany. Told in full, this story puts the US in the somewhat unflattering role of a minor supplier of raw materials, with the high-paying jobs and most investments remaining in Germany.[29]

[25] Morris, Craig. "Germany, the aluminum magnet." Renewables International. The Magazine. 27 March 2015. Accessed February 9, 2016. http://www.renewablesinternational.net/germany-the-aluminum-magnet/150/537/86553/.

[26] Patel, Tara. "France's Industrial Giants Call for Price Cap on Nuclear." Bloomberg. 17 March 2014. Accessed February 9, 2016. http://www.bloomberg.com/news/articles/2014-03-17/france-s-industrial-giants-call-for-price-cap-on-nuclear.

[27] Morris, Craig. "German renewables surcharge designed to skyrocket." Renewables International. The Magazine. 8 September 2012. Accessed February 9, 2016. http://www.renewablesinternational.net/german-renewables-surcharge-designed-to-skyrocket/150/537/56247/.

[28] Morris, Craig. "The myth of deindustrialization. Is BMW leaving Germany?" Renewables International. The Magazine. 28 October 2014. Accessed February 9, 2016. http://www.renewablesinternational.net/is-bmw-leaving-germany/150/537/82733/.

[29] Weber, Tilman, and Craig Morris. "BMW installs wind turbine on plant grounds." Renewables International. The Magazine. 18 October 2012. Accessed February 9, 2016. http://www.renewablesinternational.net/bmw-installs-wind-turbine-on-plant-grounds/150/505/57791/.

BASF, the world's largest chemicals firm, has also made headlines pertaining to German energy policy. In 2014, the New York Times wrote that the German giant was "leaning abroad" because "energy prices have jumped as a result of the government's big push for renewable energy."[30] Several mistakes are made here. First, Germany's push for renewables stems from the people, not the government. Second, the shale gas boom in the US has drastically reduced both gas and power prices in some parts of the country. One could point out that lower fossil fuel prices are not a goal of German policy; leaving carbon in the ground is. It is therefore nonsensical to ask Germany to extract its shale gas.

But the BASF story is even juicier. The company does not purchase any electricity from the German grid. It buys loads of natural gas, which it uses to make chemicals and heat for production processes. It needs so much heat, in fact, that the electricity generated in the process is almost a byproduct. BASF owns its own natural gas turbine and generates all its electricity at its headquarters in Germany.[31]

Doesn't the reliance on natural gas make it even more sensitive to gas prices? Not really—BASF buys its gas from a 100 percent subsidiary called Wintershall. If gas prices go up in Germany, revenue shifts from the parent company to the subsidiary. If they go down, the subsidiary struggles, but the parent company is more profitable. Either way, it's a wash. Corporate management at BASF has wisely and successfully made sure that German energy policy does not affect the firm's bottom line one way or the other. Nonetheless, its CEO is now moaning and groaning about the German energy transition—but it is all preemptive complaining. Mainly, BASF does not want anything to change. The firm likes things the way they are, so it started investing up to 10 billion euros at its German headquarters in 2011.[32]

---

[30] Reed, Stanley, and Melissa Eddy. "BASF, an Industrial Pillar in Germany, Leans Abroad." New York Times. 24 October 2014. Accessed February 9, 2016. http://www.nytimes.com/2014/10/25/business/international/basf-an-industrial-pillar-in-germany-leans-abroad.html?_r=1.

[31] Morris, Craig. "The myth of deindustrialization. BASF leaving Germany?" Renewables International. The Magazine. 29 October 2014. Accessed February 9, 2016. http://www.renewablesinternational.net/basf-leaving-germany/150/537/82838/.

[32] "BASF investiert Milliarden in Stammsitz." Focus. 16 September 2011. Accessed February 9, 2016. http://www.focus.de/finanzen/news/unternehmen/standort-ludwigshafen-basf-investiert-milliarden-in-stammsitz_aid:665916.html.

One repeatedly encounters warnings that Europe as a whole, if not Germany in particular, must tap its shale gas resources to compete.[33] In 2014, oil sector expert and Pulitzer Prize winner Daniel Yergin not only helped publish a study underscoring the economic benefits of shale gas for Germany, but also expressed his worries about German competitiveness for the press at the World Economic Forum in Davos, Switzerland, shortly thereafter.[34] His remarks were unfortunately far more widely reported than the Forum's own competitiveness index, which had Germany in fourth place behind only Switzerland, Singapore, and Finland—and just ahead of the United States.

Overall, the German economy looked pretty good in the years after the 2011 nuclear phaseout. In fact, one of the main complaints within the euro zone is that the Germans are too damn competitive. Whereas the US notoriously maintains a tremendous trade deficit (equivalent to more than two percent of the country's total economic output, or GDP), the Germans have a far greater trade surplus (seven percent of GDP in 2013). Even in absolute terms, the German trade surplus was the largest in the world that year at an estimated 260 billion euros, far ahead of China at just below 200 billion euros.[35]

In 2003, a new word became popular in Germany: *Exportweltmeister*, or world champion of exports. That year, the value of German exports was greater than the value of exports from any other country; 80 million Germans were not only outperforming some 320 million Americans, but also some 1300 million Chinese. And 2003 was no fluke—the situation continued for the next five years as well, ending only in the post-crisis year of 2009. The US is not having trouble competing with China; it is having trouble keeping up with Germany competing with China. The lesson has not been lost on American onlookers, either. "You

[33] Hromadko, Jan. "Yergin: Germany Must Focus on Cost-Effective Renewable Energy." The Wall Street Journal. 26 February 2014. Accessed February 9, 2016. http://www.wsj.com/news/articles/ SB10001424052702304071004579407082647574004.

[34] Eddy, Melissa. "German Energy Push Runs Into Problems." New York Times. 19 March 2014. Accessed February 9, 2016. http://www.nytimes.com/2014/03/20/business/energy-environment/ german-energy-push-runs-into-problems.html?_r=1.

[35] "Deutschland mit höherem Exportüberschuss als China." Handelsblatt. 14 January 2014. Accessed February 9, 2016. http://www.handelsblatt.com/politik/konjunktur/nachrichten/ifo-institut-deutschland-mit-hoeherem-exportueberschuss-als-china/9329150.html.

ought to study Germany," Republican politician Newt Gingrich stated in a TV debate on the economy in 2010.[36]

Granted, the EU and many economists consider Germany's trade surplus to be economically unhealthy, but let's face it—most countries would love to have Germany's problems. Its tax revenue hit a record high in 2013. In 2014, Germany's Finance Minister announced the first balanced budget since 1969.[37] That year, unemployment in Germany fell just below 5 percent according to international metrics, the lowest level since East and West Germany reunited in 1990.[38] "Countries like Germany do both capitalism and socialism better than we do," says US labor lawyer Thomas Geoghegan.[39]

The Utility Death Spiral

So much for the good news. There was one clearly negative effect of the 2011 nuclear phaseout: it was bad for the companies that owned nuclear plants. Even if you don't like nuclear power, you wouldn't want the government closing down your investments overnight, either. The phaseout of 2002 was much cleverer in that respect. The government of Chancellor Gerhard Schröder sat down with the nuclear plant owners and had them sign contracts agreeing to a phaseout after a few years of negotiations. In return for decommissioning promises, the plant owners got planning security: no new taxes or safety requirements would be implemented on nuclear reactors, and so on. Under pressure from nuclear plant owners,

---

[36] Al-Faruque, Ferdous. "'Meet the Press' Transcript — 11/14/10." Time. The Page. 14 November 2010. Accessed February 9, 2016. http://thepage.time.com/2010/11/14/meet-the-press-transcript-111410/.

[37] Thomas, Andrea. "German Government Achieves Balanced Budget Earlier Than Planned." The Wall Street Journal. 13 January 2015. Accessed February 9, 2016. http://www.wsj.com/articles/german-government-achieves-balanced-budget-earlier-than-planned-1421139601.

[38] "Germany Unemployment Rate: 4.50% for Dec 2015." YCharts. Accessed February 9, 2016. https://ycharts.com/indicators/germany_unemployment_rate.

[39] McNally, Terrence. "Why Germany Has It So Good—and Why America Is Going Down the Drain." Alternet. 13 October 2010. Accessed February 9, 2016. http://www.alternet.org/story/148501/why_germany_has_it_so_good_DOUBLEHYPHEN_and_why_america_is_going_down_the_drain.

Chancellor Merkel reneged on that phaseout in 2010, just a few months before Fukushima.

The impact of the 2011 nuclear phaseout should not, however, be confused with the general malaise utilities faced after Fukushima. The new buzzword is "utility death spiral." It stems from the rapid growth of renewable energy, not the nuclear phaseout of 2011. Germany reached its 2010 target for wind power in 2005. The solar target for 2050 was reached in 2012.[40] Renewables can be—and in Germany, have been—built faster than conventional generators need to be taken down and replaced. Critics of the Energiewende have spoken of *Wildwuchs*, the "uncontrolled growth" of renewables. In the process, the Big Four utilities in Germany—which all own nuclear—produced 16 percent less electricity in 2013 than they did in 2010. A survey taken in 2014 estimated that six percent of German households already made their own energy at home, and another 20 percent aimed to do so by the end of the decade.[41] That figure does not even include the investments that German citizens have made as members of energy cooperatives to build wind farms and biogas plants. Some utilities failed to see this trend coming; those who tried to warn them about the emerging threat of renewables were often treated like dreamers. But the chickens have come home to roost, so to speak; by 2013, around three quarters of German pig and bird farms made their own energy.[42]

In January 2015, a study entitled (in German) "The future of large energy providers" by researchers from the University of Applied Sciences in Recklinghausen detailed how the utilities were responsible for their current predicament.[43] The authors document how these firms took the

[40] Morris, Craig. "Schadenfreude about RWE? Think you would have done better?" Energypost. 10 March 2014. Accessed February 9, 2016. http://www.energypost.eu/schadenfreude-rwe-sure-done-better/.

[41] Morris, Craig. "Renewables make millions of Germans multidozenaires." Renewables International. The Magazine. 16 May 2014. Accessed February 9, 2016. http://www.renewablesinternational.net/renewables-make-millions-of-germans-multidozenaires/150/537/78900/.

[42] Morris, Craig. "70% of German pig and bird farms have photovoltaics." Renewables International. The Magazine. 9 December 2013. Accessed February 9, 2016. http://www.renewablesinternational.net/70-of-german-pig-and-bird-farms-have-pv/150/452/75232/.

[43] Bontrup, Heinz-J., and Ralf-M. Marquardt. "Die Zukunft der großen Energieversorger." Greenpeace. Hannover/Lüdinghausen, January 2015. Accessed February 9, 2016. https://www.greenpeace.de/sites/www.greenpeace.de/files/publications/zukunft-energieversorgung-studie-20150309.pdf.

windfall profits from emissions trading and excessive rent-seeking during roughly the first decade of liberalized power markets not only to make bad investments in new coal plants, but also in foreign takeovers.

For instance, Eon bit off more than it could chew throughout the 2000s. It expanded to the UK, the US, Spain, Italy, France, the Czech Republic, Slovakia, Hungary, Bulgaria, and Russia. By the time it stepped away from a major US subsidiary in 2010, Eon's net debt had grown from 6 billion euros in 2006 to 46 billion euros.[44]

Likewise, RWE also invested 9.3 billion in Dutch energy firm Essent, only to write down 4.3 billion of it by 2014.[45] RWE's other major international takeovers include the entire gas sector of the Czech Republic and American Water Works, the largest water supplier in the US.[46] Purchased for $46 per share in 2001,[47] American Water Works was sold in 2009 at a mere $21.63 per share.[48] Vattenfall was not without major missteps, either; it took over Nuon Energy of the Netherlands in 2009 for 8.5 billion euros. By 2015, the firm had been forced to write down roughly 5.7 billion of that investment.[49] Here, the words of German energy journalist Jakob Schlandt are worth quoting: "It is empirically well documented that fusions and takeovers very often have little to do with economic reason, but rather with executives who egoistically jump at the chance to bet the bank."[50]

[44] Leuschner, Udo. "Energiechronik: E.ON verkauft seine US-Tochter für 5,7 Milliarden Euro." April 2010. Accessed February 9, 2016. http://www.energie-chronik.de/100401.htm.

[45] "RWE to Write Down Another EUR 1.5bn on Dutch Essent." High Beam. 29 January 2014. Accessed February 9, 2016. https://www.highbeam.com/doc/1G1-356958576.html.

[46] Bontrup and Marquardt, *Die Zukunft der großen Energieversorger*, p. 91.

[47] "RWE Announces the Acquisition of American Water Works." RWE AG. Essen, 17 September 2001. Accessed February 9, 2016. http://www.rwe.com/web/cms/en/113648/rwe/press-news/press-release/?pmid=766.

[48] "RWE divests all shares of American Water." RWE AG. Essen, 25 November 2009. Accessed 09.02.206. http://www.rwe.com/web/cms/en/37110/rwe/press-news/press-releases/press-releases/?pmid=4004244.

[49] "Investors FAQs." Vattenfall. 23 October 2013. Accessed February 9, 2016. http://corporate.vattenfall.com/investors/investors-faqs/.

[50] Schlandt, Jakob. "Vattenfall in Deutschland: Der schwedische Blackout." Phasenprüfer. 22 September 2014. Accessed February 9, 2016. http://phasenpruefer.info/vattenfall-in-deutschland-geschichte-eines-totalausfalls/.

Not all the takeovers and mergers undertaken by big German utilities up to 2011 were bad, but this growth was not truly organic. Both RWE and Eon reported depreciation of assets in excess of investments in new replacement facilities from 2006 to 2011, meaning that the firms were actually shrinking; they were only able to report growth because of acquisitions.[51]

Furthermore, the years of great profitability were only possible because wealth was transferred from workers to shareholders. For instance, in 2012, Germany's Big Four posted a collective profit of just over 16 billion euros, whereas employing the fourth of the people laid off would have only increased expenses by just under 4.3 billion euros.[52] The study speaks of an "extreme redistribution of added value to the disadvantage of income from labor and to the advantage of income from ownership (profits, interest, rent, and leases)."[53] From 1998 to 2012, nearly a quarter of all the people employed in Germany's power sector were laid off.[54] The situation is further proof that a focus on shareholder value conflicts with community interests.

Overall, the authors of "The future of large energy providers" find that the Energiewende was not the cause of the predicament of German utilities in 2015. Rather, these firms went on an international shopping spree, which they took the liberalization of power markets to be an invitation to. Only because these companies were flooded with liquidity during the rollout of emissions trading do we now see their return to more modest profitability as a dire situation. The study uses the German phrase *Jammern auf hohem Niveau*, which could be translated as "luxury complaining," to describe the widespread concern about utility profits from 2011 to 2014.

However, as we will see in the next chapter, the utility death spiral may only just be getting started. Renewable electricity has marginalized Germany's Big Four utilities, but their share of conventional electricity has also fallen precipitously—from around 84 percent in 2010 to

---

[51] Bontrup and Marquardt, *Die Zukunft der großen Energieversorger*, p. 96.
[52] Bontrup and Marquardt, *Die Zukunft der großen Energieversorger*, p. 69.
[53] Bontrup and Marquardt, *Die Zukunft der großen Energieversorger*, p. 71.
[54] Bontrup and Marquardt, *Die Zukunft der großen Energieversorger*, p. 67.

75 percent in 2013.[55] An even better indicator is the Residual Supplier Index (RSI). It shows how often a power provider plays a dominant role in covering demand for electricity. Eon fell from an RSI of 71.8% in 2007 to 0.8% in 2012; RWE, from 93.6% to 0.8%. Within five years, these utilities were suddenly no longer indispensable for power reliability.

Shrinking corporate utilities may not be a bad thing. If you wish to prevent the redistribution of wealth from workers to shareholders, you probably want municipal utilities that primarily serve their communities. Stock corporations are required by law to increase shareholder value. It is unfair to criticize them for doing so. It is not necessary, however, to accept stock corporations as the dominant form of business enterprise. Stock corporations serve a specific purpose—just not the one people usually think of. Often, it is stated that stock markets allow undertakings to go forward with private investments when banks find the risks to be too great for low-interest loans; at the same time, the liability of investors behind corporations is limited to their respective stake. In reality, firms go public when they are already successful and venture capitalists want to cash out (think of Facebook, for instance).[56] The effect of the stock market is to allow investors to escape responsibility by becoming nearly anonymous and able to abandon ship at any time without repercussions. Venture capitalists, in contrast, need to find a buyer, as do limited-liability undertakings; abandoning ship is not so easy. And businesses embedded in their communities are neither anonymous, nor can they shift profits from workers to shareholders without hurting their reputations—and in some cases, their own bottom lines.

Energy supply is not a risky venture that requires venture capitalists anyway. During the years when German utilities expanded internationally, returns that could have been reinvested meaningfully in communities at home were paid out as dividends and spent on gambles in a global competition that the players themselves had created for each other. It's ratepayer money they are playing with.

---

[55] Bontrup and Marquardt, *Die Zukunft der großen Energieversorger*, p. 71.
[56] Ho, Karen Z. *Liquidated: An ethnography of Wall Street.* A John Hope Franklin Center book. Durham: Duke Univ. Press, 2009.

Box 13.1 *Mittelstand*, **the German dream**

Mid-size, family-owned firms with up to 499 employees and 50 million euros in revenue are called *Mittelstand* in Germany. The dream of many German entrepreneurs is to be successful in a global market niche as a *Mittelstand* firm. Even firms that become too big to be considered *Mittelstand* (such as BMW) often remain family-owned, like 95% of the companies[57] in Germany. In renewables, examples include the leading German wind turbine manufacturer Enercon and solar silicon maker Wacker.

Countries like France, the UK, and the US foster "national champions," big corporations that can compete internationally. France is the outlier here; 1% of its firms make up 97% of all exports,[58] whereas *Mittelstand* companies make up nearly a fifth of German exports. Some 1,300 of global market leaders are German "hidden champions"—midsize firms that dominate a market niche. Fewer than 400 are from the US, whereas France and the UK are each home to fewer than 100.[59]

In the UK and the US, the young entrepreneur's dream is to take a company public—make it a stock corporation. Family-owned *Mittelstand* firms are "not subject to investor pressure to reward large shareholders through practices prevalent in the United States, such as slashing wages, cutting back on worker training and research and development and buying back stock," writes the Washington Post.[60] *Mittelstand* businesspeople want protection for their family-owned firm from the short-term interests of anonymous, constantly changing shareholders. The focus can be on long-term benefits[61] when the goal is to hand down a company to

---

[57] Dostert, Elisabeth. "Erfolgreicher Mittelstand: Ein deutsches Phänomen." Süddeutsche Zeitung Online. 2 March 2014. Accessed February 11, 2016. http://www.sueddeutsche.de/wirtschaft/erfolgreicher-mittelstand-ein-deutsches-phaenomen-1.1900145.

[58] "1% des entreprises françaises concentre 97% des exportations, d'après l'Insee." Libération. 27 October 2015. Accessed February 11, 2016. http://www.liberation.fr/france/2015/10/27/1-des-entreprises-francaises-concentre-97-des-exportations-d-apres-l-insee_1409363.

[59] BMWi. "German Mittelstand: Motor der deutschen Wirtschaft. Zahlen und Fakten zu deutschen mittelständischen Unternehmen." Bundesministerium für Wirtschaft und Energie. May 2014. Accessed February 11, 2016. https://www.bmwi.de/BMWi/Redaktion/PDF/Publikationen/factbook-german-mittelstand,property=pdf,bereich=bmwi2012,sprache=de,rwb=true.pdf.

[60] Meyerson, Harold. "Germany's major export: economic optimism." The Washington Post. 24 September 2014. Accessed February 11, 2016. https://www.washingtonpost.com/opinions/harold-meyerson-germanys-major-export-is-economic-optimism/2014/09/24/d2ed77ea-4417-11e4-b47c-f5889e061e5f_story.html.

> **Box 13.1 (continued)**
>
> future generations. In such cases, the brand becomes crucial; if bad business practices have tarnished the name, the legacy for the family is hurt. *Mittelstand* firms are therefore members of their communities.
>
> German companies offer on-the-job training as an alternative to university degrees, and 85 percent of these young people train at *Mittelstand* firms. As Trade Minister, Ian Livingstone called for a "Brittelstand" in the UK, and Americans like labor expert Robert Reich[62] have pointed to Germany's "Dual System" (academic courses combined with hands-on training) as a reason why Germany is "a world leader in precision manufacturing." Such skills are needed to develop new technologies for the energy transition. By creating numerous mid-size firms instead of a small number of large corporations, Germany avoids having strong incumbents that will resist changes that hurt their own bottom lines. The *Mittelstand* thus fosters the very kind of fundamental transition that the Energiewende entails.

What's more, this international expansion itself reveals how much the energy transition's success depends on new players. In 2014, the CEO of RWE admitted, "We got started with renewables late, maybe too late."[63] Yet, RWE is one of the biggest investors in wind power in the UK; Eon, one of the biggest in the US and the UK. In both cases, these investments do not conflict with their own corporate assets in Germany; the firms were otherwise small newcomers on these foreign markets. These German utilities have yet to make such significant investments in renewables at home, where they have more stranded assets. So if you want a fast transition, you want a larger number of smaller companies.

---

[61] Marsh, Sarah. "Insight: The Mittelstand—one German product that may not be exportable." Reuters. 14 November 2012. Accessed February 11, 2016. http://www.reuters.com/article/us-germany-mittelstand-idUSBRE8AD0KV20121114.

[62] Reich, Robert. "Affirmative Action in College Admission for the Rich." 22 March 2015. Accessed February 11, 2016. http://robertreich.org/post/114356426465.

[63] Sorge, Nils-Viktor. "Ein Konzernchef auf dem Weg in die in Kaltreserve." Manager Magazin. 4 March 2014. Accessed February 9, 2016. http://www.manager-magazin.de/unternehmen/energie/portrait-peter-terium-auf-bilanz-pk-rwe-a-956831.html.

Indeed, for many experienced renewable energy proponents in Germany, the utility death spiral is a sign that the Energiewende is working, not failing. Because Germany (along with Denmark) is the only country with considerable energy democracy,[64] foreign reports sometimes do not suspect that low utility profits might be a cause for celebration among those calling for greater energy democracy and less corporate control. For instance, in 2013 the *Financial Post* misconstrued the fall in profits at German utilities as an indication that the Energiewende "is getting ever nearer to destroying the business model of companies meant to invest in renewables, namely the big sector leaders such as Eon and RWE."[65] The Germans who have faced opposition from RWE and Eon in their own community projects will be interested to hear that these two firms are the ones who should be driving the energy transition. Such people view these utilities as obstacles.

The Emergence of "Energy Poverty" in the Elections of 2013

By the time the election campaign of 2013 got underway, Germany had not only gotten through two winters without any major power outages due to the Energiewende; the country had even helped prevent a blackout in France during the cold spell at the beginning of February 2012,

---

[64] EESC. "Changing the Future of Energy: Civil Society as a Main Player in Renewable Energy Generation." European Economic and Social Committee. Final Report. January 2015. Accessed February 10, 2016. http://www.eesc.europa.eu/resources/docs/eesc-2014-04780-00-04-tcd-tra-en.docx.

"The potential for civic investment in the energy transition has only been partially realised, with substantial differences between the EU [member states]. Denmark and Germany are probably the best examples of the transition being largely carried out by civil society." (p. 9).

[65] Steitz, Christoph, and Vera Eckert. "Germany grapples with switch to renewables." Financial Post. 21 January 2013. Accessed February 9, 2016. http://business.financialpost.com/news/energy/germany-grapples-with-switch-to-renewables?__lsa=adae-ce25.

when power lines were maxed out from Germany to France.[66] It should have been a time of celebration for the Energiewende. It wasn't.

Instead, the first-ever massive campaign against renewable energy was undertaken, orchestrated by the Institut Neue Soziale Marktwirtschaft (INSM), a pro-business initiative financed by the Employers' Associations of the Metal and Electronics Industry. As such, the initiative represents the interests of the hiring class, not low-income households which might have trouble paying power bills. But instead of arguing that higher electricity prices were hurting their businesses, the employers found their bleeding hearts and began writing about "energy poverty."

The goal was to get the public to resent the high cost of renewable electricity in the run-up to the election. Never mind that Germany's largest charity organization Caritas focused on getting more low-income households involved in "an energy mix that make sense economically and ecologically."[67] In Frankfurt, Caritas built a new Passive House office complex along with a day care center because "nobody freezes in a Passive House building."[68] Instead of complaining about "energy poverty," Caritas did something about it—by offering some 20,000 energy audits annually for free to poor households so they could learn how to reduce energy consumption.[69] The focus was always on all types of energy, not just electricity; heating bills are typically higher than power bills in Germany.

The concern about "energy poverty" was generally limited to groups that oppose renewables. Everyone else was concerned about general poverty. In addition to Caritas, other entities that usually stand up for the

---

[66] Morris, Craig. "German power exports to France increasing." Renewables International. The Magazine. 6 February 2012. Accessed February 9, 2016. http://www.renewablesinternational.net/german-power-exports-to-france-increasing/150/537/33036/.

[67] Raible, Helga. "Energiewende selbst gemacht." Caritas. 13/2013. Accessed February 9, 2016. http://www.caritas.de/neue-caritas/heftarchiv/jahrgang2013/artikel/energiewende-selbst-gemacht.

[68] Sprondel, Hartmut. "Wer im Passivhaus sitzt, muss nicht frieren." Caritas. 13/2013. Accessed February 9, 2016. http://www.caritas.de/neue-caritas/heftarchiv/jahrgang2013/artikel/wer-im-passivhaus-sitzt-muss-nicht-frier.

[69] Sans, Reiner. "Stromspar-Check Plus beugt Energiearmut vor." Caritas. 02/2013. Accessed February 9, 2016. http://www.caritas.de/neue-caritas/heftarchiv/jahrgang2013/artikel/stromspar-check-plus-beugt-energiearmut-?searchterm=energiearmut.

interests of the poor remained staunch proponents of the Energiewende. For instance, the Rosa Luxemburg Foundation is the political think tank of the Left Party, which is the farthest to the left among parties in the Bundestag. It calls for such things as higher welfare payments, higher minimum wages, and more progressive taxation. In 2013, it produced a study entitled "Poverty risks from the Energiewende," and the subtitle was telling: "myths, lies, and arguments."[70]

The Rosa Luxemburg Foundation addressed the figure of 312,000 households that had had their power cut off in 2011, as reported by the German Network Agency. The statistic became well-known after it appeared in an article published by *Der Spiegel*, which was doing its utmost to discredit the Energiewende at the time. Though taken outside Germany for a mainstream, credible news weekly with leftist leanings, it had in fact long been the only major news outlet that regularly gave a platform to climate change deniers, as German journalist Gerd Rosenkranz remembers.[71] In 2004, he was one of two staff members who left *Der Spiegel* in protest when a terrible hatchet job against wind power was published after their more neutral article had been suppressed.

Der Spiegel did not investigate whether the number was rising significantly. The Rosa Luxemburg Foundation found that the number had remained practically stable since 2006 although retail rates had risen by around a quarter.

And it gets worse. *Der Spiegel* claimed that electricity was becoming a "luxury item" in Germany because power bills might rise to "90 euros a month," at which point Germans who could not afford electricity would

---

[70] Morris, Craig. "Impact of Energiewende on the poor." Renewables International. The Magazine. 25 June 2013. Accessed February 9, 2016. http://www.renewablesinternational.net/impact-of-energiewende-on-the-poor/150/537/68361/; and Pomrehn, Wolfgang. "Armutsrisiko Energiewende? Mythen, Lügen, Argumente." Rosa Luxemburg Stiftung. Luxemburg Argumente Nr 4. Berlin, March 2013. Accessed February 9, 2016. http://rosalux.de/fileadmin/rls_uploads/pdfs/Argumente/lux_argu_Armutsrisiko_dt.pdf.

[71] Conversation with the authors.

return to "the Stone Age."[72] (As electricity only became widely available at the beginning of the twentieth century, the Stone Age apparently lasted until the nineteenth century for *Der Spiegel*.) Since 90 euros was equivalent to around 110 dollars in 2013, the claim should have seemed unusual to Americans, for whom such a monthly power bill is the average, not remarkable.

But instead of wondering why *Der Spiegel's* coverage of the Energiewende was so obviously bad, the international press uncritically repeated the German paper's claims, partly because the magazine was the only German news outlet publishing regularly in English at the time. Although comparisons with other countries might have seemed obvious, none of the reports attempted to make any. For the US, no statistics are even available at the national level for disconnections due to nonpayment of power bills, for instance. Instead, one finds utility-by-utility reports, which then need to be extrapolated for a comparison. 312,000 disconnections is equivalent to around 0.7 percent of Germany's 40 million households. In 2012, Seattle City Light disconnected 1.7 percent of the households it served for nonpayment.[73] Likewise, the California Public Utilities Commission had to step in when utilities began shutting off 70,000 customers a month—five times the German level—in 2009.[74] In Australia, the frequency of household power cut-offs is also easily twice as great as in Germany.[75]

---

[72] "How Electricity Became a Luxury Good." Der Spiegel Online. 9 April 2013. Accessed February 9, 2016. http://www.spiegel.de/international/germany/high-costs-and-errors-of-german-transition-to-renewable-energy-a-920288-druck.html.

[73] Burkhalter, Aaron. "A startling disconnect." Real Change. 18 September 2013. Accessed February 9, 2016. http://realchangenews.org/2013/09/18/startling-disconnect; and Morris, Craig. "What the New York Times does not tell you. Poverty an issue for social policy, not energy policy." Renewables International. The Magazine. 20 September 2013. Accessed February 9, 2016. http://www.renewablesinternational.net/poverty-an-issue-for-social-policy-not-energy-policy/150/537/73112/.

[74] "Help from the Southern California Gas Company." Accessed February 9, 2016. http://www.utilitybillassistance.com/html/california_laws_and_rules_that.html.

[75] Bita, Natasha. "Ban on power cut-offs for bills under $300." The Australian. 11 February 2012. Accessed February 9, 2016. http://www.theaustralian.com.au/news/nation/ban-on-power-cut-offs-for-bills-under-300/story-e6frg6nf-1226268212153.

Furthermore, no press outlet focused on a comparison of the charges households faced when their power is cut off. In Germany, the average fee came in at 32 euros in 2011.[76] In Seattle, the reconnection fee was $164.

Only the UK has good statistics for energy poverty—because it also has a definition: a household is considered "energy poor" if it spends more than 10 percent of its income on energy.[77] Of course, energy includes electricity, heating oil or gas, and gasoline for your car. In 2011, roughly 9 percent of households in England alone (not the entire UK) were found to be suffering from energy poverty.[78] Germany only collects statistics for heat and electricity because gasoline consumption is not metered by household (it is estimated in England). In 2011, the average low-income household in Germany spent 6.2 percent of its net budget on electricity and natural gas.[79] In the US, a low-income household with a net annual budget of around 15,000 dollars would have spent nearly 12 percent on residential energy (excluding fuel for cars) in 2014—nearly twice as much as in Germany.[80] In 2015, the UK's Association for the Conservation of Energy published a survey of energy poverty in 16 EU countries. The UK had fallen to fourteenth place in 2013 at 9 percent of households living in energy poverty. Germany was fourth at 3.6 percent.[81]

[76] "Monitoringbericht 2012." Bundesnetzagentur für Elektrizität, Gas, Telekommunikation, Post und Eisenbahnen. Bonn, 5 February 2013. Accessed February 9, 2016. http://www.bundesnetzagentur.de/SharedDocs/Downloads/DE/Allgemeines/Bundesnetzagentur/Publikationen/Berichte/2012/MonitoringBericht2012.pdf?__blob=publicationFile, p. 15.

[77] Morris, Craig. "Germany and UK: Energy poverty still hard to compare." Renewables International. The Magazine. 25 February 2014. Accessed February 9, 2016. http://www.renewablesinternational.net/energy-poverty-still-hard-to-compare/150/537/77095/.

[78] "Fuel Poverty Report—Updated August 2013." Department of Energy & Climate Change. London, 2013. Accessed February 9, 2016. https://www.gov.uk/government/uploads/system/uploads/attachment_data/file/226985/fuel_poverty_report_2013.pdf.

[79] BMUB. "Antwort der Bundesregierung auf die Kleine Anfrage der Abgeordneten Bärbel Höhn, Christian Kühn, Oliver Krischer, Markus Kurth, Nicole Maisch, Julia Verlinden und der Fraktion." Bundesministerium für Umwelt, Naturschutz, Bau und Reaktorsicherheit. Berlin, 20 January 2014. Accessed February 9, 2016. http://www.baerbel-hoehn.de/fileadmin/media/MdB/baerbelhoehn_de/www_baerbelhoehn_de/Fotos/Kleine_Anfrage_Heizkosten_01.pdf.

[80] "Energy Cost Impacts on American Families, 2001-2014." American Coalition for Clean Coal Electricity. February 2014. Accessed February 9, 2016. http://americaspower.org/sites/default/files/Trisko_2014_1.pdf.

[81] Morris, Craig. "Who is the "cold man of Europe"?" Energy Transition. The German Energiewende. 11 December 2015. Accessed February 9, 2016. http://energytransition.de/2015/12/who-is-the-cold-man-of-europe/.

To understand why poor households in Germany do not oppose high electricity prices more, it helps to understand two things: the level of welfare payments and the freedom to make your own energy. Energy poverty is a subset of poverty. If you have more poverty, you have more energy poverty. German welfare payments depend on particular circumstances, but a family of four was likely to receive roughly 24,000 euros in 2013. At the exchange rate of that year, the amount was equivalent to around USD 32,500. At the time, the poverty level for a four-person household in the US came in at just above $23,000 annually, so German welfare payments for a family of that size were nearly 50 percent above the US poverty line.[82]

Second, poor households can indeed become involved in the energy transition. *Der Spiegel* displays an incredible double standard when it writes:[83]

> … renewable energy subsidies redistribute money from the poor to the more affluent, like when someone living in small rental apartment subsidizes a homeowner's roof-mounted solar panels through his electricity bill.

The rich have also invested in energy corporations more than the poor, benefiting from dividends in the process. Yet, *Der Spiegel* does not mention how the stock market redistributes wealth from bottom to top (such as with low capital gains taxes). British journalist George Monbiot wrote in 2015 that the only people who benefit from solar investments in the UK are those "rich enough to carry the costs of installation,"[84] but how does that differ from other investments? Furthermore, *Der Spiegel* seems to think that you have to own a home to invest in renewable energy.

[82] Morris, Craig. "What the New York Times does not tell you. Poverty an issue for social policy, not energy policy." Renewables International. The Magazine. 20 September 2013. Accessed February 9, 2016. http://www.renewablesinternational.net/poverty-an-issue-for-social-policy-not-energy-policy/150/537/73112/.

[83] "How Electricity Became a Luxury Good." Der Spiegel Online. 9 April 2013. Accessed February 9, 2016. http://www.spiegel.de/international/germany/high-costs-and-errors-of-german-transition-to-renewable-energy-a-920288-druck.html.

[84] Monbiot, George. "With this attack on community energy the big six win out over 'big society'." The Guardian. 23 January 2015. Accessed February 9, 2016. http://www.theguardian.com/environment/georgemonbiot/2015/jan/23/community-energy-companies-big-six-big-society?CMP=share_btn_tw.

In reality, community energy projects allow citizens to get involved in solar projects on school buildings, local wind farms, and biogas units. A single share in such cooperatives costs less than 500 euros in two-thirds of German cooperatives—with the minimum amount being even less than 100 euros in some cases. As the head of Germany's BSW-Solar puts it, "Energy cooperatives democratize energy supply in Germany and allow everyone to benefit from the energy transition even if they do not own their own home."[85]

Furthermore, the discussion about "energy poverty" conflates citizens with consumers. In fact, the energy sector should ban the word "consumer" from its vocabulary because it's outdated. In the future, citizens will both consume and produce energy.

In addition to the INSM and *Der Spiegel*, the German government also jumped on the energy poverty bandwagon. Above all, Environmental Minister Peter Altmaier played a prominent role in this respect. At the beginning of 2013, his off-the-cuff estimation that the Energiewende might cost a trillion euros went viral. Spread across the 37 years from then until 2050 (the full time frame for Germany's energy transition), that figure works out to be less than 30 billion per year, but the press failed to make that calculation. If they had, they could also have reported that Germany already spent 24 billion on renewables in 2014, so getting five times more renewable energy will apparently not even cost a quarter more. Environmental taxation experts at Green Budget Germany discovered that there was no calculation at all—no formal study—to critique when they set out to investigate the number; they also pointed out that offsets of conventional fuel were not subtracted from the amount, nor were the lower costs of environmental and human health impacts. "No one ever thinks to multiply 90 billion euros in fossil fuel costs by 30 years," leading German economist Claudia Kemfert also complained.[86] In the end, the tax experts found that the Energiewende would lead to

[85] Morris, Craig. "Cooperatives drive renewables." Renewables International. The Magazine. 24 July 2012. Accessed February 9, 2016. http://www.renewablesinternational.net/cooperatives-drive-renewables/150/537/39736/.

[86] Kemfert, Claudia. *Kampf um Strom: Mythen, Macht und Monopole.* Hamburg: Murmann, 2013.

net benefits, not net expenditures, by mid-century.[87] So did a Fraunhofer study in 2015.[88]

Another estimate was floated at the beginning of 2012, this time by Siemens executive Michael Süss. Once again, there was no actual investigation behind the number, though that small fact did not prevent it from being covered uncritically by Reuters.[89] However, Süss only referred to the cost of replacing nuclear power by 2030. His time frame was therefore half as long as Altmaier's and covered only 6 percent of German energy consumption; yet, Süss's number was not 3 percent of Altmaier's, but 170 percent: 1.7 trillion euros. Although a handful of top journalists worked on the Reuters report, none of them saw it fit to do a simple calculation to see what Süss thought replacing nuclear would cost by the kilowatt-hour: around 1.65 euros, easily 15 times more than solar currently costs—and nearly 20 times greater than the most expensive onshore wind power in Germany.[90] The number was ludicrous.

In late 2012, Minister Altmaier's proposal to "put the brakes on power prices" began to dominate the Energiewende discussion in the run-up to the elections that fall. By April of 2013, the idea of some kind of political limit on retail rates had been buried once and for all, but it gave political legitimacy to the orchestrated cost concerns vociferously propagated elsewhere.[91] It was possible to focus on the rising price of renewables because the misnamed "renewable energy surcharge" was skyrocketing. In 2012, it was 3.59 cents per kilowatt-hour, but it jumped to 5.28 cents in 2013.

[87] Morris, Craig. "Criticism of Altmaier's trillion euro price tag." Renewables International. The Magazine. 12 March 2013. Accessed February 9, 2016. http://www.renewablesinternational.net/criticism-of-altmaiers-trillion-euro-price-tag/150/537/61170/.

[88] Morris, Craig. "What will the Energiewende cost?" Energy Transition. The German Energiewende. 18 November 2015. Accessed February 9, 2016. http://energytransition.de/2015/11/energiewende-cost/.

[89] Steitz, Christoph. "Siemens puts cost of nuclear exit at 1.7 trillion euros." Reuters. 17 January 2012. Accessed February 9, 2016. http://www.reuters.com/article/us-siemens-energy-idUSTRE80G10920120117.

[90] Morris, Craig. "Do the math! 1.7 trillion euros for German nuclear phaseout?" Renewables International. The Magazine. 18 December 2014. Accessed February 9, 2016. http://www.renewablesinternational.net/17-trillion-euros-for-german-nuclear-phaseout/150/537/84285/.

[91] "Strompreisbremse ist endgültig gescheitert." Handelsblatt. 20 April 2013. Accessed February 9, 2016. http://www.handelsblatt.com/politik/deutschland/energiekosten-strompreisbremse-ist-endgueltig-gescheitert/8096864.html.

The surcharge primarily grew not because more was being spent on renewable electricity. On the contrary, feed-in tariff payments remained essentially flat from 2012 to 2013. In contrast, wholesale prices had begun plummeting. Because the surcharge covers the difference between feed-in tariff payments and revenue from sales on the wholesale power exchange, lower wholesale prices lead to a higher surcharge. Furthermore, the government had increased industry exemptions to the surcharge dramatically. Pushing the total cost burden onto a smaller ratepayer base naturally inflated the surcharge for those who are not exempt: households and small firms. In 2006, Chancellor Merkel's first full year in office, 492 industrial firms were exempt;[92] by 2013, that number had risen to 2000.[93]

Energy experts who supported renewable energy struggled to explain to the public why something called the "renewable energy surcharge" was nonetheless not a good price tag for the Energiewende, but the explanation was complicated.[94] The main sound bite that got through to the public was that renewable power was making German power prices more expensive.

In reality, however, Germans were never up in arms about the price tag of the Energiewende. To date, there has not been a single demonstration against the cost of renewable electricity. There have, however, been repeated demonstrations for the energy transition—by citizens who want to protect their right to make their own energy.

In fact, on the only TV debate held between Chancellor Merkel and her challenger from the Social Democrats for the elections of 2013, the Energiewende was not even discussed.[95] As in the past, there were far

---

[92] Morris, Craig. "Merkel coalition runs wild with industry exceptions." Renewables International. The Magazine. 4 September 2013. Accessed February 9, 2016. http://www.renewablesinternational.net/merkel-coalition-runs-wild-with-industry-exceptions/150/537/72535/.

[93] Morris, Craig. "German industry exemptions for 2015." Renewables International. The Magazine. 21 October 2014. Accessed February 9, 2016. http://www.renewablesinternational.net/german-industry-exemptions-for-2015/150/537/82558/.

[94] "The right price tag for the Energiewende.: Interview with Felix Matthes and Hermann Falk." Renewables International. The Magazine. 19 November 2013. Accessed February 8, 2016. http://www.renewablesinternational.net/the-right-price-tag-for-the-energiewende/150/522/74670/.

[95] Morris, Craig. "Chancellor Merkel begins third term in office. "Energy transition is most urgent problem"." Renewables International. The Magazine. 18 December 2013. Accessed February 9, 2016. http://www.renewablesinternational.net/energy-transition-is-most-urgent-problem/150/537/75634/.

more pressing matters on the table: the Greek crisis, the future of the euro, and the working poor in Germany. These three items are related; Germany is so successful in international trade partly because wages have not kept up with productivity gains. Poverty—not just "energy poverty"—was on the agenda. After the 2013 elections, Germany adopted its first-ever minimum wage. Up to then, wage negotiations had not been set by the government, but left up to labor unions and employers.

The biggest change from the 2013 elections was the disappearance of the Free Democrats from Parliament. The FDP, the closest thing Germany had to a libertarian party, was the staunchest political opposition to the Renewable Energy Act. Their criticism of renewables was not the reason they lost so many votes, however. Rather, party leaders were seen to be ill-fit for real politics. The nearly 70-year-old party chairman was involved in a scandal at the beginning of 2013, when he made advances toward a female journalist on the job. More importantly, the party's top politician in the coalition—Foreign Minister Guido Westerwelle—was one of the least liked politicians in the country. In 2009, the party had joined the coalition with a promise to lower taxes and reduce the size of government. On that platform, the FDP managed to get around 15 percent of the vote that year, but after the elections it turned out that an overwhelming majority of the German public opposed lower taxation, particularly during the discussion about governmental budgets ballooning as they bailed out banks.[96] The FDP managed to get lower taxes for hotels implemented, which only brought about accusations of pandering to their own clientele of ritzy hotel chains.

With the FDP gone from Parliament and the majority left of center (Social Democrats, Greens, and the Left Party) unable to bridge differences to form a coalition, the only feasible constellation was now a grand coalition between the CDU and the SPD under a reelected Chancellor Merkel—exactly the coalition the country had from 2005 to 2009. It could have been a return to the politics of those years. In fact, the new coalition made the most drastic changes to renewable energy policy since 2000.

---

[96] Morris, Craig. "58% of Germans oppose lower taxes." 9 January 2010. Accessed February 9, 2016. http://notesfromotherside.blogspot.de/2010/01/58-of-germans-oppose-lower-taxes.html.

Pushing Back the People: EEG Amendments of 2014

Up to 2013, responsibility for Energiewende tasks had been spread across three ministries: economics, environment, and transport and buildings. This arrangement had one main benefit: it created an open debate. For instance, if the Economics Ministry produced a study criticizing the cost of renewable electricity, the Environmental Ministry could respond with its own findings and vice versa. Because all this discussion was held in public with published documents, citizens and civil society could inform themselves and participate in the debate. Not having a single energy ministry fostered a public debate that was part of the Energiewende's learning curve.

There was also a downside. In 2014, the German Accounting Office complained that conducting multiple, competing studies on similar topics was inefficient in terms of cost. Furthermore, it all led to a lack of proper coordination.[97] Merkel's new grand coalition seemed to agree. After taking office in 2013, it rearranged mandates, putting practically all the Energiewende within the Economics Ministry—with former Environmental Minister Sigmar Gabriel at the helm. Now, if the Environmental Ministry did not like something the Economics Ministry was proposing, it would be told not to comment on matters outside its mandate (which still includes climate change).

Greater control of the Energiewende was indeed the main goal of the policy changes adopted in 2014, but other goals were officially stated. When the grand collation revised the EEG that year, three objectives were given:

- maintaining the growth corridors for renewables;
- keeping costs in check; and
- and maintaining the "diversity of actors."

---

[97] Morris, Craig. "German Audit Office says Energiewende too expensive." Energy Transition. The German Energiewende. 1 September 2014. Accessed February 9, 2016. http://energytransition. de/2014/09/german-audit-office-says-energiewende-too-expensive/.

The first objective referred to the targets for new installations. The EEG had always specified a goal of "at least" a certain percentage of renewable electricity or energy, not ceilings. But a limit was imposed in 2014 for biogas units. No more than 100 megawatts annually could be installed with feed-in tariffs. To understand how little that is, consider that four time as much will eventually be needed annually to replace old capacity.

In contrast, not much changed for solar. It had grown in excess of all expectations at around 7.5 gigawatts annually from 2010 to 2012. In those years, investors rushed to complete projects before the announced cuts to feed-in tariffs took effect. The government had therefore already implemented a target corridor for solar of 3.0 gigawatts annually. If more were newly built, scheduled feed-in tariff reductions would be increased; if less was built, these reductions for new arrays would slow down or be suspended temporarily. In 2013, that change brought the market down to 3.3 gigawatts, followed by a mere 1.9 gigawatts in 2014—far below the corridor, which was also reduced for photovoltaics that year to 2.5 gigawatts.

There had never been a corridor for wind power, but one was added in 2014, also at 2.5 gigawatts. The immediate result was a run on the market, with 4.75 gigawatts being installed that year—nearly 50 percent more than the previous record year of 2012. Wind farm planners had rushed to complete projects before financial support dropped further. But because far more than 3 gigawatts was installed, feed-in tariffs were reduced more than scheduled for wind power in 2015; the sector had brought about the negative outcome it was trying to escape. But the rush may continue until 2017 because what the sector fears the most is the switch from feed-in tariffs to auctions.

Internationally, auctions seem to have produced less expensive green electricity than German feed-in tariffs have, which brings us to the second objective in the 2014 EEG amendments: keeping costs in check. And indeed, the nominal rates obtained, say, in auctions for wind power in Brazil have been roughly half as low as feed-in tariffs for wind power in Germany. Policymakers tend to put this outcome down to the policy design itself; in auctions, the government or power sector experts set the volume to be built, and market participants compete on price. The lowest bids win, so obviously auctions lead to the lowest cost by their very design.

Or do they? A closer look tells a different story. To begin with, most countries only just started building a lot of wind turbines after 2010, so the first projects being auctioned are going up in the best locations in those countries. In contrast, Germany has been building wind farms at a strong level since the late 1990s; the best sites are already taken. When we compare auctions abroad to German feed-in tariffs in the wind sector, we are not only comparing policies, but also the best foreign sites to average sites in a country with modest potential.

In 2015, the International Renewable Energy Agency (IRENA) produced an overview of wind farms around the world, showing that the best sites in Brazil and China attained capacity factors of close to 50 percent—compared with less than 20 percent for the entire German wind turbine fleet onshore.[98] In other words, a turbine installed in a good Brazilian location would produce roughly 1.5 times more electricity than it would in the average German site. Brazilian wind power should therefore cost only 40 percent as much as German wind power does—a level that has not yet quite been reached. German feed-in tariffs produce very low prices with consideration of the country's wind potential.

The situation is similar for photovoltaics. In 2014, auctions in Dubai produced a record low price for solar power of six USD cents per kilowatt-hour. On the other hand, Dubai has twice as much sunlight as German does. Solar power in Dubai should therefore be half as expensive as in Germany—but it isn't. The lowest feed-in tariff for solar power from new arrays at the time came in at 8.72 euro cents per kilowatt-hour. When adjusted for the exchange rate and sunlight conditions, new German solar power was 10 percent cheaper than the widely celebrated record low price in Dubai in 2014.[99]

In 2015, the first two pilot auctions for photovoltaics revealed that lower prices would not result from the policy change. The first round produced a price of 8.93 cents, clearly above the lowest feed-in tariff.

[98] Morris, Craig. "World Energy Council (WEC) survey finds Energiewende not model for the world." Energy Transition. The German Energiewende. 10 April 2015. Accessed February 9, 2016. http://energytransition.de/2015/04/wec-survey-finds-energiewende-no-model/.

[99] Morris, Craig. "Solar in Dubai reaches record low price." Renewables International. The Magazine. 27 November 2014. Accessed February 9, 2016. http://www.renewablesinternational.net/solar-in-dubai-reaches-record-low-price/150/452/83692/.

Granted, the second round reached a new low price of 8.49 cents, but those systems have two years to go into operation. By that time, scheduled decreases would have brought down feed-in tariffs to 8.43 cents—hardly a difference.[100]

In addition to wind overshooting its corridor, photovoltaics continues to shrink in 2015, and biogas will not even hit its already extremely low target. Furthermore, wind is to switch over to auctions as well in 2017. At that point, overshooting the growth target will no longer be possible; the market will either meet or fall short of the target. The switch to auctions will have one main outcome: preventing a faster energy transition than planned. If the growth corridors are not working well and auctions will not bring down the cost, how is the goal of "keeping costs in check" going?

If you ask the government, quite well. In 2015, the German Economics Ministry pointed out that power production prices—conventional and renewable—had fallen for the first time in years thanks in part to "last year's amendments." Consumer advocates are not so certain, pointing out that fossil fuel prices are down globally, not as a result of German energy policy pertaining to a different type of energy: renewables.[101] And in 2014, the same ministry explained that the goal of the amendments was to bring the average feed-in tariff down to 12 cents—the level it was already at for new renewable electricity.[102] Essentially, the government is taking credit for the status quo.

Finally, there is the third objective of the 2014 EEG: a "diversity of actors." Under feed-in tariffs, any project deemed worthy could go forward. Under auctions, only the cheapest can be built; everyone else is told no. Like all other investors in the energy sector, a citizen-driven community cooperative spends time and money planning a project and obtaining

---

[100] Morris, Craig. "8.49 cents for second round of photovoltaics auctions." Renewables International. The Magazine. 2 September 2015. Accessed February 9, 2016. http://www.renewablesinternational.net/849-cents-for-second-round-of-pv-auctions/150/452/89859/.

[101] Morris, Craig. "German government claims responsibility for lower power prices." Renewables International. The Magazine. 30 June 2015. Accessed February 9, 2016. http://www.renewablesinternational.net/german-government-claims-responsibility-for-lower-power-prices/150/537/88509/.

[102] Morris, Craig. "Cutting feed-in tariffs "by a third"" Renewables International. The Magazine. 10 April 2014. Accessed February 9, 2016. http://www.renewablesinternational.net/cutting-feed-in-tariffs-by-a-third/150/537/78125/.

a permit. Under feed-in tariffs, they do so knowing they will build if the feasibility study shows a good possibility of turning a profit. Under auctions, they are likely to be told they cannot build even if they want, regardless of the planning and permit. In fact, the first two pilot auctions for photovoltaics held in Germany both produced several times more losers than winners—an outcome the government itself spun as a sign of success, calling the auctions "oversubscribed," and hence a sign of great interest in the market.[103] In reality, more than three-quarters of bidders went empty-handed and were left sitting on their lost time and money.

Under those circumstances, community projects are unlikely to participate in auctions at all, and if they do once, they are unlikely to try a second time if they lose the first time. Auctions are a policy that mainly large corporations with deep pockets are happy with; they have the liquidity to absorb the occasional losing bid, and they will be around long enough to pass on these losses in future bids, once weaker bidders have been weeded out. As a result, auctions produce higher prices than necessary because companies later price in lost bids from the past.

The German government itself, it should be noted, is not especially pleased about community projects being shut out of auctions. Whenever they get a chance, officials reiterate how important citizen participation in the Energiewende is; without public support, the energy transition will fail. But at present, they have trouble admitting that community cooperatives are being shut out. In a parliamentary query in 2015, the Greens asked the government what was happening with the "diversity of actors," and the answer was an exercise in sophistry: "there is no legal definition of 'great diversity of actors.'"[104] That reply can be turned around: the government has set goals for itself that it has not even clearly defined.

If the three main policy objectives of the 2014 EEG are not clearly being met, we come back to our analysis of what the actual goal is: greater

---

[103] Morris, Craig. "Second German photovoltaics auctions over." Renewables International. The Magazine. 7 August 2015. Accessed February 9, 2016. http://www.renewablesinternational.net/second-german-pv-auctions-over/150/452/89358/.

[104] "Kleine Anfrage der Abgeordneten Dr. Julia Verlinden, Oliver Krischer, Christian Kühn (Tübingen),Annalena Baerbock, Matthias Gastel, Bärbel Höhn, Steffi Lemke, Peter Meiwald, Markus Tressel, Dr. Valerie Wilms und der Fraktion Bündnis 90/Die Grünen: Auswirkungen der Novelle der Erneuerbare-Energien-Gesetzes 2014." Drucksache 18/5774. Berlin, 13 August 2015. Accessed February 9, 2016. http://dip21.bundestag.de/dip21/btd/18/057/1805774.pdf.

central control of the Energiewende. Though never openly stated, this goal has some justification. Photovoltaics was indeed built at an unsustainably high level from 2010 to 2012. Most estimates for the amount of photovoltaics needed toward 80 percent renewable electricity in Germany estimate that around 100 GW will be needed. At a solar panel service life of 25 years, 7.5 gigawatts annually is nearly twice that much at almost 190 GW. To make things worse, photovoltaics was still relatively expensive at the time. Unfortunately, the current state of affairs means that Germany built lots of photovoltaics when it was expensive but is building less now that it is cheap.

But "central control" does not mean that Berlin decides everything. Top policymakers want to be taken out of the line of fire. The Network Agency handles the auctions; the government merely sets the volume. All the potential pitfalls of auctions—failure to build, excessively high prices, market consolidation, and so on—can then be passed off as market failures. Under feed-in tariffs, policymakers take more of the blame themselves.

There is some wiggle room for FITs going forward. The European Commission has set forth a de minimis rule. Projects smaller than six megawatts or six wind turbines up to 18 megawatts do not have to take part in auctions, which the European Commission otherwise calls for. Originally, the Commission only wanted to allow for exemptions up to 3 megawatts or three turbines, but the limit was doubled to accommodate Germany. Strangely, the German government now does not wish to make use of this leeway, which would allow a wide range of community projects to move ahead.[105] Indeed, even in 2014 there were signs that Berlin was using the intervention from Brussels in order to appear to be forced to make a change that it wanted in reality, though it did not want to admit such a thing.[106]

---

[105] Morris, Craig. "Upcoming auctions for photovoltaics: a legal perspective." Renewables International. The Magazine. 19 December 2014. Accessed February 9, 2016. http://www.renewablesinternational.net/upcoming-auctions-for-pv-a-legal-perspective/150/537/84338/.

[106] Morris, Craig. "Brussels and Berlin settle dispute until 2017." Renewables International. The Magazine. 10 July 2014. Accessed February 9, 2016. http://www.renewablesinternational.net/brussels-and-berlin-settle-dispute-until-2017/150/537/80210/.

In July 2015, the Economics Minister made its current position clear in a policy paper.[107] It explains that 6 megawatts/turbines is far too great a number because a lot of project developers who are clearly not themselves community projects work within that wind farm size. Furthermore, the Ministry fears that developers might begin gaming the system by artificially splitting up larger projects into six-turbine subprojects.

A simple definition of what constitutes *Bürgerenergie* (citizen energy) would solve the problem, of course. The government could, for instance, stipulate that such projects are exempt if they have a certain percentage of citizen ownership, and voting rights are by person, not by the number of shares. Bidders can also be limited to one de minimis project a year. Other details can also be tweaked, but the specifics are not the point here. The point is that the government's renewable energy policy increasingly seems designed to shut out the very citizen and community groups that have sustained the energy transition for at least the past 25 years. At present, this policy failure can still be excused as a temporary flaw. If it is not fixed soon, however, one can only conclude it is intentional.

Ironically, this pushback against energy democracy is occurring with staunch proponents of the Energiewende in office. Before Fukushima, the debate largely revolved around two camps: those against and those in favor of the energy transition. After Fukushima, the opponents dispersed; they still exist, but they no longer form a coherent voice in the political debate. Now, two schools of thought within the Energiewende shape the discussion: those who focus on cleaning up polluting industry, and those who focus on bringing about new players with renewables, including citizens and communities. This rift within Energiewende proponents was always there, but it has just now come to the foreground. As we will see in the next chapter, the future success of Germany's energy transition hinges largely on these two groups both reaching their goals.

---

[107] BMWi. "Ausschreibungen für die Förderung von Erneuerbare-Energien-Anlagen. Eckpunktepapier." Bundesministerium für Wirtschaft und Energie. Berlin, July 2015. Accessed February 9, 2016. https://www.bmwi.de/BMWi/Redaktion/PDF/Publikationen/ausschreibungen-foerderung-erneuerbare-energien-anlage,property=pdf,bereich=bmwi2012,sprache=de,rwb=true.pdf.

The Energiewende is a work in progress and will remain one. Mistakes have been made, and there are more to come. Numerous metaphors have been used for the Energiewende, and our book opened with a mention of "Germany's Man on the Moon." Perhaps we should close this chapter— the last one on the past—with another common meme indicating the seriousness of the task. It's not the best metaphor; the energy transition should rejuvenate Germany, not give the elderly a few more years of life. But the metaphor's popularity among top politicians and energy sector experts reveals their own nervousness: the Energiewende as "open-heart surgery."[108]

---

[108] "Tweet of Peter Altmaier." Twitter. 11 November 2014. Accessed February 9, 2016. https://twitter.com/peteraltmaier/status/532316313062875137.

14

Will the Energiewende Succeed?

If the Energiewende is a football game, it has just entered the second half if measured from the publication of the 1980 *Energiewende*; 2016 is 36 years later, with only 34 years left until 2050, when the transition's targets have to be met. And if we say the actual transition did not really begin until the Feed-in Act of 1991 or maybe even the Renewable Energy Act of 2000, then we are only in the first half. The game is far from over. It is therefore much too early to say that the Energiewende is succeeding or failing. But the game has gone fairly well up to now; victory is not unlikely.

But whether we are in the first or the second half, one thing is certain: there is no halftime. Germany's energy transition will continue without any break. Electricity is still delivered every second at the exact level of demand. Businesses continue to grow, entrepreneurs continue to invest. Technologies are being developed at a rapid pace; we don't know exactly what will be available and affordable by 2050. And most of us continue to go about our daily lives with two major wishes: continued prosperity along with a bit of peace and quiet.

Here, we see that the Energiewende's success can partly be measured in terms of things hard to capture as a number. Happiness is elusive math-

© The Editor(s) (if applicable) and The Author(s) 2016
C. Morris, A. Jungjohann, *Energy Democracy*,
DOI 10.1007/978-3-319-31891-2_14

ematically. Yet, if the energy transition doesn't make people happier, then what good is it?

As we have seen above, the Energiewende has numerous impacts outside of the official targets. Public involvement means public (self-) education. The Germans are teaching themselves and each other about energy and politics, thereby developing skills that will be useful in the future. And of course, the transition allows people to make their own energy, both as individuals and members of committees. If Germany reaches its official targets for 2050 but does so primarily with utility projects, sidelining public participation in the process, then the original spirit of the Energiewende—the driving force since the 1970s—will have been lost. The environment would still be cleaner and carbon emissions still reduced to a sustainable level, but corporations too big to fail would run the show. These firms would use their money and influence the shape of the public sphere—politics, the media, society, and so on—to suit their bottom line. Citizens would be spoken of primarily as consumers. For the authors, this outcome would be a huge missed opportunity.

What about the official targets? By 2050, Germany plans to cut its primary energy consumption in half. By that time, renewables are to make up at least 60 percent of energy consumption. If consumption remained the same, renewables would only make up 30 percent—compared to around 12 percent in 2014. The transition will be the easiest in the power sector, where renewables are to cover at least 80 percent of supply. Obviously, 79 percent renewable power would not be seen as a true failure, but remember the importance of the wording: "at least." More is also okay. But is this really possible?

Cutting primary energy in half may sound fanciful, but no magic will be needed. First, we need to understand the difference between primary and final energy. Primary energy is input; final energy, output. A lump of coal is primary energy; the electricity that comes out of the plant, final energy. Gasoline you put in your car's tank is primary energy; the motive force that reaches the wheels, final energy.

An old coal plant has an efficiency of around 33 percent. In other words, you put three units of coal in for every unit of electricity you get out. Internal combustion engines (ICEs) in cars are even worse at closer to 20 percent efficiency—five parts gasoline in for every unit of motive force that moves your car forward.

If we replace the old coal plant with a technology that transfers all its primary energy input into final energy output (such as wind and solar do), we cut primary energy use by 67 percent. Primary energy is an expression of how much of a finite resource we remove from the environment. No matter how many solar panels we install, we cannot "use up the Sun" faster. Likewise, wind turbines affect local wind conditions, but the impact on the total amount of wind on the planet will be hard to measure even in a future with many times more wind farms.[1]

Battery-powered electric vehicles have efficiencies around 90 percent. Passive House architecture optimizes building orientation and uses excellent insulation in order to cut demand for heat by 75 percent below the level of the German building code in 2015. And Passive House essentially becomes the new standard throughout Europe when the EU's requirement for Nearly Zero-Energy Buildings takes effect around 2020.

In other words, the German goal of cutting primary energy consumption in half is actually only another way of saying the country will have Passive House architecture and cars charged with electricity from wind and solar. The goal for primary energy efficiency is not a separate target, but another way of expressing the same thing.

So what does Germany need to do to get Passive House in new buildings and refurbishments? How can it promote not only electric mobility, but also walkable cities and cycling? And is 80 percent renewable electricity really feasible when almost all of it will be fluctuating wind and solar power?

Most of the Energiewende research and political position papers focus on the next steps that need to be taken: which grid lines need to go where, how does what policy need to be changed, and so on. This discussion is quite well-documented, even in English, such as at websites of Clean Energy Wire, Agora Energiewende, EnergyTransition.de, and the German economics ministry BMWi. To provide a new angle, we investigate the future not from today's viewpoint, but from two other perspec-

---

[1] Two methods are used: the physical content method (which the IEA uses) and the substitution method (which BP uses).

tives: the first Energiewende study of 1980; and looking back from 2050, when the goals are to be met.

Energiewende Studies from the 1980s: A Clear Political Agenda

In 1980, three researchers from the newly founded Öko-Institut published a book entitled "Energiewende: growth and prosperity without petroleum and uranium." The inclusion of "growth and prosperity" is crucial; amidst a debate about whether our lifestyles have to change fundamentally, the authors argued that sacrifice would not be necessary. This debate continues to be held today: Will people be able to fly as often in 2011 as they do today? In focusing on "growth and prosperity," the 1980 book does not specifically try to address this fundamentalist Green position as much as it answers the more exaggerated criticism from skeptics that the energy transition is a return to the open fires of cavemen; on the contrary, the authors write, people will perceive the convenience of multimodal mobility and the greater comfort from Passive House architecture as progress—not to mention cleaner air without coal emissions. But something is missing from the 1980 book's title: coal.

In reality, it's not missing at all. The most advanced scenario in the book would have Germany getting nearly half of its energy from renewables, with around 55 percent still coming from "domestic coal." Germany was a major producer of both hard coal and lignite at the time. In the wake of two oil crises, the main concern was the reliability of energy supply. The plan included the liquefaction of coal to produce a fuel that could replace gasoline and diesel, roughly 98 percent of which the country has to import.[2] Natural gas was also somewhat problematic; large amounts had recently been discovered in the North Sea, but Germany still imported around a third of this fuel from the Soviet Union—a politically tense situation. In contrast, Germany has the largest lignite resources in the

---

[2] BMWi. "Energiestatistiken." Bundesministerium für Wirtschaft und Energie. Accessed February 9, 2016. https://www.bmwi.de/BMWi/Redaktion/PDF/E/energiestatistiken-energiegewinnung-energieverbrauch,property=pdf,bereich=bmwi2012,sprache=de,rwb=true.pdf.

world—enough to last at least another century or two, depending on the estimate, at current levels of consumption (roughly a quarter of power supply).

Nuclear was not considered an option on the basis of cost: "20 years (and 20 billion deutsche marks in tax money) later, Germany now has a small fleet of nuclear plants that cover only 10 percent of power supply and only two percent of final energy consumption…We would be just as well off (or not) without any nuclear power." The share of nuclear energy in final energy never rose significantly above 6 percent in Germany, half the level of renewables at present. Furthermore, Germany has always imported basically 100 percent of its uranium, so the focus on domestic energy supply ruled out nuclear power.[3]

The book also focused on efficiency. Here, the authors found nuclear to be a bad option because the reactors had already become huge and, for safety reasons, were built far away from centers of consumption. As a result, significant amounts of waste heat could not be recovered. Instead, they praised the much higher efficiencies of fossil-fired cogeneration units, which were much smaller and could be connected to district heat networks. The efficiency of a large nuclear reactor is around 40 percent; that of a cogeneration unit, easily twice as great. The authors also knew about Passive House architecture before it was even called such. Their overview of disparate projects worldwide revealed, however, that Germany was not yet a leader; at the time, Switzerland and the UK seemed the furthest ahead with energy-efficient buildings.

What about renewables? photovoltaics was still an application for satellites in the 1970s, but the authors assessment hits the bull's-eye: "Notably, we have not included photovoltaics in our calculation, although prices have already begun dropping so fast that we might need to expect solar cells to play a role, especially in distributed power supply." They were equally prescient about wind power, speaking of future three-megawatt wind turbines with rotor diameters of around 110 meters, which we had by around 2000.

Why did it take so long? The authors already understood that the monopoly market structures in the German energy sector at the time

---

[3] BMWi. *Energiestatistiken.*

would unnecessarily slow down these trends. They therefore formulated an unambiguous political agenda: "The current monopolist structures on energy markets have to be done away with." Distributed renewables would replace central power stations, and distributed biomass units would be "hard for large corporations to monopolize." But the authors also understood the limits of sustainable biomass; they reminded readers that there would eventually be competition between food and energy crops.[4]

Germany's failure to rein in its own coal sector is easier to understand against this historic backdrop. Indeed, the French do not have so much nuclear power because they were more concerned about global warming than everyone else in the 1970s; rather, like the Germans, they lack significant oil resources—but unlike the Germans, they also have practically no coal. No country has ever left affordable coal in the ground; the US and the UK partly transition from coal to gas when gas became cheaper. Still, does the recommendation of 55 percent energy from domestic coal not reveal a blind spot in the Energiewende debate from the outset?

In an interview from 2013, Florentin Krause, one of the book's coauthors, called this interpretation a misreading: "None of our scenarios suggest increasing coal consumption."[5] Rather, overall energy consumption would have been cut almost in half, with nearly half of the remaining energy being renewable, so carbon emissions would have been reduced by around three quarters in the most progressive scenario. Keep in mind that such proposals were widely derided as preposterous at the time.

Furthermore, Krause believes he and his colleagues were far ahead of the curve in even talking about carbon emissions: "To my knowledge, our work was the first scenario study for Germany that incorporated the concern over carbon emissions. This was 12 years before the 1992 UN climate conference in Rio de Janeiro. We were a lonely voice at the time."

---

[4] Morris, Craig. "Looking back at the Energiewende 1980—Time for a Coal Phaseout." Energy Transition. The German Energiewende. 30 March 2013. Accessed February 9, 2016. http://energytransition.de/2013/03/time-for-a-coal-phaseout/.

[5] Morris, Craig. ""Efficiency lacks a loud lobby": An interview with Florentin Krause." Energy Transition. The German Energiewende. 17 April 2013. Accessed February 9, 2016. http://energytransition.de/2013/04/an-interview-with-florentin-krause/.

Unfortunately, the views expressed in the 1980 Energiewende book hardly impacted policymakers, at least not immediately. In fact, the opposite can be stated: had the recommendations in the 1980 Energiewende book been heeded, Germany's energy transition would not be primarily an electricity transition, but would also have paid equal attention to heat and transport. The book discussed options for electric mobility and bio-fuels "where electric drives are less practical," Krause remembers. And his main criticism of current policy is that too little attention is paid to efficiency. The book contained a scenario investigating efficiency gains not only in homes, cars, and appliances, but also in industry.

Conventional German energy experts derided Öko-Institut's most progressive proposals along with those of Heidelberg-based IFEU—another independent research group newly founded to produce alternative scientific investigations for anti-nuclear protesters—as unscientific, if not outright science fiction.[6] The idea that energy consumption could be cut in half without slowing down economic growth was quite controversial at the time, but by the end of the decade this decoupling was already happening. By the 1990s, the camp of pro-Energiewende energy experts in Germany was gaining respect because its predictions had clearly turned out to be more accurate than those of conventional experts. But this widespread respect was hard-earned and a long time coming. In 1985, Öko-Institut produced a follow-up Energiewende book, its main question being why nothing had happened.

The answer in the 1985 book was that a quick energy transition was not in the interest of incumbent energy corporations. Amazingly, the authors implicitly anticipate the current utility death spiral. Kilowatt-hour prices drop as we move from the smallest consumers (households) toward the largest (industrial firms). The rationale generally given is that bulk purchases make unit prices cheaper, but the authors convincingly prove that the reason is different. In the first half of the 20th century, industrial firms often made their own electricity and heat. In order to expand, large utilities had to underbid the price of on-site generation for these firms. The utility recovered its profits from captive consumers: households and small

---

[6] Pröstler, Leo. "Angriffe gegen das Institut." *Öko-Mitteilungen*, no. 5 (82): 29–32. Accessed February 10, 2016.

businesses. In the 1980s, these parties had few options to the retail rate, though the first wind power generators did provide electricity at around seven pfennigs per kilowatt-hour, a fraction of the retail rate. However, even those early turbines often had a rated capacity of 55 kilowatts, a size that would overwhelm a normal household.

Entitled "The Energiewende is possible," the 1985 book thus found that small power consumers cross-subsidize large power consumers in Germany—a situation that persists even today, only that now households and small businesses have the option of installing solar, increasingly with battery storage.

Funded with donations from the public (as was the 1980 book), the authors of *Energiewende is possible* could speak their minds freely. Not surprisingly, some of their proposals are uncompromising. For instance, they took the idea of distributed energy to its logical conclusion in calling for "one power plant per company."[7] They explained how it was "impossible for investments to lead to losses" in the energy sector. Because the sector was regulated, profit margins were protected; after all, the government cannot force a company to do business at a loss. In return, margins must not be excessively high lest consumers be gouged.

But the problem with this arrangement, the authors found, was that German regulators were not as savvy as the utilities they monitored. The officials focused on the prices, not on whether projects need to be built. Power providers told regulators that everything proposed was needed to prevent blackouts. The regulators then reviewed the impact on rates without ever properly questioning the specter of blackouts.[8]

The result was overbuilding. Grid experts estimate that a country needs roughly a dispatchable generation capacity exceeding peak power demand by around 10 to 15 percent. When power markets were liberalized in Germany at the end of the 1990s, this reserve was closer to 25 percent. Back in the early 1980s, it was around 40 percent. Under liberalized market conditions, such overcapacity would have been unprofitable

---

[7] Morris, Craig. ""One power plant per company"" Energy Transition. The German Energiewende. 19 March 2015. Accessed February 9, 2016. http://energytransition.de/2015/03/one-power-plant-per-company/.

[8] Morris, Craig. ""Impossible for investments to lead to losses"" Energy Transition. The German Energiewende. 18 March 2015. Accessed February 9, 2016. http://energytransition.de/2015/03/impossible-for-investments-to-lead-to-losses/#utility.

because these power plants simply would not have been able to run often enough for a lack of power demand. Under regulated monopoly conditions, however, the firms simply passed on their costs at rates approved by regulators who didn't know better.

Even environmental campaigns could be passed on to consumers in this manner. For instance, in the mid-1980s RWE was proud of being behind the "largest and most complex environmental protection program ever undertaken by a single company." The price tag was 9 billion deutsche marks. The authors were not impressed: "Essentially, the firm is saying that it will charge an additional 9 billion marks to its customers, and in return it will also receive even larger profits because the guaranteed profit margin is a percentage of overall revenue. And the company even expects applause."[9] In other words, power providers at the time focused on increasing revenue; it was the only way to increase profits.

The authors found similar problems with grid lines. In the post-war era, Germany had increasingly built central power stations further away from cities, thereby requiring the grid to be expanded. Also, power plants grew bigger. In 1964, a decree made it difficult for plants smaller than 300 MW to be built. By 1984, 80 percent of the country's power plants were larger than 200 MW, compared to only 54 percent in 1962.[10]

Today, there is much debate over the amount of grid investments needed for the Energiewende. The grid has practically not been expanded at all for Germany to go from almost 0 percent non-hydro renewable electricity in 1990 to around 30 percent in 2015. Much of this capacity is spread across the country in small units and does not need long power lines.

A number of conclusions can be drawn from these early studies for our discussion about the future:

- cogeneration has not made much progress, nor has efficiency;
- too little is still being done to promote renewables in heat and transport;

---

[9] Morris, Craig. *"Impossible for investments to lead to losses"*.

[10] Morris, Craig. *""Energy wasted by design""* Energy Transition. The German Energiewende. 17 March 2015. Accessed February 9, 2016. http://energytransition.de/2015/03/energy-wasted-by-design/.

- the debate on grid expansions revolves around two competing visions: one of distributed energy supply close to points of consumption; the other, a focus on the best wind and solar resources, with additional power lines bringing this electricity to consumers elsewhere;
- focal points have changed—the nuclear phaseout remains central, but combating climate change has replaced national energy independence as a main objective; and
- a fast energy transition will require new players; incumbents will be the losers, so they will fight for their lives.

Finally, the 1985 book reminded its readers of ordo-liberalism. The authors complain that the debate at the time revolved around free markets versus government regulation. They point out that free markets do not come about when government stops intervening. On the contrary, oligopolies then form; firms with the most market power push out small competitors. The book calls for a free market but makes it clear that it will require governmental intervention to prevent the market from failing. In other words, the state brings about a free market by continually intervening where necessary.

It is hard to imagine the call for "one power plant per company" being made by pro-Energiewende researchers today. Such statements are a comment on the overall political system and the market. The book investigates what society needs. It was only possible because members of society—private citizens as donors—funded the publication. As coauthor Dieter Seifried remembers, Öko-Institut was largely funded with donations at the time.

To paraphrase Margaret Thatcher, society does not contract a lot of research studies. German energy researchers now focus more on topics that moneyed interests—companies, industry organizations, governmental ministries, and NGOs—want investigated. Today, the Öko-Institut still has donors, but they provided only 230,000 euros of support in 2013[11]—less than 2 percent of the Institute's total budget of around 13 million.[12]

---

[11] "Around the World: Annual report of the Oeko-Institut 2013." Berlin, April 2014. Accessed February 9, 2016. http://www.oeko.de/oekodoc/2024/2014-609-en.pdf.

[12] "Webpage of the Öko-Institut." Accessed February 9, 2016. http://www.oeko.de/en/the-institute/.

The Crucial Role of Citizens and Communities

The Energiewende absolutely still needs the research done by such institutes. But in the overall volume of material produced, studies focusing on the big picture—what is best for society—have become few and far between. The early pro-Energiewende researchers not only provided answers, but also formulated the questions themselves. Today, these researchers provide answers to questions formulated and paid for by others: industry associations, governmental ministries, and so on. The need for future work from such sources might color the answers given. German climate and energy policy researcher Oliver Geden speaks of "policy-based evidence-making" (instead of evidence-based policymaking).[13] Questions about how large energy corporations should be are hardly ever asked these days.

One is most likely to find these investigations in studies funded by NGOs.[14] The role of citizens and communities in Germany's energy transition is therefore generally dealt with from the perspective of policymakers, industry, and enterprises. We thus find repeated emphasis on citizen participation as a crucial instrument toward increasing acceptance of, say, new power lines. This view relegates citizens to a group that needs to be brought in line with big projects; the Energiewende is seen here as something properly handled from the top down. It wasn't always this way.

For policymakers and utilities in Germany, the need to overcome NIMBYism is the reason why they want communities and citizens involved in the energy sector. This view does not take account of the main reasons why citizens want input. Citizens call for "greater local self-determination and citizen engagement" along with ecological concerns, as Patrick Graichen (now the head of Berlin-based think tank Agora Energiewende) explains in his history of the Power Rebels in Schönau.

---

[13] Geden, Oliver. "Policy: Climate advisers must maintain integrity." Nature. 6 May 2015. Accessed February 9, 2016. http://www.nature.com/news/policy-climate-advisers-must-maintain-integrity-1.17468.

[14] Such as Bontrup, Heinz-J., and Ralf-M. Marquardt. "Die Zukunft der großen Energieversorger." Greenpeace. Hannover/Lüdinghausen, January 2015. Accessed February 9, 2016. https://www.greenpeace.de/sites/www.greenpeace.de/files/publications/zukunft-energieversorgung-studie-20150309.pdf.

People get together in local fire brigades and sports clubs, for instance, in order to get to know each other and create communities for themselves.[15]

Practically all the Energiewende's shortcomings stem from this dilemma: citizens and communities (new players) have been on the move; the big players, less so—and the German government now seems keen on limiting the role of communities even as industry is let off the hook. The Volkswagen scandal unfolding in 2015 is a story of corporate cheating, but also one of governmental watchdogs benevolently turning a blind eye for the preceding decade. And the German government continues to protect its industry today; in late 2015, the German Environment Ministry rejected stricter pollution limits, specifically for ammonia and methane.[16] Clearly, industry is not a driver for a cleaner environment or an energy transition—the public is. Only because public will became law around the usual ministerial channels with both the 1991 Feed-in Law and the 2000 EEG did the Energiewende get off the ground.

Of course, we should not expect citizen cooperatives to start manufacturing electric vehicles; communities can't do everything. But electric cars are not a silver bullet, either, and communities can do something even more important by creating pedestrian zones, bike lanes, and affordable, convenient public transportation—in other words, by reducing people's need for cars. Just don't expect corporate carmakers to welcome *that* transition.

The role that electric cars will play is overstated because people often think some new technology is needed. Actually, walking and cycling are the best option both in terms of the climate and personal health. Electric cars can help where there is urban sprawl, but proper urban planning can prevent the sprawl, thereby reducing the need for electric cars. The most progress will therefore still come from communities, not car companies.

Attempts to get industry involved in the global energy transition have not fared well up to 2016. Many of those who ostensibly supported the

---

[15] Graichen, Patrick. *Kommunale Energiepolitik und die Umweltbewegung: Eine Public-Choice-Analyse der "Stromrebellen" von Schönau.* 1. Aufl. Mannheimer Beiträge zur politischen Soziologie und positiven politischen Theorie 7. Frankfurt am Main: Campus-Verl., 2003. Univ., Diss. DOUBLEHYPHENHeidelberg, 2002.

[16] "Luftreinhaltung: Bundesregierung legt Veto im EU-Umweltrat ein." Windkraft Journal. 31 December 2015. Accessed February 9, 2016. http://www.windkraft-journal.de/2015/12/31/luftreinhaltung-bundesregierung-legt-veto-im-eu-umweltrat-ein/76900.

call for a carbon price did so knowing that any real-world arrangement reached would remain toothless. Here, it is worth quoting at length Tom Burke, head of the UK's energy analyst agency E3G, from 2015:

> Changing your business model is no simple task…Its first product is a decision to buy time to think about how to deal with the collision between their business model and a safe climate. Hence the call for a carbon price. The intent is to create the impression of an industry in favour of urgent action whilst actually slowing that action down…
>
> The call for a carbon price is a shield with which to defend themselves from calls for faster change. If we are not decarbonising fast enough, they will argue, it is not their fault. If only governments were brave enough to put the carbon price up higher and faster, they will lament, we would get there sooner.

Granted, progress has appeared on the horizon. Global coal consumption dropped in 2015 for the first time in decades, and the growth of carbon emissions slowed down in 2015, albeit at a level too high to prevent significant climate change. Obviously, we still need to work with industry. But we must remember that they are not really the ones calling for dramatic, fast change. A survey taken for the German government in the summer of 2015 revealed that only 5 percent of the public want coal power 30 years from now.[17]

For Germany's energy transition to succeed, citizen and community involvement with renewables and efficiency must be protected, and far more success needs to be made in cleaning up the big players.

The Two Camps Within the Energiewende Today

Until Fukushima, there were broadly two camps within the Energiewende: those for it and those against it. After Fukushima, the camp of opponents dispersed; they still exist, but they no longer constitute a single unified

---

[17] Schlandt, Jakob. "Interne Regierungs-Umfrage: Bürger wollen Kohle-Ausstieg." Phasenprüfer. 28 September 2015. Accessed February 9, 2016. http://phasenpruefer.info/buerger-wollen-kohle-aus/.

force speaking outright. Nowadays, the discussion is being held between two somewhat diverging pro-Energiewende worldviews that were always there but have only recently come to the foreground.

On the one hand, we have those who want to focus on cleaning up the polluters. This group argues that the most progress is to be made by solving the biggest problems, particularly carbon emissions from industry and the conventional energy sector. This **Polluters Pay camp** supports renewables but has sometimes argued that a focus on renewable energy alone will not move us fast enough. In the worst case, they have sometimes doubted that renewables can grow quickly enough.

The other school of thought holds that industry and conventional energy firms do not want to clean up their act, nor do they want a rapid energy transition, so quick change will only come from new players. As our history of the Energiewende shows, the **New Players camp** has had the most success up to now, so you might expect this camp to be directing energy policy today.

They are not. Two of the most prominent representatives of the New Players camp are Hans-Josef Fell and Hermann Scheer. Fell retired from the Bundestag in 2013, and Scheer passed away in 2010 just before reaching retirement age.

Of course, numerous organizations represent the New Players camp nonetheless today. The most prominent of them is arguably German wind power association BWE. Whereas the US (American Wind Energy Association) and EU equivalent (European Wind Energy Association) primarily represent the interests of wind turbine manufacturers, Germany's BWE primarily represents investors—who, historically, have been farmers and community cooperatives. The organization has openly expressed its support for onshore wind, where citizen energy cooperatives are highly active, as opposed to offshore wind, where the main investors are large utilities. The wind turbine manufacturers perceived a need to provide greater support for offshore wind and therefore created a separate Offshore Wind Energy Foundation.

The BWE is also behind a monthly magazine called *Neue Energie* (New Energy). Although the BWE only represents wind turbine owners and not investors in other types of renewables, the magazine nonetheless has a history of covering all sources of renewable energy and related issues. In doing so, the monthly reflects the interest of the citizens who read it; after all, they are not only interested in wind power, but in the Energiewende. The authors know of no other monthly industry journal in any country that covers all renewables; all other publications mainly cover their own industry, usually strictly from the viewpoint of the organization's board.

The German Renewable Energy Federation, the BEE, is another relatively old organization worth mentioning from the New Players camp. It closely monitors the current pushback against community energy, as does the more recently founded Citizen Energy Alliance (BEEn). Finally, the *Bundesverband Neue Energiewirtschaft* (BNE) represents new energy suppliers; it was founded in 2002 shortly after liberalization of the power market. Apparently, its members do not feel that the utility lobby group BDEW represent its interests equally alongside those of larger utilities.

A clash between Scheer and Öko-Institut in the early 2000s illustrates well the early existence of these two pro-Energiewende camps (New Players and Polluter Pays)—and how governmental bodies contracting studies were able to formulate questions that changed the debate. In 2001, the state of Baden-Württemberg—which back then had a government critical of renewables—asked Öko-Institut and German Aerospace Center DLR to investigate what the implementation of a green certificate system similar to the one used in the UK might look like in Germany.[18] The question came as an affront to Scheer in only the second year of the Renewable Energy Act (EEG), which used feed-in tariffs.

The paper states that it investigates a harmonization of energy policies within the EU, a charge that would return with a vengeance in 2013, when the EU finally managed to get the German government to abandon feed-in tariffs and roll out auctions, a process that will be completed in 2017. The paper used a phrase that would later become a buzzword (*marktnah* or literally "close to the market" but meaning "exposed to

---

[18] Timpe, Christof, Heidi Bergmann, Uwe Klann, Ole Langniß, Joachim Nitsch, Martin Cames, and Jan-Peter Voß. "Umsetzungsaspekte eines Quotenmodells für Strom aus erneuerbaren Energien. Kurzfassung." Im Auftrag des Ministeriums für Umwelt und Verkehr Baden-Württemberg. Öko-Institut and DLR. Freiburg, Stuttgart, Heidelberg, August 2001. Accessed February 9, 2016. http://www.dlr.de/tt/Portaldata/41/Resources/dokumente/institut/system/publications/Quote_BaWue_Kurz_final.pdf.

wholesale power prices"): "renewables are to be more exposed to the market so that no further financial support is needed." (It also explicitly called for policy support for the co-firing of biomass in coal plants, an option that is anathema today in Germany.)

Scheer promptly wrote the paper's main author a complaint, and the author attempted to assure the politician that the paper did not call Scheer's favorite policy into question—but to no avail. Scheer faced severe criticism already in Berlin, and the notion that a policy that was taking renewables nowhere fast in the UK might be better was too much for him to stomach. And in reality, Scheer's criticism was unfair; the Öko-Institut wasn't saying anything Scheer hadn't basically said before. When calling for feed-in tariffs in the 1990s, he had written that only five years of policy support would be needed, with the market taking over afterward.[19] Now that he had his policy, he was eating those words as he called for indefinite extensions.

Other conflicts also flared up between Scheer and pro-Energiewende experts at the time. In 2003, economics institute DIW and the Öko-Institut investigated the need for new conventional power plants to replace the old ones. The government had recently proposed a new target, which would be adopted in 2004: 20 percent renewable electricity by 2020. The authors wrote: "Even if...the 2020 target is adopted, a lot of electricity will still have to come from other sources."[20] They did not doubt that the renewable power target could be met; one of the authors, Felix Matthes of Öko-Institut, had co-authored a study in 1991 with Dieter Seifried, which forecast the level of wind and solar power generation in 2010 within 10 percent of what actually occurred that year.[21] But in the 2003 paper, Matthes called for new conventional plant projects, mainly gas turbines, to replace upcoming decommissions of old coal plants. In other words, their focus was on cleaning up the polluters.

---

[19] Scheer, Hermann. *Sonnen-Strategie: Politik ohne Alternative.* Serie Piper Bd. 2135. München: Piper, 1998, p. 215.

[20] "Wochenbericht des DIW Berlin 48/03: Energiepolitik und Energiewirtschaft vor großen Herausforderungen." Accessed February 9, 2016. http://www.diw.de/sixcms/detail.php/284434.

[21] Morris, Craig. "Halftime for the Energiewende." Renewables International. http://www.renewablesinternational.net/halftime-for-the-energiewende/150/537/86407/.

To be fair, the assumed need for new conventional plants seemed obvious at the time and was widely held.[22] Nuclear made up around 30 percent of electricity supply but would disappear by around 2022 under the phaseout agreement of 2002. If renewables grew to 20 percent, power from fossil fuel would actually have to increase by 10 percentage points to fill the gap left by nuclear.

Scheer saw things differently. He believed that renewables would grow even more quickly and began warning utilities not to invest in new coal and gas plants. He was a lone voice with this opinion, but he turned out to be right; in 2015, Germany already had 33 percent renewable electricity (as a share of demand, so excluding power exports), so renewables equaled 30 percent nuclear power seven years before the phaseout was even over. The 2020 target has since been increased to 35 percent, and even this higher target will likely be surpassed in 2016.

The conflict between the New Players camp around Scheer and the Öko-Institut smoldered on behind the scenes until December 2004, when Scheer went public in the German press, accusing the Öko-Institut of supporting a policy that would create more work for researchers like the paper's authors. As he pointed out, feed-in tariffs do not require expert reviews, whereas certificates of origin do.[23] He saw the Institute's proposals as a job creation attempt for itself.

The institute defended itself in a press release entitled "The EEG is the right instrument," but the first sentence was telling: "The Öko-Institut believes the EEG is the right instrument for the midterm…"[24] Implicitly, "midterm" meant that feed-in tariffs would have to go, probably sooner than later. The press release argued that certificates could be combined with feed-in tariffs and listed two points of criticism. First, feed-in tariffs had yet to demonstrate that they could actually bring down the cost of a

---

[22] Jungjohann, Arne, and Craig Morris. "The German Coal Conundrum." Unpublished manuscript, last modified February 6, 2016. https://us.boell.org/sites/default/files/german-coal-conundrum.pdf.

[23] Scheer, Hermann. "Kommerzieller Kurzschluss." TAZ. 13 December 2004. Accessed February 9, 2016. http://www.taz.de/1/archiv/?id=archivseite&dig=2004/12/13/a0184.

[24] "„Das EEG ist das richtige Instrument": Das Öko-Institut weist Vorwürfe des SPD-Bundestagsabgeordneten Scheer zurück." Öko-Institut. Freiburg, 15 December 2004. Accessed February 9, 2016. http://www.oeko.de/fileadmin/wikipedia/041215_eeg.pdf.

kilowatt-hour of renewable electricity, and second, the EEG did not take account of European power trading.

The institute would certainly not express the first concern today. Through the deployment it triggered, German feed-in tariffs brought down the cost of both wind and solar power more than anyone, including Hermann Scheer, ever thought possible. The second item, however, is the one that brought down German feed-in tariffs in 2013. It wasn't so much that FITs conflict with EU law; the European Court of Justice (ECJ) has repeatedly ruled in favor of them, in fact.[25] Rather, politicians in Brussels simply continued to claim there was a conflict (despite the ECJ's repeated rulings otherwise), and eventually Berlin used this European intervention as an excuse to toss out FITs, whose shortcomings are considered a failure of government, in favor of auctions, whose shortcomings are considered a failure of markets. "We will have millions of units to produce wind and solar power. It is adventurous to think the government could steer them. The market will be needed to arrive at efficiency," Undersecretary Baake has said.

Baake is a member of the Green Party, making him an unusual top official in a grand coalition of Christian Democrats and Social Democrats. His appointment to this post was a testimony of the wide respect he has throughout the political spectrum; it was also a tactical move by Energy Minister Gabriel, who made it harder for Greens to criticize the government's energy policy.

Baake is occasionally referred to as a co-author of the Renewable Energy Act (EEG) of 2000.[26] In fact, he had little to do with it. Hans-Josef Fell remembers him being skeptical of the law for similar reasons as those held by the Öko-Institut.[27] At the time, Baake worked as Undersecretary at the Environmental Ministry; as described in previous chapters, the EEG was one of the exceptional laws that came from

---

[25] Morris, Craig. "ECJ rules in favor of national payments for renewables." Renewables International. The Magazine. 1 July 2014. Accessed February 9, 2016. http://www.renewablesinternational.net/ecj-rules-in-favor-of-national-payments-for-renewables/150/537/80006/.

[26] Mihm, Andreas. "Grünen-Vordenker will Ökostrom-Subventionen deckeln." F.A.Z., 10 October 2013. Accessed February 9, 2016. http://www.faz.net/aktuell/wirtschaft/wirtschaftspolitik/energiewende-gruenen-vordenker-will-oekostrom-subventionen-deckeln-12610839.html.

[27] Personal correspondence.

Parliament, not from ministry officials. Baake was instrumental in negotiating the nuclear phaseout with industry, and he otherwise dealt with such issues as emissions trading and recycling. In other words, Baake is a good example of someone from the Polluters Pay camp.

In 2012, a new think tank was created with Baake at the helm: Agora Energiewende.[28] By 2016, Agora had become quite influential in shaping the Energiewende discourse in Berlin. When Baake left to become Energy Undersecretary in 2013, his assistant Patrick Graichen became director of the think tank. Like numerous other staff members at the think tank, Graichen is quite familiar with the community renewable energy movement in Germany; he wrote his dissertation on the Power Rebels from Schönau. The think tank's Council, however, consists of only a single person representing renewable energy: an executive from a wind farm development firm.[29] The solar sector is no longer represented after Photon's Philippe Welter stepped down in March 2013 to focus on his firm's insolvency proceedings; he has not been replaced with anyone from the solar sector as of November 2015.

Otherwise, the Council includes, among others, two executives from energy-intensive industry (a chemicals firm and an aluminum producer), two members of the conventional power sector (utilities association BDEW and Eon), numerous politicians (German and European), the head of Germany's Network Agency and an executive from a grid operator, two labor union heads, and two representatives of municipal utilities. Not all these people have historically opposed the Energiewende. For instance, one of the municipal utility representatives is Holger Krawinkel, who published a book in 1991 to introduce Denmark's energy transition to the Germans.[30] In the decades since, he supported the energy transition as a top consumer advocate. Overall, Agora Energiewende seems to focus on bringing together representatives of the old camp against the Energiewende with those who have historically supported it; some staff members, like Gerd Rosenkranz, are long-time supporters of renewables.

---

[28] See about Agora: http://www.agora-energiewende.de/de/ueber-uns/agora-energiewende/.

[29] See about the Council of Agora: http://www.agora-energiewende.de/en/about-us/council-of-the-agora/.

[30] Krawinkel, Holger. *Für eine neue Energiepolitik: Was die Bundesrepublik Deutschland von Dänemark lernen kann.* Orig.-Ausg. Fischer alternativ. Frankfurt am Main: Fischer, 1991.

But with its focus, the think tank thus fits into the Polluters Pay camp better than into the New Players camp.

There has thus been a major shift in policy steering from the New Players camp toward the Polluters Pay folks. While the New Players group had great success boosting renewables, an argument can be made that more input from the other camp is now needed because more progress is needed getting conventional energy firms and industry on board. The trick will be keeping the involvement of citizens and communities—the groups that actually want the transition—in the process.

Three Energiewende Scenarios for 2050

At present, policymakers are focusing on the next steps for the midterm, such as grid upgrades and power market design. Up to now, renewable power generators have been able to sell electricity at a guaranteed floor price (feed-in tariffs) regardless of prices on the exchange. Unlike conventional power, renewables have priority grid access. This policy is now derided as "produce and forget," meaning that renewable power producers did not need to worry about their impact on the grid. In 2014, however, 1 percent of green electricity was curtailed because the grid could not absorb it, and that figure is rising.[31] That level, however, is relatively low in an international comparison; around 3.5 percent of wind power was curtailed in fiscal 2013 to 2014 in the UK, for instance.[32]

In addition, unscheduled power flows (loop flows) are passing from northern Germany to southern Germany via neighboring countries to both the East and the West. Some power lines are obviously needed to mitigate these loop flows, but some proponents of community energy

---

[31] "Zahlen, Daten und Informationen zum EEG." Bundesnetzagentur. 9 November 2015. Accessed February 9, 2016. http://www.bundesnetzagentur.de/DE/Sachgebiete/ElektrizitaetundGas/Unternehmen_Institutionen/ErneuerbareEnergien/ZahlenDatenInformationen/zahlenunddaten-node.html.

[32] "Winter Outlook 2014/15." National Grid. Warwick, October 2014. Accessed February 9, 2016. http://investors.nationalgrid.com/~/media/Files/N/National-Grid-IR/reports/Winter%20Outlook%20Report%202014.pdf.

charge that these new transmission lines will primarily protect baseload coal power from having to ramp down.

In redesigning the power market, Energy Undersecretary Baake says the old approach of "produce and forget" simply won't work going forward.[33] But having renewable power generators assume greater responsibility for grid stability also comes at a cost. In 2014, a group of economists wrote to the European Commission, which began the pushing for less "produce and forget" in all member states in 2013. They pointed out that uncertainty about revenue (not knowing how much of the power you generate can actually be sold) will increase risk and hence the cost of capital.[34] Because wind and solar have no fuel costs, installation costs occur almost entirely upfront, making interest rates far more important than they are for energy sources with fuel costs. The economists also argued that policy redesigns should focus on the market for flexibility options—everything that reacts to fluctuating solar and wind, such as dispatchable generators, storage, and demand shifting—instead of burdening wind farms and solar arrays unnecessarily.

Wind and solar can help stabilize the grid to some extent. For instance, they are already required to provide reactive power when connected to the medium-voltage grid or higher. It would also be possible to require them to run constantly at 90 percent of current output so that they could provide an additional 10 percent of electricity when there is a shortfall. (In general, ramping down is less of a problem than ramping up for conventional generators.) The question is then how renewable electricity generators would receive compensation in such a scheme.

Another question is how much of what renewable source will be needed. We now look at what goal attainment might look like in 2050 and try to draw conclusions working backward until today.

A number of scenarios for a renewables-based energy system have been published for 2050. For international onlookers, the first thing to note

[33] "German renewables sector urged to step up investments." EurActive. 12 September 2014. Accessed February 9, 2016. http://www.euractiv.com/sections/energy/german-renewables-sector-urged-step-investments-308404.

[34] "Economic analysis: Negative prices from priority for renewables." Renewables International. The Magazine. 21 September 2014. Accessed February 9, 2016. http://www.renewablesinternational.net/negative-prices-from-priority-for-renewables/150/537/81835/.

is that several studies have found that very high penetration levels of renewable energy, specifically green electricity, are not only possible, but also affordable—usually, however, only with consideration of the external environmental and health care costs. In fact, German researchers have produced so many such scenarios that they are often asked to author such investigations for other countries as well. For instance, in 2003 German and Japanese researchers produced "Energy Rich Japan," which investigated a 100 percent renewable energy scenario for the country.[35] German energy analyst Sven Teske and experts from the German Aerospace Center DLR, who also helped produce previous editions of the *Leitstudie* for the German government, are also behind Greenpeace's *Energy (R)evolution* study. Updated every few years, Greenpeace's publication investigates a global scenario—and has proven more accurate over the past decade than any other global forecast, such as the International Energy Agency's World Energy Outlook. In the following, we focus on the main three scenario investigations for 2050 published before this book went to press.[36]

"The Costs of the Energiewende" (ISE)

In November 2015,[37] Fraunhofer ISE published an overview investigating a carbon emission reduction of 80 to 95 percent, the German government's official target range for 2050. There was bad news for campaigners

---

[35] "Energy Rich Japan, Report." Accessed February 9, 2016. http://www.energyrichjapan.info/en/welcome.html.

[36] A handful of similar scenarios were also published in previous years and are worth mentioning in particular: "Wege zur 100% erneuerbaren Stromversorgung: Kurzfassung für Entscheidungsträger." Sachverständigenrat für Umweltfragen (SRU). Accessed February 9, 2016. http://www.umweltrat. de/SharedDocs/Downloads/DE/02_Sondergutachten/2011_Sondergutachten_100Prozent_ Erneuerbare_KurzfassungEntscheid.pdf?__blob=publicationFile; Nitsch, Joachim. "Lead Study 2008. Further development of the "Strategy to increase the use of renewable energies" within the context of the current climate protection goals of Germany and Europe." BMU. Stuttgart, October, 2008. Accessed February 11, 2016. http://www.bmub.bund.de/fileadmin/bmu-import/files/english/pdf/application/pdf/leitstudie2008_en.pdf (and follow-up studies); and Gerhardt, Norman, Fabian Sandau, Britta Zimmermann, Carsten Pape, and Bofinger, Stefan, Hoffmann, Clemens. "Geschäftsmodell Energiewende. Eine Antwort auf das "Die-Kosten-der-Energiewende"-Argument." Fraunhofer-Institut für Windenergie und Energiesystemtechnik (IWES). Kassel, January 2014. Accessed February 9, 2016. https://www.fraunhofer.de/content/dam/zv/de/forschungsthemen/energie/Studie_Energiewende_Fraunhofer-IWES_20140-01-21.pdf.

[37] "Was kostet die Energiewende? Wege zur Transformation des deutschen Energiesystems bis 2050." Fraunhofer ISE. Presseinformation 28/15. 5 November 2015. Accessed February 9, 2016.

Table 14.1 Installed capacity needed in 2050 in GW for percent share of renewables in power generation

	80%	85%	90%
Onshore wind	147	168	204
Offshore wind	24	33	42
Photovoltaics	122	166	290

Source: Fraunhofer ISE, "Was kostet die Energiewende? Wege zur Transformation des deutschen Energiesystems bis 2050."

carrying the "100 percent renewable" banner: the study found that the installed capacity needed for numerous technologies would double in order to move from an 80 percent emissions reduction to 90 percent. This finding was not surprising; economists speak of the Pareto Principle, which states that 20 percent of causes produce 80 percent of effects (Table 14.1).

The Fraunhofer ISE study investigates seven scenarios for the 80 percent CO_2 reduction; the table above only takes the one with the lowest installed capacity in order to stress the maximum difference. The amount of installed capacity nearly doubles from around 290 GW to around 540 GW. Peak power demand is currently around only 80 GW. But the main finding was that even an 85 percent carbon reduction was affordable under realistic assumptions.

An 85 percent reduction would also require 157 GW of solar thermal for low-temperature heat in buildings and industry. Excess electricity would be stored as heat. Surprisingly, very little electricity would be curtailed at only two percent. Interestingly, biomass is not emphasized much in the text (a sign of how bioenergy growth is being deemphasized in the German debate), though it would still make up around 14 percent of total energy supply in this scenario. Finally, roughly two times more natural gas and oil would be used than raw biomass, but no coal at all would be consumed. Without a coal phase-out, "a more than 80 percent reduction will be very hard to reach," the study finds.

---

https://www.ise.fraunhofer.de/de/presse-und-medien/presseinformationen/presseinformationen-2015/was-kostet-die-energiewende-2013-wege-zur-transformation-des-deutschen-energiesystems-bis-2050.

The remaining fossil fuel would be a combination of natural gas and oil. The latter would still make up 73 percent of liquid fuel consumption, with natural gas spread primarily across cogeneration and power consumption (Table 14.2).

Otherwise, the 85 percent scenario would require 76 GW of electrolysis units, which would produce hydrogen from excess electricity. Battery storage is given in gigawatt-hours, not gigawatts, and comes in at 81 GWh in cars and 74 GWh in stationary systems. The amount of hydrogen that is "methanized" (meaning that carbon is added to the hydrogen to produce synthetic methane) is significant and roughly equivalent to the amount of hydrogen produced via electrolysis and consumed directly—in other words, roughly half of the hydrogen would be converted into methane.

When hydrogen is produced from excess electricity and used to generate electricity again, roughly half of the energy is lost in the process, and methanation adds to the cost even further. The study does not, however, expect the need for power-to-fuel to become significant until around 2030—and synthetic methane not until around 2045.

German net power exports are currently skyrocketing; they were above 60 TWh or around 10 percent of power production in 2015. The Fraunhofer study found that the situation would be very much the reverse in a scenario with 85 percent carbon reduction, with power imports rising to around 100 TWh starting in 2023 (the first year without nuclear power), roughly a sixth of total German power consumption at present. Nonetheless, energy imports would fall dramatically in the process from the current level of around 85 percent to roughly 25 percent. On the

Table 14.2 Installed capacity in GW required for various levels of carbon emission reductions by 2050

	Minus 80%	Minus 85%	Minus 90%
Hard coal	7.6	0.0	0.0
Soft coal	2.9	0.0	0.0
CCGT gas	38.1	37.8	49.6
Gas turbine	27.4	33.9	59.0
CHP	16.8	13.9	24.4

Source: Fraunhofer ISE, "Was kostet die Energiewende? Wege zur Transformation des deutschen Energiesystems bis 2050."

other hand, if neighboring countries also decide to follow a similar path dependent on imports, where will the power come from?

The researchers found that even the 85 percent scenario would probably be less expensive than business as usual if carbon has a price of around 100 euros per ton and fossil fuel prices increase by 2 percent per annum. Other externalities, such as health care, were not taken into consideration, however. That carbon price may seem extremely high given the low level in 2015 of less than 10 euros, but the figure is given with a discount rate up to 2050, putting it within realistic reach.

"Power Reliability in the Energiewende" (IWES)[38]

In 2014, another Fraunhofer institute, IWES, published a study investigating how power reliability could be guaranteed even with 100 percent renewable electricity (remember, the official goal is only 80 percent).

Though the study does not use the word *Dunkelflaute*, it helps to understand the study better if we learn the word. Dunkel means "dark"; solar is simply not available half the time, and solar power production is significant for only around six hours a day even when the sun is shining. Flaute is "doldrums"—when the wind is not blowing. So the "dark doldrums" are times when solar and wind power is not available in sufficient amounts. How can this gap be covered?

Many people stress power storage. The assumption is that batteries, for instance, will store excess electricity and make it available again during the "dark doldrums." But the study points out that batteries will mainly be used to stabilize the grid. In other words, the batteries will not store large amounts of power for much later, but instead small amounts of power, releasing those small amounts shortly thereafter.

In other words, batteries will provide "ancillary services" to stabilize the hertz frequency, as opposed to storing kilowatt-hours for later. Whereas only around 150 GWh of electricity would come from batteries in the

---

[38] "Sichere Stromversorgung mit 100% Erneuerbaren Energien ist möglich." Kombikraftwerk 2. Accessed February 9, 2016. http://www.kombikraftwerk.de/start.html.

ISE study, where they are largely used to store electricity for mobility, IWES puts the figure at 2700 GWh in this different scenario focusing on the power sector.

The two studies are thus hard to compare; IWES focuses on reliability in the power sector (preventing blackouts) with no account taken of cost, whereas ISE investigates costs for electricity, motor fuels, and heat. In the IWES analysis, around 290 GW of renewable capacity would be needed (see chart below).

For a fictitious future January with 100 percent renewable electricity, power demand would roughly be at the level of today, ranging normally from 40 to 80 GW. The biggest difference would be power storage options, which kick in to harvest excess power production, thereby bringing total power consumption above 100 GW regularly. A significant amount would be power-to-gas, an option that did not become significant in the ISE study until 2045. Nonetheless, a significant amount of electricity produced would still be lost.

"Space Requirements for the Energiewende" (BBR)[39]

Finally, the German Environmental Ministry's Construction and Spatial Planning Agency (BBR) published a report in 2015 investigating the space requirements for an all-green power supply. The assessment was based on the two categories of "taboo zones" and "suitable zones," as defined by the German Environmental Agency (UBA). In other words, the evaluation of feasibility took account of areas where renewable energy installations are completely ruled out, say, for conservation reasons.

The study found that Germany could install 125 gigawatts of wind turbines on only 1.7 percent of the country—on land, not including offshore wind farms. Likewise, 143 gigawatts of photovoltaics could be installed on 0.9 percent of the country. Both of these numbers are smaller than what IWES

---

[39] Bundesministerium für Verkehr und digitale Infrastruktur. "Räumlich differenzierte Flächenpotentiale für erneuerbare Energien in Deutschland." BMVI-Online-Publikation, Nr. 08/2015. BBSR, August 2015. Accessed February 9, 2016. http://www.bbsr.bund.de/BBSR/DE/ Veroeffentlichungen/BMVI/BMVIOnline/2015/DL_BMVI_Online_08_15.pdf?__blob=publica tionFile&v=2.

estimated necessary for a reliable power supply. To have 100 percent renewable electricity, Germany would only need to devote 2.6 percent of its area to wind farms and solar arrays. Farming can continue, however, in wind farms, and solar can be put on built structures; the land can have a dual use.

To put that number into context, roads alone currently cover just over four percent of Germany. The Lusatia coalfields cover 0.2 percent of Germany; the coalfields on the Rhine near Cologne, 0.1 percent. The area of groundwater impacted by lignite mining alone in eastern and central Germany are estimated at over one percent of the total area in Germany. Settled areas (including roads) take up 13 percent; rivers and lakes, 2.4 percent. Other areas (a category that includes Germany's tremendous coalfields) make up 1.7 percent of the country. Since Germany won't need those coalfields anymore, we could subtract them from the equation. In other words, the space that Germany would need to reach its targets for wind and solar are close to what the country now lists under "other."

The study found that "even if a third of all suitable roofs are not used for solar because the building owners simply don't want to, the technically suitable roof potential is enough to install 65 GW by 2032 … without any need for new ground-mounted arrays at all." In other words, Germany could get around 10 percent of its electricity from photovoltaics using only two thirds of suitable existing rooftops.

About 52 percent of the country is currently used for farming, followed by 30 percent forest. For bioenergy, the study puts the potential within the power sector at around 52 TWh annually—hardly more than the 42.8 TWh of electricity generated from biomass in 2014. In other words, no major biomass growth is needed.

What Is Needed to Shift to a Renewable Energy Economy?

Taken together, the three Energiewende scenarios provide valuable insights into what is needed for Germany to shift to a renewable energy economy:

- There is no single master plan—no set number of solar panels, wind turbines, and so on that Germany needs to install by 2050. Estimates vary, and that's good because we don't know what technology will look like in 2030, much less 2050. But the estimates also overlap roughly, so there is a consensus.

- A high level of renewable energy is affordable, would increase energy security and is technically possible within German spatial limitations.
- Massive overbuilding of solar and wind will be necessary; the total installed capacity in every scenario exceeds peak power demand severalfold.
- Germany does not need (and does not plan) to ramp up energy from biomass significantly from the current level.
- Significant storage will eventually be necessary, but power curtailment (losses) might still also be significant.
- No money will be made on wholesale power markets in the future; profits will shift to ancillary services (backup power, frequency stabilization, etc.).
- A focus on all energy sectors, not just electricity, indicates that electrification of mobility and heat will be crucial for the overall goal of 60 percent renewable energy in all sectors (including transportation and heat).

The amount of renewable electricity that will eventually be generated will thus exceed both demand and the volume that can be stored. Requiring solar and wind—which will make up the bulk of power production in all scenarios for Germany—to react to market prices, however, will be of limited value.

Due to the "cannibalization effect," wind and solar price themselves out of the market; when a lot of wind and solar power is available, electricity prices are automatically low on the power exchange, where the merit order lines up dispatchable power plants by price, so that the market will take the cheapest unit. The merit order has come to be conflated with the market, but it does not represent the value of the electricity generated. For instance, if solar and wind push dispatchable units down to a point where they can switch off, power prices in the merit order reached zero or go negative. Germany may have 70 GW of electricity at that moment from solar and wind, but no money can be made on the exchange when the merit order puts the price at zero. In return, prices rise when there is very little solar and wind—and hence, great demand for dispatchable electricity. But solar arrays and wind turbines cannot react to those prices; they react to the weather. So if Germany really wants all that wind and solar, it should simply pay for it without calculating its effects on the residual power load. The idea that wind and solar could someday "do without financial support" (meaning without feed-in tariffs), as the Öko-Institut argued back in 2001 and Hermann Scheer himself argued in the 1990s, was misleading all along. If we want very high levels of wind and solar, payment still needs to come from a reliable source. The kind of floor price provided by feed-in tariffs is an obvious option.

In addition to low capital costs for wind and solar, there would be much stronger price signals for flexibility options—the things that can react: dispatchable plants (conventional, but also biomass), storage units of all kinds, and demand management. The more the financing of solar and wind is independent of the spot power market, the more power prices will fluctuate, creating strong incentives for demand shifting, power storage, and so on.

The central power exchange—the EEX in Leipzig—clearly does not send all the price signals we need. The merit order is not the only shortcoming. EEX prices are not only nationwide, but also currently include Austria, and the push for an Energy Union in the EU with more international power trading will further erode price differences between international grids. More power trading is good, in that countries like Switzerland and Norway, with tremendous adjustable hydropower, can serve as a gigantic battery for countries like Germany, which will have to make do with fluctuating solar and wind for their energy transition. But below this gigantic continental grid that is being built, there will still be regional bottlenecks.

Grid hubs with their own price signals for power shortages and overloads are therefore needed, primarily on medium-voltage grids and lower. If such signals were sent out, households in a particular region could begin heating water with electricity when there is local excess renewable power. The Enera project in an area in northwest Germany which already has more than 100 percent renewable power net is one such example.[40]

---

[40] Budke, Jan. "Verteilnetz und Übertragungsnetz an einen Tisch bringen." 19 January 2016. Accessed February 9, 2016. http://energie-vernetzen.de/index.html.

Just under 400,000 people live in the model region, only 5 percent of the German population. At that size, Germany would need roughly 20 such hubs. When residents and businesses in the Enera region begin to store power as heat in times of great wind power production, the situation might be considerably different in Bavaria to the southeast. The EEX currently does not provide any such regional signals. Unfortunately, there is too little talk about the need for distributed power hubs and power-to-heat at present.

In addition, we need to pay attention to more than the power sector. In particular, efficiency and the sectors of heat and transport require immediate and sustained attention. Passive House architecture will fortunately become more or less the standard for new buildings around 2020 throughout the EU with zero-energy buildings, but renovation rates need to be increased. Furthermore, Denmark has banned oil-fired heaters entirely; Germany could follow this example. Incentives need to be provided for the storage of excess electricity as heat.

Furthermore, technological breakthroughs are always welcome, but none of the scenarios described above require anything that doesn't already work today. In fact, some of the greatest gains will be made by going low-tech. For instance, electric cars will play an important role in the future, but we should focus primarily on creating walkable cities where cars are not needed at all. Greater urban density would also allow more consumers to be hooked up to district heat networks, which are beneficial in numerous ways. To name just two, they allow waste heat from power generators to be recovered usefully, and they could be bidirectional, thereby solving the problem of limited heat storage capacity in small, distributed units. The grid was the great facilitator for solar power; district heat networks could be for renewable heat.

Finally, if all these conclusions were heeded, the result would be far more community involvement. Communities will not build electric cars, but they will promote public transportation, bike paths, and walkable urban areas. A more straightforward pricing mechanism for wind and solar, such as feed-in tariffs under the de minimus rule described above, would further allow community projects to continue to go forward in the energy sector.

Likewise, the EU is pushing for competitive auctions whenever municipalities sell 20-year commissions for utility services. The goal is to provide consumers with the lowest prices—but citizens are not just consumers. The EU should allow municipalities to take over their own local gas and power grids if a majority of citizens wish that to happen. Even if this option does not produce the lowest cost, people simply are sometimes willing to pay slightly more for a better service; in this case, that could include revenue flowing back into public services rather than being paid out to out-of-town investors. If communities are strengthened in this way, they will have more money to invest in making their belt environment walkable and bikable, for instance.

Energy Democracy at the Crossroads

At present, the EU is pushing for a harmonization of energy policy among member states. The German government seems to be taking the opportunity to make fundamental changes. All this is happening under the banner of the free market, with Undersecretary Baake stating in 2014: "The Energiewende was never meant to be a renationalisation of policy, I'm sorry if we gave that impression. It must be implemented within a European context."[41]

In reality, though it would be impolitic of Baake to say so, the charge of Germans going it alone for their energy policy is unfair; all EU member states have always done their own thing in the energy sector. Brussels has pushed for harmonization since at least the 1990s, but all member states have fought back, some (such as France) more than others. The Treaty of Lisbon from 2009 specifically gives the EU Commission, the European Parliament, and member states shared responsibility for energy policy because member states fought to keep control.

But energy democracy is not a term widely understood outside Germany and Denmark. If these changes are implemented, community

---

[41] "German renewables sector urged to step up investments." EurActive. 12 September 2014. Accessed February 9, 2016. http://www.euractiv.com/sections/energy/german-renewables-sector-urged-step-investments-308404.

projects will have to compete with international energy corporations. In the worst case, the German public may wake up in, say, 2030 to find that the large companies entrusted with the transition have convinced politicians that 50 percent renewable electricity and 30 percent renewable energy is enough; the 2050 targets could be abandoned—just as Denmark's new government is now calling its 2050 targets into question.[42] Already, German economists with ties to conventional power providers are demonstrating that there is no business case for more than 50 percent renewable electricity based largely on wind and solar.[43] But that is a microeconomic view. Civil society will have to remind politicians that there will still be macroeconomic benefits beyond what is good for some companies.

The German public originally called for energy democracy in the 1970s because they felt that their politicians and energy experts were pushing them around. They created their own alternative group of energy experts, who have since become respected in the mainstream. They developed modern turbines under the noses of Germany's best engineers, and they bought enough solar panels when the technology was still prohibitively expensive in order to enable the mass production that the world now benefits from.

Over the decades, climate change has taken center stage in the Energiewende alongside the nuclear phaseout. Today, climate campaigners in Germany are not even always aware that the goal of lower carbon emissions was not the original ambition; energy democracy was.[44] In the worst case, without energy democracy, climate targets are not attainable; civil society is calling for action toward an energy transition, whereas there will be winners and losers in the business world if one is achieved,

---

[42] Morris, Craig. "Are we losing Denmark next?" Energy Transition. The German Energiewende. 30 September 2015. Accessed February 9, 2016. http://energytransition.de/2015/09/are-we-losing-denmark-next/.

[43] Morris, Craig. "No business case for lots of wind and solar." Energy Transition. The German Energiewende. 15 July 2015. Accessed February 9, 2016. http://energytransition.de/2015/07/no-business-case-for-lots-of-wind-and-solar/.

[44] Morris, Craig. "German environmentalists speak of "coal transition"." Renewables International. The Magazine. 22 August 2013. Accessed February 9, 2016. http://www.renewablesinternational.net/german-environmentalists-speak-of-coal-transition/150/537/72199/.

so some big businesses will always fight back. In the best case, without energy democracy, new corporations—Google, Tesla, Siemens, and so on—may handle the transition for us so that climate targets are indeed met in the end. In 2050, we would then wake up in a climate that might still be livable; in return, we would have a business environment in which SMEs do not focus on serving their local communities, but on "going public" and making a fortune cashing out on the stock exchange. It will be a world in which citizens will be thought of only as consumers. One in which the companies that saved us from climate change will be too big to fail.

Technocrats would tell voters that proposed policies would provide the cheapest solutions. Alternatives to those solutions would come from outside a small coterie of experts and not be taken seriously. Democracy would devolve into a system in which moneyed interests manage political majorities—instead of being seen as a way of resolving natural conflicts of interest through compromise.

Energy experts, as Bloomberg New Energy Finance CEO Michael Liebreich pointed out in 2015, are used to thinking of the energy sector in terms of climate change and environmental pollution. "There is, however, a third level on which the struggle between defenders of clean and fossil energy must be understood, and that is in terms of the social structures in which we want to live."[45] In explaining why he is committed to renewables, Liebreich writes of the fossil fuel sector:

> ...gorging on subsidies and externalising substantial costs onto others, helping it to generate massive profits, of which it spends a meaningful proportion on lobbying, physical protection, community compensation and even—as history has repeatedly shown—downright bribery. These actions, in turn, help protect political access and entrenched economic advantage. This is as much a part of the old energy system as any pipeline or power station.

---

[45] Liebreich, Michael. "Liebreich: An Energy Sector Transformed Must Be An Energy Sector Reformed." Bloomberg New Energy Finance. 18 August 2015. Accessed February 9, 2016. http://about.bnef.com/blog/liebreich-energy-sector-transformed-must-energy-sector-reformed/.

The year 2015 marked the midpoint of Germany's energy transition, 35 years after the original Energiewende book of 1980 and 35 years before the ultimate targets for 2050. In most other countries, the window of opportunity for an energy transition is only now opening. But not everyone is being told that this window is briefly open to something other than clean energy and a stable climate: energy democracy.

15

Act Now or Be Left Out

The energy transition represents a one-time window of opportunity to democratize the energy sector. Once utilities have built giant renewable power plants, the market will be closed to community projects.

The Energiewende is a broad mix of energy and climate policies. There is no master plan. It is an iterative process—work in progress. Mistakes have been made, and more are to come. It is Germany's Man on the Moon project—but that successful American project started off with Apollo 1, which burned up on the launch pad. The Energiewende may yet have its Apollo 1. And like America, Germany might learn from its mistakes and march on.

The Energiewende movement began in the mid-1970s as a grassroots protest against large infrastructure projects. The first one was against a nuclear plant in a farming region—but also against plans to attract industry, specifically a lead production plant. The German anti-nuclear movement came out of this protest, but locals at the time mainly opposed industrialization; they did not want large power plants (nuclear or otherwise) attracting other polluting big business. Citizens felt that profits were being privatized and risks socialized. Awareness of and concern about radioactivity gradually came later, as did the understanding of

© The Editor(s) (if applicable) and The Author(s) 2016
C. Morris, A. Jungjohann, *Energy Democracy*,
DOI 10.1007/978-3-319-31891-2_15

climate change. The grassroots Energiewende movement predates public awareness of global warming.

The people who said "no" to nuclear knew they had to say "yes" to something else. They chose renewables and efficiency. From the beginning, Energiewende proponents understood that the corporate market structures in the German energy sector would unnecessarily slow down the transition to renewables. Distributed renewables needed to replace central power stations, and revenue from the energy sector would be devoted to efficiency projects as well. The breakthrough was political: the Germans got their government to do what it wanted even when it hurt big energy companies. The Energiewende began as a fight for energy democracy.

In 1980, the first book entitled *Energiewende* was published, and others followed in the 1980s—but little progress was made otherwise. The incumbent utilities and their allies in the government watered down or completely blocked most initiatives for renewables. Still, Energiewende proponents succeeded with three policy breakthroughs:

- In 1990, an unlikely coalition of parliamentary backbenchers from the conservative Christian Union and the Greens drafted what became the **1991 Feed-in Act**. It barely passed in the late night, final parliamentary session of a unifying Germany. The law mainly got wind power and small hydropower moving.
- In 1998, the Germans elected the first completely new coalition without any party from the previous one. Circumventing the Ministry of Economics, parliamentarians from Social Democrats and the Greens then passed the **Renewable Energy Act of 2000,** which laid the foundations for a boom in renewables growth.
- In 2011, the nuclear accident at Fukushima led the then Chancellor Merkel to adopt the Energiewende as her own term. The **2011 nuclear phaseout law**, supported in a cross-party-consensus, put an end to the remaining support for nuclear within her party and sealed the fate of nuclear power in Germany once and for all.

For civil-society groups, there is a lesson to draw from these policy breakthroughs. They all happened under rare circumstances. All of them were based on years of preparation. The groundwork for progress must therefore be laid in advance. There will be many years of frustrating standstill, but windows of opportunity eventually open up, and all that preparation suddenly pays off. The important thing is to recognize those opportunities on the horizon—and remember, during the most frustrating hours when they are nowhere to be seen, that they will eventually come.

These policy breakthroughs must not be confused with silver bullets. As our history shows, the Energiewende could have never happened at a stroke of a pen. The policy in place today was built up step by step over decades through successive and interrelated policy changes. None came as a fundamental surprise or was revolutionary at the time passing. It was hard work, a fight all along, sometimes also playing defense. But building up policy bricks, one after the other, consolidating and improving them over time helped build up momentum in the long run.

At times, however, the Energiewende simply got lucky. In 2002, a flood in eastern Germany, the US invasion of Iraq, and a bumbling Bavarian politician inadvertently kept the Energiewende going.

Otherwise, the history of the Energiewende contains a number of surprises. Repeatedly, renewables have grown much faster than expected. The 2004 goal was at least 20 percent renewable power by 2020. Germany reached that level in 2011, so the target was raised to 35 percent by 2020. Even that target will be reached ahead of schedule, possibly in 2016; in 2015, renewable electricity met 33 percent of demand. German utilities long doubted that solar and wind would make much of a difference. No utility today will repeat this mistake.

Billionaires quoted in the media on clean energy usually stress the need for breakthroughs. But innovation has often come from deployment more than R&D. Furthermore, Germany's biggest energy firms have been of little help in the energy transition. Newcomers made the difference. While breakthroughs would help, few are needed. Most of the tech we will need in 2050 already exists, and prices will come down through deployment.

Most countries have better renewable energy potential than Germany does; none face Germany's high legacy solar costs. The transition will be cheaper and easier outside Germany. A high level of renewables is nonetheless affordable in Germany, would increase energy security, and is possible within German space restrictions. Solar and wind will need to be overbuilt; in every scenario for 2050, installed capacity exceeds peak power demand by far. But Germany does not need (or plan) to ramp up biomass significantly from the current level; the focus for bioenergy is now on waste. The electrification of mobility and heat will be crucial for the overall goal of 60 percent renewable energy. In the power sector, profits will no longer be made mainly on wholesale power markets but also on ancillary services (backup power, grid stabilization, etc.).

Large shares of wind and solar power will require flexible backup generation capacity. Baseload has to go, and nuclear is the least flexible type of baseload. The more nuclear you have, the harder it is to integrate wind and solar. The early retirement of existing conventional power plants precedes the need for storage. The former is not a technical challenge, but a financial and political one—like the nuclear phaseout.

To date, there has not been a single demonstration against the cost of renewables. There have, however, been repeated demonstrations *for* the energy transition—by citizens who want to protect their right to make their own energy. Except for the nuclear phaseout after Fukushima, energy was never in the foreground of German politics. Issues like migration, the financial crisis, and globalization were. The word "Energiewende" was not even mentioned in the 2013 TV election debate between Chancellor Merkel and her challenger. As a generational project, the Energiewende is, however, constantly in the background.

The Energiewende is a political response to societal demands, not something the energy market brought about. Relying on the energy experts who don't want the Energiewende in the first place is therefore a recipe for failure. The public, not the corporate energy world or industry, wants the energy transition, so citizens must remain involved.

Yet, studies focusing on what is best for society have become rare in Germany, such as the Prometheus study of September 2015.[1] The early pro-Energiewende researchers not only provided answers, but also

---

[1] Walk, Heike, Melanie Müller, and Dieter Rucht. „Prometheus. Menschen in sozialen Transformationen am Beispiel der Energiewende." September 2015. Accessed February 12, 2016. https://protestinstitut.files.wordpress.com/2016/02/ipb_prometheus-studie.pdf.

formulated questions themselves. We need communities—and progressives engaging more with classical conservatives who want to strengthen communities, not just libertarians who stress individual rights.

Lately, the German government's renewables policy increasingly seems designed to shut out the very citizen and community groups that sustained the energy transition for the past 25 years. At present, this policy failure can still be excused as an oversight. If it is not fixed soon, however, one can only conclude it is intentional. If Germany reaches its official targets for 2050 with utility projects, sidelining public participation in the process, the original spirit of the Energiewende will have been lost. The environment would still be cleaner and carbon emissions still reduced to a sustainable level, but corporations too big to fail would run the show. These firms would use their money and influence the shape of the public sphere—politics, the media, society, and so on—to suit their bottom line.

To make more people aware of the reasons why the focus of the energy transition should be on energy democracy, a Bill of Rights would be useful. Numerous groups have come up with such proposals. For instance, the UK's Fuel Poverty Action has published its own Energy Bill of Rights, which largely focuses on protecting the poor, though it does mention active citizen participation in community energy.[2] The Brussels-based chapter of the Rosa Luxemburg Foundation also published a document entitled "Strategies of energy democracy" in February 2016.[3] An update of the Foundation's previous work, the paper provides an overview of international voices on energy democracy and adds labor unions to the debate. The Prometheus study mentioned above also points out the numerous reasons why citizens become involved in energy cooperatives aside from profits: community building, personal skill development, political engagement, and so on. People do not always need financial incentives; they sometimes volunteer at local schools, fire brigade, and so

---

[2] Fuel Poverty Action. *Energy Bill of Rights*. May 2014. Assessed on February 12, 2016. http://www.fuelpovertyaction.org.uk/home-alternative/energy-bill-of-rights-2/.

[3] Angel, James. "Strategies of Energy Democracy." Rosa Luxemburg Foundation, Brussels Office. February 2016. Assessed February 12, 2016. http://rosalux-europa.info/userfiles/file/EnergyDemocracy-UK.pdf.

on.[4] As Anna Leidreiter of the World Energy Council puts it, "renewable energy is about more than climate protection. It's about power to the people and making the world a more equal place!"[5]

Too often, the discussion about the energy transition focusses only on affordability and carbon emissions. These are, no doubt, important criteria to consider. But so are civil rights in the energy sector. Based on the lessons from this history of the Energiewende, the authors would therefore like to end with a proposal for a Bill of Rights for the energy transition. The list is based on the lessons we draw from the Energiewende's history and from other such proposals, like the ones listed above. It is intended merely as a contribution to the international discussion, not as definitive. Criticism is welcome; in fact, the more, the merrier. Our intent is to encourage an international discussion on energy democracy—ultimately, to move civil rights into the foreground of the energy transition debate. In the end, it is up to civil society to call for energy democracy with or without an actual Bill of Rights for the energy transition. Our draft below is merely an invitation for you, our readers, to comment on— and come up with your own. Instead of disparate groups calling, say, for the right to put solar on their roofs in one state, we should call for the national civil right to do everything the energy transition requires (not just residential solar). If this book helps start a global conversation about ownership within the energy sector and about the energy sector within democracy, the authors will consider their endeavor to have been successful (Box 15.1).

Box 15.1 Bill of Rights for the Energy Transition

1. **You have the right to make and sell your own energy**—and receive a fair price for it. Renewable energy, especially sun and wind, are public goods like air or water. Everyone shares a right to use them; no one has the right to monopolize access to them.

---

[4] Morris, Craig. "The co-benefits of community energy." Energy Transition. The German Energiewende. 8 October 2015. Accessed February 12, 2016. http://energytransition.de/2015/10/the-co-benefits-of-community-energy/.

[5] Personal correspondence with Craig Morris.

Box 15.1 (continued)

2. **In a democracy, people are citizens first, consumers second.** Energy is not just a commodity, but the basis for modern life. Cost is important, but so are benefits to citizens, which go beyond low prices.
3. **Energy prices must reflect all costs, including those for health care and environmental impacts.** These costs must be must be internalized, not passed on to society. In particular, nuclear reactor operators—not the public—must pay the full cost of insurance for catastrophic accidents.
4. **The risks of energy projects must be borne by those who profit.** If profits are private, so should be the risks.
5. **People have a right to a minimum amount of affordable clean energy services** (not merely an amount of energy). But the focus must remain on energy efficiency.
6. **Communities affected by energy projects have a right of refusal.** But with rights come responsibilities. Saying no is not enough. Communities that reject projects must say what they want and how they plan to get it.
7. **Information on energy projects should be publicly accessible.** A Freedom of Information Act for energy is needed. An informed citizenry is essential to democracy.
8. **Community projects must have priority in policy and in land use planning.** They include, but are not limited to, grid infrastructure, storage, wind farms, solar arrays on shared private or publicly owned roofs, bioenergy facilities running on local biomass feedstock, walkable cities with good bike paths, and efficiency projects.
9. **Minority opinions must be taken into account; the winner does not take all.** Negotiations should be held to seek a consensus, not produce a tyranny of the majority.
10. **The needs of future generations must be taken into account** when undertaking the energy transition.

Index

© The Editor(s) (if applicable) and The Author(s) 2016
C. Morris, A. Jungjohann, *Energy Democracy*,
DOI 10.1007/978-3-319-31891-2

Printed by Books on Demand, Germany